The Reshaping of America

The Reshaping of America

Social Consequences of the Changing Economy

EDITED BY

D. Stanley Eitzen
Colorado State University

Maxine Baca Zinn
University of Michigan–Flint

PRENTICE HALL
Englewood Cliffs, New Jersey 07632

LIBRARY OF CONGRESS
Library of Congress Cataloging-in-Publication Data

The Reshaping of America / D. Stanley Eitzen, Maxine Baca Zinn, editors.
p. cm.
Bibliography: p.
ISBN 0-13-774506-0
1. High technology industries—United States. 2. Information technology—United States. 3. Industrial management—United States. 4. United States—Economic conditions—1981- 5. United States—Social conditions—1980- I. Eitzen, D. Stanley. II. Zinn, Maxine Baca.
HC110.H53R47 1989
306′.0973—dc19 88-19137
CIP

Editorial/production supervision and interior design: Joy Moore/Jeanne Genz
Cover design: Ben Santora
Cover art: *Estate* by Robert Rauschenberg, 1963; courtesy of Leo Castelli Gallery, New York City
Manufacturing buyer: Peter Havens

Printed in the United States of America
10 9 8 7 6 5 4 3 2 1

ISBN 0-13-774506-0

Prentice-Hall International (UK) Limited, *London*
Prentice-Hall of Australia Pty. Limited, *Sydney*
Prentice-Hall Canada Inc., *Toronto*
Prentice-Hall Hispanoamericana, S.A., *Mexico*
Prentice-Hall of India Private Limited, *New Delhi*
Prentice-Hall of Japan, Inc., *Tokyo*
Simon & Schuster Asia Pte. Ltd., *Singapore*
Editora Prentice-Hall do Brasil, Ltda., *Rio de Janeiro*

To
Michael Harrington
whose work, ideas, and life
have informed and inspired
critical thought and action
among countless scholars

Contents

Preface

We are in the midst of a momentous turning point in history. American society is being transformed by the convergence of four powerful forces: microelectronic technology, the globalization of the economy, the swift movement of capital, and the shift from an economy based on the manufacture of goods to one based on information and services.

This book is organized to elaborate on the nature and the contours of the changes taking place and their consequences for society, organizations, communities, and individuals. The Introduction describes the forces that are reshaping ing America. Parts I through V focus on the consequences of the structural transformation for the economy itself; how work is being redesigned and jobs redistributed; how inequalities by class, race, and gender are being exacerbated; how the societal institutions of the family, education, and polity are being altered; and the spatial reorganization of communities, cities, and regions as some areas flourish under the new conditions, while others recede. Part VI examines various policy alternatives to meet the challenges of these significant societal changes.

The selection of articles for inclusion was determined by their power to describe, explain, and analyze the various facets of change under investigation. We have edited some of these articles to improve clarity, avoid repetition, and eliminate materials not central to our analysis. We have deleted many of the footnotes and some tables to shorten some of the articles so as not to burden the reader with excessive scholarly minutia. Those readers interested in an in-depth analysis are invited to read the original articles and to explore the references listed at the conclusion of many of the articles included in this anthology.

The story of the reshaping of America is unfolding around us. This transformation will continue to affect us as individuals and us collectively as a society for decades. Our hope is that the reading of the materials reprinted here will enhance the understanding of the social forces at work, increase analytical skills, and heighten awareness of what the problems are and what we might do collectively to solve them.

D. Stanley Eitzen
Maxine Baca Zinn

The Reshaping of America

Introduction
THE FORCES RESHAPING AMERICA

D. STANLEY EITZEN AND MAXINE BACA ZINN

> We are living in truly extraordinary times. Massive changes since the end of World War II are impacting our way of life . . . in ways heretofore unmeasured. Governing these changes—underlying them, and basic to them—is the economic factor. Economics—easily the least understood of the great forces of change—has literally changed our world (Robert J. Buckley, retired chairman, Allegheny International, Inc., in a speech delivered before the economics department, University of Pittsburgh, January 29, 1987).

THE GREAT TRANSFORMATIONS IN HISTORY

There have been two fundamental turning points in human history, and we are in the midst of the third. The Neolithic revolution began about 8,000 B.C., marking the transition from nomadic pastoral life, where the animal and vegetable sources of food were hunted and gathered, to life in settlements based on agriculture. During this phase of human existence cities were built, tools were created and used, language, numbers, and other symbols became more sophisticated, and mining and metal working were developed.

The second fundamental change, the industrial revolution, began in Great Britain in the 1780s. While the Neolithic agricultural revolution took almost ten thousand years to run its course, the second lasted but two hundred years. And the industrial revolution really involved three distinct but related technological revolutions, each of which brought fundamental changes to the economy, the relationship of people and work, family organization, and the like. The first phase lasted about sixty years in Great Britain and involved primarily the application of steam power to textiles, mining, manufacturing, and transportation. The second phase in the industrial revolution occurred between 1860 and 1910 largely in the United States, Great Britain, and Germany. It was marked by a significant cluster of inventions and discoveries—the use of oil and electricity as energy sources for industry and transportation, the telephone and telegraph, motor cars and airplanes, and the first plastics. The final stage of the industrial revolution is still in progress and involves the major technological breakthroughs of atomic fission and fusion, supersonic aircraft and missiles, television, biotechnology, and computers. This third phase of the industrial revolution is giving way to a new era. Whereas employment throughout the industrial revolution was dominated by manufacturing, the new era is characterized by a shift in employment toward service occupations and the collection, storage, and dissemination of information (Jones 1982; Zinn and Eitzen, 1987:76–77).

Each of these surges in invention and technological growth had major implications for the economy and other societal institutions and for individuals and families within society. The objective of this book is to examine the consequences of the transformation occurring now. To set the stage for this inquiry, the next two sections describe the forces transforming American society and the special conditions that are making this turning point in history different, much different, from the previous two.

THE INTERRELATED FORCES TRANSFORMING AMERICA

Several powerful forces are converging in the United States that are transforming its economy, redesigning and redistributing jobs, exacerbating inequality, reorganizing cities and regions, and profoundly affecting families and individuals. These forces are (1) technological breakthroughs in microelectronics, (2) the globalization of the economy, (3) capital flight, and (4) the shift from an economy based on the manufacture of goods to one based on information and services.

The New Technologies Based on Microelectronics

The computer chip is the technology transforming America toward a service/information economy. Micro-based systems of information allow for the storage, manipulation, and retrival of data with speed and accuracy unknown just a few years ago. Computer transactions are measured in multiples of picoseconds (pico means one trillionth). Information can be sent instantaneously via communications satelite throughout the world in microseconds. Parallel processing with supercomputers gives machines the ability to reason and make judgements. Computer-aided design (CAD) permits engineers to design and modify an incredible array of products in three dimensions very quickly. Computer-aided manufacturing (CAM) is replacing traditional manufacturing with conventional machines and workers. The problem with these industrial robots is that while they increase productivity, they displace rather than create jobs (Oxford Analytica, 1986:232). Moreover, the use of robotics will replace the higher-paid semiskilled workers, not the unskilled manual workers. "Thus, the continuing use of robots will result in a minimal increase in the number of skilled positions, a significant reduction of the middle layer of semi-skilled positions and a substantial increase in the number of manual workers at the base of the skill hierarchy. The overall effect is a deskilling of the workforce" (Dassbach, 1986:59).

Globalization of the Economy

Because of the size of the domestic market, the relative insulation of the Pacific and Atlantic Oceans, and superior technological expertise, the American economy throughout most of this century has been relatively free from the competitive pressures from abroad. This has changed dramatically since about 1970. The United States, once the world's industrial giant, has lost its premier status.

> In 1960, the United States led the world in per capita gross domestic product (GNP less income from abroad) by a comfortable margin. In 1980, the United States stood eleventh among nonsocialist countries, and the leader outpaced the United States by 39 percent. U. S. steel production accounted for 26 percent of the world's total output in 1960, while Japan's was just over 6 percent. By 1980 Japan was at 15.5 percent and the United States at only 14.0 percent. In 1967, the States built more cars and trucks than the rest of the nonsocialist world combined. By 1982, the United States was third behind Japan and Europe. (Dolbeare, 1986:62)

Examined another way, again comparing the twenty years from 1962 to 1982, automobile imports to the United States rose from 1.9 percent to 22.8 percent; in consumer electronics from 18.5 to 42.1 percent; communications and electronic equipment up from 1.3 percent to 13 percent; textiles from 4.0 to 15.4 percent; and shoes and leather from 5.3 to 43.5 percent (Dolbeare, 1986:63). Put in a closer time frame, in 1982 some 10.8 percent of all goods and services purchased in the United States were imported. By 1987, it had increased to 13.6 percent (Karmin 1987).

This competition from abroad means reduced profits to American corporations. Their typical response has been to cut costs by demanding concessions from workers, reducing the labor costs as the semiconductor industry did when sixty-five thousand workers were laid off over a two-year period because of the intense competition with Japan (Koepp, 1986). Many corporations in the hardest hit areas such as steel just shut down plants completely, throwing thousands of employees out of work and the communities in which they were located into difficult dislocations. Another strategy employed by American corporations was to compete as strongly as possible. Massive investment has occurred in the automobile industry (General Motors spent $40 billion from 1980 to 1984), for example, but these monies were spent on labor-saving devices, which, of course, did little to help workers (Oxford Analytica, 1986: 211–212).

Capital Flight

Private businesses in their search for profit make crucial investment decisions. The term *capital flight* refers to the investment choices that involve the movement of corporate monies from one investment to another. This takes several forms: (1) investment overseas, (2) plant relocation within the United States, and (3) mergers. While these investment decisions may be positive for the recipients of the move, they also take investment away (disinvestment) from others (workers and their families, communities, and suppliers).

Overseas Investment by U. S. firms American multinational corporations have invested heavily in foreign countries. In 1984 the direct investment by U. S. based firms in foreign countries amounted to $233.4 billion, up from $192 billion in 1980, and $16 billion in 1950 (Vernon 1986). "The total overseas output of American multinational corporations is now larger than the gross domestic product of every country in the world except the United States and the Soviet Union" (Bluestone and Harrison, 1982:42). The foreign revenues of the largest U. S. based multinationals in 1985, for example, were Exxon, 68.1 percent of their total revenues; Mobil, 57.2 percent; Texaco, 47.2 percent; and I. B. M., 43 percent (*Forbes,* 1986:207).

Corporate capital is invested overseas because manufacturing overseas is profitable, mainly because of cheap and nonunionized labor and the relative lack of the kind of regulations found in the United States that the companies believe to be excessive and expensive on such matters as pollution and worker safety. The primary reason, though, is greater profit from lower wages. For example, there are over eleven hundred U. S. owned plants—owned by corporations such as Ford, General Motors, RCA, Zenith, and Westinghouse—located in Northern Mexico close to the U. S. border (these plants are called maquiladoras). U. S. corporations are allowed to ship raw materials, components, equipment, and machinery to Mexico duty free. They are delivered to factories in Mexico and then assembled by laborers working for 75 cents an hour (1987 wages). The finished products are then exported back to the United States with duty paid only on the value added. Obviously, the corporations profit greatly from such an arrangement.

Relocation of Businesses Corporate administrators may decide to move their business to another locality. Such decisions involve what is called plant migration or more pejoratively, "runaway shops." The decision may be to move the plant to Mexico, as we have seen, to the Caribbean (all baseballs for major league baseball, for example, are manufactured in Haiti), or to the Far East where many U. S. plants involved in textiles, electronics assembly, and other labor-intensive industries are located.

Capital is also moved within the United States as corporations shut down operations in one locality and start up elsewhere. Profit is the motivation for investment in a new place and disinvestment in another. Corporations move their plants into communities and regions where wages are lower, unions are weaker of nonexistent, and the business climate more receptive (that is, there are lower taxes and greater government subsidies to the business community).

Regardless of whether plants are moved within the United States or to foreign countries, there are consequences to individuals and communities. Plant closures are devastating. Workers in the affected plants are suddenly unemployed and so, too, may be many in the affected communities whose jobs were directly and indirectly tied to that plant (such as transportation, supplies, and services). Also, real estate, banking, schools, and other areas are adversely affected. The local governments can no longer provide the same level of services because of a lower tax base. The recipient communities benefit from the increase in jobs, greater tax revenues, and the image of growth and progress. The boom communities, however, often cannot meet the greater demand for new roads, sewage treatment, schools, hospitals, recreation facilities, and housing that the new plants engender.

Mergers Another type of capital flight occurs when corporations use their capital to purchase companies in related or unrelated enterprises rather than to expand and modernize their plants. In 1981 there were 2,314 mergers at a cost of $82.6 billion. In 1986 there were 4,022 mergers costing $190.0 billion. Some of the largest mergers during this period were Standard Oil of California's purchase of Gulf Oil for $13.3 billion, Texaco's buyout of Getty Oil for $10.1 billion, and General Motors' buying of Hughes Aircraft for $5.2 billion. Perhaps the most interesting merger occurred when U. S. Steel, which was having trouble competing with the modern steel plants in Japan and South Korea, paid $6 billion to acquire Marathon Oil. Obviously, the company could have used this money to rebuild its antiquated plants, but it chose, rather, to buy a profitable company in an unrelated industry.

This trend toward megamergers has at least three negative consequences: (1) it increases the centralization of capital, which reduces competition and raises prices for consumers; (2) it increases the power of the huge organizations over workers, unions, and governments; and (3) megamergers diminish the number of jobs.

From Manufacturing to Services

Manufacturing, the backbone of the American economy in this century, is no longer dominant. In 1947 employment in the service sector of the economy reached 50 percent, and now it is over 70 percent. Whereas people mostly worked at producing goods, now they tend to be doing work in offices, banking, insurance, retailing, health care, education, custodial work, restaurant work, security, and transportation.

Overall manufacturing has remained relatively stable, but within this sector about twenty industries have experienced steady declines since 1975. These "sunset" industries (for example, steel, tires, shoes) have declined in both output and employment (Kutscher and Personick 1986). Over fifteen hundred plants in these industries have closed permanently since 1975. And, literally, millions of jobs have been lost that will not be replaced. The Labor Department reported, for example, that 2.3 million manufacturing jobs disappeared from January 1980 through the end of 1984 and that 90 percent of these jobs most likely will never return (reported in Noble 1985). Moreover, the data suggest that this trend will continue. Peter Drucker, the esteemed management expert, has predicted that by the year 2015 as many as fifteen million manufacturing jobs will have evaporated and only about 5 to 10 percent of the work force will be holding manufacturing jobs—roughly equivalent to the proportion now engaged in agriculture (cited in Hart, 1983:11).

The "sunrise" industries, which are characterized by increased output and employment, are involved in the production of "high tech" products (computers, communications equipment, medical instruments, fiber optics, bioengineering, and robotics). These industries are creating many new and exciting products. The employees are typically highly skilled, but "only a very small number of very highly trained workers are needed to provide the technical breakthroughs critical to industry—less than 10 percent of all jobs through 1995. The rest of the jobs will be routine, and new labor-saving technologies will cause workers to be 'de-skilled' " (Oxford Analytica, 1986:252). The production workers in high tech, unlike those in heavy industry, tend to be nonunionized and to be paid relatively low wages and benefits.

The third category of manufacturing involves those industries that have gained in output but have lost employment (for example, food processing, metal products, industrial machinery, and automobiles). The source of this seeming incongruity (high productivity with a loss in employment) is automation, a topic we will discuss shortly.

To repeat, the manufacturing sector of the economy overall has been relatively stable. But this stability masks a massive redistribution of job opportunities with declines in some industries (steel, for example, has lost jobs steadily since 1975 and is projected to drop another 20 percent between 1984 and 1995) and increases in the high-tech industries. But the redistribution of jobs and opportunities is not only within the manufacturing sector. The major change is in the growth of the service sector of the economy.

From 1970 to 1985 some thirty million jobs were created. Most of these were in the service sector—from professional and administrative to clerical and service workers. About one half of these jobs were "bad" jobs in the sense that they involved few skills, were poorly paid, had little responsibility attached to

Table 1 What Jobs Will Be Expanding Most

Job Category	*Growth in Jobs 1982–1995*
Building custodians	779,000
Cashiers	744,000
Secretaries	719,000
Office clerks	696,000
Salesclerks	685,000
Registered nurses	642,000
Waiters, waitresses	562,000
Teachers (elementary)	511,000
Truckdrivers	425,000
Nursing aides, orderlies	423,000
Accountants, auditors	344,000
Auto mechanics	324,000
Blue-collar supervisors	319,000
Kitchen helpers	305,000
Guards, doorkeepers	300,000

SOURCE: U. S. Department of Labor Statistics, "Employment Projections for 1995," Bulletin No. 2147 (March 1984):43.

them, and provided poor job security. They have been typically filled by teenagers, women, racial minorities, and the poorly educated. Oxford Analytica, a group of scholars from Oxford, England, commissioned by three major American corporations to project trends in the United States, predicted that "the likely distribution between 'good' and 'bad' jobs in the future will not be an improvement on that of the past decade. Most of the new jobs will be service jobs and many of those will be 'bad' jobs with attendant problems of poor payment" (1986:251). Table 1 provides job projections from the U. S. Department of Labor that reveal the likelihood of most new jobs being in the "bad" category of the service sector.

AGE OF DISCONTINUITY

Every new era poses new problems of adjustment, but this one differs from the agricultural and industrial eras. The earlier transformations were gradual enough for adaptation to take place over several decades, but conditions are significantly different now. The rate of change is phenomenal and unprecedented. There is a global economy now in which communication is instantaneous and capital is incredibly mobile. The potential work force in the United States is expanding (women and minorities) at the very time when technology combines with other forces to reduce jobs. These factors, which are discussed in this section, result in considerable discontinuity and disequilibrium (Jones, 1982:4–39).

Technology

There are several aspects regarding technology that make this era different than the preceding ones. First, new technologies, some quite revolutionary, are being developed rapidly. These have potentials that can be positive and negative and many that are unforeseen. Lasers and fiber optics, for example, have a wide variety of applications as components in telecommunications systems, information processing, and entertainment (three-dimensional television). Fiber optics will increase the capacity of communications networks and make them more efficient. These technologies will generate considerable investment, render other technologies obsolete, generate new jobs, and destroy others.

Biotechnology is just beginning to effect changes in a wide range of industries—foods, agriculture, fuel, pharmaceuticals, chemicals, waste treatment, and natural resource recovery. The changes occurring in this field are rapid and far reaching. This is a field of tremendous growth potential (Oxford Analytica, 1986:233–234). Its future consequences for the economy and jobs are unknown.

Of greatest significance for our purposes are the technological changes directly affecting work and workers. For the most part, the new technologies, especially those related to computers, are reducing jobs and wages. The cost of technology in the microchip field has fallen dramatically relative to the cost of labor. "Despite inflation and the rising cost of resources, the price of each unit of performance in micro-technology is 100,000 times cheaper than it was in 1960. 'Miniaturization' has destroyed the historic relationship between the cost of labor and the cost of technology, permitting exponential growth with insignificant labor output" (Jones, 1982:36). The declining cost of computing power is illustrated by what has happened in personal computers. "A 1977 state-of-the-art personal computer processed 100,000 machine instructions per second and came with 64 kilobytes of main memory and 160 kilobytes of disk storage. A 1987 machine not only costs much less, but also has 20 times the processing power, 20 times the main memory and 500 times the disk storage capacity. The price of storing a single digital unit of data in a memory chip fell from one-tenth of a cent in 1976 to one-thousandth of a cent in 1986, and it will keep dropping by about 35 percent per annum" (Forester, 1987:2).

The capacity of microchips staggers the imagination. Microchips increased their capacity to store information ten thousand times between 1971 and 1980, and the pace continues. We now have "superchips" with one million transitors per chip. "Before the end of the century gigabit—one billion components per chip—integration may be possible" (Forester, 1987:2). The incredible storage and speed capabilities of modern computers is almost beyond comprehension. "Imagine . . . two computers conversing with each other over a period. They are then asked by a human being what they are talking about, and in the time he takes to pose the question, the two computers have exchanged more words than the sum total of all the words exchanged by human beings since *Homo sapiens* first appeared on earth 2 or 3 million years ago" (Simons, 1985:165).

Microprocessors, of course, are programmable, making robots possible. Robots can, for example, cut materials, weld, paint, and assemble. These robots, unlike human workers, do not get bored or tired, go on strike, require cost of living increases, or bicker among themselves. Not surprisingly, corporate managers are purchasing robots to replace workers at an accelerating rate. General Motors, for example, had fifteen hundred robots at the end of 1982 and projects to have twenty-one thousand industrial robots working its assembly lines by 1990 (Dassbach 1986). Robots are drastically altering the workplace. For example, robots perform more than 98 percent of the spot welding on Ford's Taurus and Sable cars. And at a General Dynamics plant in Fort Worth, Texas, one robot drills 550 holes in the vertical tail fins of an F-16 plane in three hours, a job that takes three workers eight hours to do (Bock, 1987:46). The result of this trend, of course, is fewer and fewer industrial jobs and for the workers who remain, less and less leverage in bargaining with management. A University of Michigan study has predicted that robots will displace 4.3 percent of the industrial work force—roughly 200,000 jobs—by 1990. Robotics manufacturing, meanwhile, will create only about 44,500 jobs by 1995 (*New York Times* 1986).

Robotics are not just mechanical menials. They can be equipped with a multitude of sensors for vision, touch, and proximity. They can react to these sensory inputs and adapt. Even more incredible, they can be linked (computer-integrated manufacturing or CIM) into one comprehensive, integrated manufacturing system. In short, robots and computer-controlled machines will work together to plan, schedule, transport, control inventory, and perform many, if not most, of the manufacturing operations that used to require human skills.

The elimination of jobs because of superautomation is not limited to industrial factories. Offices are increasingly electronic. Engineers and architects now draw three-dimensional designs, then update, test, and store them almost instantaneously. In agriculture, there are robot fruit pickers and sheepshearers, computerized irrigation systems that use sensors to calculate water and fertilizer needs in different parts of a field, and automated chicken houses. Retail stores, banks, and brokerage houses use on-line transaction processing to obtain instant information and to conduct transactions. Laser scanning and bar codes are transforming the physical handling of goods by retailers and wholesale distributors. As a final example of technological change affecting jobs, there is the widespread use of televisions, telephones, and personal computers for the purposes of home banking and shopping.

Today's technology is different from that occurring at other stages in history. We now have "smart machines." As Jones has put it: "Computers can be programmed to parallel human mental processes—including the exercise of judgment and intuition. Machines have never before posed a challenge to mental (or white-collar) work: no steam engine could play chess, no slide rule could converse" (Jones, 1982:37). Throughout history technological changes have tended to increase overall productivity, and although there were considerable dislocations to workers in some fields, employment actually increased overall. In the past the introduction of machines extended the capacity of the labor force. The machines of the past (for example, sewing machines, typewriters, telephones, and motor vehicles) were designed to have one operator for each machine. Not only were they designed to have at least one operator per machine, they required other workers to make, sell, and maintain them. Now, however, there has been a significant shift toward labor-displacing technology in which machines are designed to reduce, if not eliminate, human labor. "Much technological innovation in the past was 'labor-complementing'—it extended the capacity of the existing labor force, and the machines themselves changed the nature of work. *But there has been a significant shift to 'labor-displacing' technology where low-cost machines are specifically intended to reduce, if not eliminate, labor inputs*" (Jones, 1982:39; emphasis added).

The effects of technology on social organization are not random or aimless. The costs and benefits are not shared equally by all members of society. For this reason, "there is no such thing as technological change pure and simple" (Smith, 1987:7). Technology is shaped by the prevailing social, demographic, and political conditions. Technology is part of the social relations of production that can assume the particular forms given it by the most powerful and forceful people in society (Noble, 1977:xxii).

The Global Economy

The profound transformation in the international economy is another factor making this time in history different from the previous transitions. For thousands of years caravans have moved across the land, and ships have sailed along the winds and currents. People explored, conquered, and exploited other peoples and their resources; but the pace was slow, and the interaction among different nations muted by time and space. Now, however, there is instant communication around the globe, and transportation anywhere is only a day away rather than six months or more as was once the case. These dramatic advances in communication and transportation have at least two important ramifications. First, technological advances now diffuse rapidly. During the industrial revolution technological change was quite slow (although rapid compared to previous epochs). The first team-powered cotton mill in the United States, for example, began in 1847, sixty-three years after its adoption in Britain. Now technological breakthroughs in one country rapidly spread to other nations.

The second consequence of advances in transportation and communications is the enhanced mobility of capital. As Harrison and Bluestone put it: "The ability to move managers and key components at nearly the speed of sound by jet and to move money and the information needed to coordinate production at nearly the speed of light enables capital, as never before, to go anywhere in the world" (*Working Papers,* 1983:43). Corporations of almost any size now make plans regarding raw materials, workers, and markets across national boundaries. "The globe is the unit of planning" (Coates, 1986:539). Corporations seek cheap labor, which means a transfer of jobs—millions of jobs—to other nations. Americans now buy more foreign goods than foreigners buy ours, driving down the number of American jobs and the wages of American workers. America's economic, political, and even military power, supreme since World War II, is now seriously threatened by the globalization of the economy.

Demography

Three population facts—declining overall growth, disproportionate growth by racial minorities, and much greater participation by women in the work force—combine to make this time in economic history different from other stages. Let us examine each of these demographic trends briefly.

A major source of economic growth during the industrial revolution has been rapid population growth in the technologically advanced countries. Construction, the provision of services, and the like provided a significant source of employment especially in the rapidly growing urban centers. Since about 1970, however, population growth (births minus deaths) has stabilized in some countries and actually declined in others. This fact provides a further impetus for corporations to seek markets and workers in other countries.

While the fertility rate for whites has declined significantly in the United States, the rate for racial minorities has remained about the same. This imbalance, combined with the massive influx of legal and illegal immigrants, has led to the disproportionate growth of racial minorities. Throughout American history racial minorities have been denied equality in education and jobs. They have been relegated to the semiskilled and unskilled jobs in agriculture, in the mines, in construction, in the packing plants, and on the production lines. This bias continues today at a time when these are the declining and obsolete jobs, made unnecessary in a society increasingly committed to knowledge, high technology, and education. This puts immigrants at a distinct disadvantage—more so than at any time in history—and so, too, are the racial minorities already here but locked in poverty by institutional racism, blighted neighborhoods with inferior services (especially education), and fewer and fewer jobs.

Another major demographic trend contributing to the uniqueness of the present economic transformation is the entry of women into the work force on a massive scale. In 1950 about one third of the women held jobs outside the home, now more than two thirds do, and the proportion is growing. This change is directly related to shifts in the economy, especially the expansion of the service sector. Two thirds of the jobs created since 1975 have been taken by women. In fact, it is precisely this great increase in women's paid work that has been a primary factor in the rise of service jobs. These changes, combined with the economic requirements of families (high cost of housing, education, and the like), have made women's labor force participation a permanent phenomenon and have contributed to their greater autonomy.

There are two related problems resulting from the permanent addition of many millions of women to the work force. First, this influx of workers has and is occurring at the very time that the number of jobs in the relatively high-pay industrial sector is declining. Second, women, for the most part, are competing against each other for the expanding number of jobs at the low-paying end of the service/information sector. During the 1970s the service sector of the economy grew by fourteen million jobs and three fourths of these were filled by women new to the labor force. These jobs tended to be temporary, part-time, and low paying. They were not associated, typically, with skill and achievement but rather tended to require obedience, dependence, and repetitiveness. In short, although there are exceptions, most working women have jobs consistent with traditional gender expectations.

These factors—microelectronic innovation, the global economy, and demography—combine to make the present turning point in history unique. The old answers no longer apply, yet few economists and other social scientists are confronting this reality. The thinking of most "has been shaped by the axioms of the Industrial Revolution, and they are unable to entertain the possibility of radical discontinuity with the past. . . . Attempts to apply old remedies to a new situation are as futile as attempting to use a Ptolemaic world-view to interpret an Einsteinian universe" (Jones, 1982:39).

The purpose of this book is to examine the current transformation of the economy, focusing on the consequences, sometimes wrenching, for categories of workers, families, minorities, communities, and the society. There are few other topics as important as these for social scientists, policy makers, and others to ponder, understand, and solve.

REFERENCES

BLUESTONE, BARRY and BENNETT HARRISON. 1982. *The Deindustrialization of America.* New York: Basic Books.

BOCK, GORDON. 1987. "Limping Along in Robot Land." *Time* (July 13):46–47.

BUCKLEY, ROBERT J. 1987. "Economic Armageddon." *Vital Speeches of the Day* 53 (March 15):347–349.

COATES, JOSEPH F. 1986. "Twenty-Twenty Vision." *Vital Speeches of the Day* 52 (June 15):536–540.

DASSBACH, CARL H. A. 1986. "Industrial Robots in The American Automobile Industry." *Insurgent Sociologist* 13 (Summer):53–61.

DOLBEARE, KENNETH M. 1986. *Democracy at Risk: The Politics of Economic Renewal,* revised edition. Chatham, N.J.: Chatham House.

Forbes. 1986. "The 100 Largest Multinationals." *Forbes* (July 28):207.

FORESTER, TOM. 1987. *High-Tech Society: The Story of the Information Technology Revolution.* Cambridge, Mass.: The MIT Press.

HART, GARY. 1983. "Investing in People for the

Information Age." *The Futurist* 17 (February):10–14.

JONES, BARRY. 1982. *Sleepers, Wake! Technology and the Future of Work.* New York: Oxford University Press.

KARMIN, MONROE W. 1987. "Will the U. S. Stay Number One?" *U. S. News & World Report* (February 2):18–22.

KOEPP, STEPHEN. 1986. "Feeling the Crunch from Foreign Chips." *Time* (October 27):72–73.

KUTSCHER, RONALD E. and VALERIE A. PERSONICK. 1986. "Deindustrialization and the Shift to Services." *Monthly Labor Review* (June):3–13.

New York Times. 1986. "Rise of the Robots." *The New York Times* (January 5):3,1.

NOBLE, DAVID. F. 1977. *America by Design: Science, Technology and the Rise of Corporate Capitalism.* New York: Alfred A. Knopf.

NOBLE, KENNETH B. 1985. "U. S. Study Details Employment Shift." *The New York Times* (June 8):1,9.

NOBLE, KENNETH B. 1986. "America's Service Economy Begins to Blossom—Overseas." *The New York Times* (December 14):E5.

OXFORD ANALYTICA. 1986. *America in Perspective: The Social, Economic, Political, Fiscal, and Psychological Trends that Will Shape American Society for the Next Ten Years and Beyond.* Boston: Houghton Mifflin.

POWELL, WALTER W. 1987. "Explaining Technological Change." *American Journal of Sociology* 93 (July):185–197.

SIMONS, GEOFF. 1985. *Silicon Shock: The Menace of the Computer.* Oxford: Basil Blackwell.

SMITH, JOAN. 1987. "Terminal Illnesses." *The Woman's Review of Books* 14 (April):6–7.

VERNON, RAYMOND. 1986. "Multinationals Are Mushrooming." *Challenge* 29 (May–June).

Working Papers. 1983. "An Interview with Bennett Harrison and Barry Bluestone." *Working Papers for a New Society* 10 (January–February):43–51.

ZINN, MAXINE BACA and D. STANLEY EITZEN. 1987. *Diversity in American Families.* New York: Harper & Row.

PART I
The Consequences of the Structural Transformation for the Economy

American society is being fundamentally transformed, as noted in the Introduction, by four interrelated factors: (1) new technologies, (2) global economic interdependence, (3) capital flight, and (4) the dominance of the information and service sectors over basic industries. Since each of these forces is economic, it is clear that the current metamorphosis is at base an economic transformation. This presents some analytical confusion because both the sources of change (the independent variables) and the consequences of change (the dependent variables) are economic phenomena. While the Introduction discussed the economic forces behind the structural transformation, this section focuses on the consequences of the dramatic changes occurring now in the economy.

The American economy has shifted from an economy based primarily on the manufacturing of products to one based on knowledge and services. As Peter Drucker, the esteemed professor of management, has argued: "Knowledge has become the central 'factor of production' in an advanced, developed economy. . . . Knowledge has actually become the 'primary' industry, the industry that supplies to the economy the essential and central resources of production" (1969).

This change from an economy based on industrial production to one based on knowledge is very significant. Whatever the productive force in society is, it affects the mode of production, which determines the way people make their living—which, in turn, affects the way they relate to one another. In short, the fundamental shift from one type of product (manufacturing) to another (knowledge) changes the production, distribution, consumption, and relational patterns in society (Gurley 1978).

The economic transformation is affecting the relative importance of capital. When the United States was an industrial society, the key resource was capital. A lot of people may have known how to build a manufacturing plant, but they were limited because they did not have the money to build and operate one. Now with knowledge being the key resource, those with the specialized brainpower to make and utilize the technological breakthroughs are needed along with the capital to create wealth and power (Bell 1973). Some have claimed that we have moved from a labor-intensive economy (agriculture), to a capital-intensive one (manufacturing), now to one that might be labeled brain intensive (Naisbitt, 1984:6).

The global economy coupled with the ease of capital movement has increased the power of the largest multinational corporations. Their economic power may even supersede the political power of individual nations, which are dependent on the multinationals for the necessary resources, products, and services.

The shift away from manufacturing has meant a dramatic decline in the demand for unskilled and semiskilled workers. Further, the power of organized labor is diminished, and the gap between those with specialized knowledge and training and the relatively uneducated has increased. A fundamental twofold question facing society now and in the foreseeable future is, Will there be enough work in society and will the workers be paid enough so they can afford the consumption of goods and services produced, thus sustaining domestic economic growth (Harrington 1987)? These issues about work and inequality, although fundamental to the economy, are the special topics of Parts II and III.

What are the effects of the transformation on American capitalism itself? Well over one hundred years ago Karl Marx argued that capitalism contains the seeds of its own destruction. Among the contradictions of capitalism that appear especially germane under present conditions is that capitalists in their pursuit of profit make rational decisions that turn the masses into an organized political class bent on a more equitable economic system. The merging of companies now tends to eliminate jobs, both blue collar and white collar. Investing in robotics reduces employment. Plant closings and transferring the operations overseas may aid the capitalists but clearly not their workers. The threats to move or to reduce the work force coerces worker concessions and increases their alienation. Lower wages brings the very real contradiction of increasing corporate profits in the short run but leaving workers without enough compensation to consume the goods and services produced, which reduces corporate profits and brings economic chaos in the long run. What will be the result of these and other economic contradictions? Will capitalism bent on profits without considering the human consequences prevail? And, if so, at what costs? Are the conditions of the current transformation the necessary ones to bring about the demise of capitalism and the rise of socialism, as Marx prophesied?

The changing economic conditions brought about by the new technologies, global interdependence, capital flight, and the dominance of the knowledge/service sector raise some some fundamental issues that must be addressed by the polity and citizenry. Should the new global capitalism be a self-regulatory market economy or one regulated and directed by individual governments and/or multinational agreements? The issue is whether to protect domestic products by restricting imports through embargos and high tariffs. Should the United States government subsidize its basic industries so that they can compete in global markets? What is the proper role of government toward the surplus population (the unskilled and uneducated)? What is the responsibility of corporations toward the displaced employees and communities resulting from corporate decisions to locate elsewhere? The societal responses to these issues clearly have economic repercussions, especially regarding whether society moves closer to a capitalist or socialist economy.

The economic consequences of the structural transformation are considered throughout this book. This section examines five issues: (1) the nature of the economic transformation; (2) the relationship between the new technologies and economic productivity; (3) the current phase of corporate restructuring to be competitive in the global marketplace; (4) the consequences of plant closures on workers and communities; and (5) the fundamental contradiction between capital and community.

REFERENCES

BELL, DANIEL. 1973. *The Coming of Post-Industrial Society.* New York: Basic Books.

DRUCKER, PETER. 1969. *The Age of Discontinuity.* New York: Harper & Row.

GURLEY, JOHN G. 1978. "The Materialist Conception of History," in Richard C. Edwards, Michael Reich, and Thomas E. Weisskopf (eds.). *The Capitalist System,* second edition. Englewood Cliffs, N.J.: Prentice-Hall, pp. 43–49.

HARRINGTON, MICHAEL. 1987. *The Next Left: The History of a Future.* New York: Henry Holt.

NAISBITT, JOHN. 1984. *Megatrends.* New York: Warner Books.

1 *The Nature of the Economic Transformation*

KENNETH M. DOLBEARE

An economic transformation may seem to share some of the characteristics of general decline in that some industries may be starved for investment capital and particularly vulnerable to foreign competition. But the reasons that lie behind the decline of certain industries during a transformation are rooted in some revolutionary technological development and other changes associated with it. And the latter simultaneously lead to the rise of new industries to replace the older ones in a nation's economy. In other words, *structural* change in the nature of the economy itself is involved.

The U. S. economy is undergoing just such structural change today, in three basic ways. *What* we produce is shifting from heavy industry and primary metals to high technology and services. *How* we produce is shifting from labor-intensive and high-volume practices to highly automated and more flexible methods. *Who* does the work is changing also, with many assembly tasks shifted to Third World countries. In the United States, more people are needed in service and technical jobs and fewer in the familiar blue-collar mass production occupations. If not entirely irresistible, the forces making for this structural change are at least very powerful. If the United States does not make changes necessary to keep up with foreign competition, it risks losing the profits and jobs it needs to sustain its economy. We must run very fast, in effect, in order to stay where we are.

Source: Kenneth M. Dolbeare, *Democracy at Risk: The Politics of Economic Renewal,* revised edition (Chatham, N. J.: Chatham House, 1986), pp. 69–73.

It is easy to exaggerate the role of a single factor in this transformation, and to end with dire predictions about the eclipse of the role of human intelligence in production or other work. But it is probably true that among the many new materials and applications involved, the microprocessor or semiconductor chip—a tiny silicon wafer with a myriad of electronic circuits printed on it—lies at the heart of our economic revolution. One Nobel Prize-winning economist was quoted in 1982 as ranking the chip just behind the wheel and equal to the steam engine among history's technological thresholds. It is at once the replacement for giant computers and cadres of scientists and engineers, making possible both manufacturing and other tasks with less material, lower energy consumption, greater reliability, higher productivity—and far fewer workers. We shall someday produce many new products and be able to *do* much more, much of it at lower costs. But along the way there will be profound upheaval for millions of people.

The so-called sunset industries are often actually chains of industries, from producers of raw materials to fabricators of components to manufacturers of finished products. Steel, whose precipitous decline in share of the world's market was cited earlier, is a good example. The U. S. steel industry in 1982 had a production capacity 10 percent lower than it did in 1975, and was still operating at only 38 percent of that capacity. A total of 200,000 jobs, both blue and white collar, had been eliminated since 1965, mostly in the late 1970s and early 1980s. Steel companies were rapidly diversifying out of steel production instead of modernizing their plants. Bethlehem Steel, for example, was only acting in ways typical of many of the older industries when it declared its intent to sell still-profitable plants as a way to consolidate its operations and free capital for new and more profitable ventures.

The problem facing the steel industry was partly intense pressure from the automated competition of Japanese and German companies. This pressure was so heavy that the industry and its principal union went together to the national government in search of protective legislation. A good share of the industry's problems, however, stemmed from the sharp decline in automobile sales experienced by American manufacturers in the late 1970s and early 1980s. Even more came from the fact that the materials going into cars were changing in character. The "downsizing" of American-made cars and trucks came relatively late, but it arrived with a vengeance. The steel content of the average Ford car, for example, was cut by 31 percent between the years 1977 and 1982. In 1976, the U. S. steel industry sent about 25 percent of its total output into automobile production, but by 1981 this figure was 15 percent of a much lower total. Much of the impetus for this change came from the national government's efforts to deal with the oil shortages of the mid-1970s. In the words of a leading business journal:

> Congress, in legislating fuel-consumption standards for American cars, committed itself to a radically different industrial structure. You downsized the

American car to save gas, and you wound up downsizing not only the U. S. auto industry itself but a substantial part of the industrial base that supported it. You saved gasoline. And you destroyed blue-collar jobs.

Other industries also were affected by the downsizing of American cars, as well as by the sharp inroads made by foreign imports in total sales. Table 1.1 gives some idea of the scope and economic meaning of weight reductions accomplished by Ford as it moved to meet fuel efficiency standards set for 1985 models. Not only steel but also iron (and behind them both, coal) and rubber were seriously affected by the changes, as attested by a wave of plant closings in the Midwest.

These data make clear that the unemployment and borderline bankruptcy problems of the industries centered on automobile production will not go away with economic recovery. Indeed, recovery is likely to make the sunset nature of these industries (and the jobs they used to provide) even more obvious. A good proportion of the remaining American automobile production is likely to be accomplished through fewer plants but much more fully automated methods than ever before.

This combination of energy-induced downsizing, the substitution of newly developed materials, and automated production processes is evident in other industries as well. Housing, for example, has turned toward construction of smaller residences and increased use of lighter, prebuilt materials. Some of

Table 1.1 The Changing Ingredients of the Average Ford Car

Material	*Pounds (Dry Weight)*		
	1977	1982	1985
Plastics	165	224	225
Aluminum	110	133	135
High strength steel	105	252	270
Cold rolled steel	820	510	490
Hot rolled steel	1419	864	760
Cast iron	620	352	315
Glass	93	74	67
Rubber (synthetic, natural)	180	129	120
Sound deadeners	85	46	42
Copper and brass	35	32	25
Lead	29	28	25
Zinc die castings	34	10	10
Other	65	50	49
TOTAL WEIGHT	3760	2704	2533

SOURCE: Reprinted by permission of *Forbes* Magazine, November 22, 1982. © Forbes Inc., 1982.

this was intended to improve fuel consumption, and some was the result of changes in interest rates and the multiplication of investment opportunities alternative to the traditional home mortgage. But the effect on the lumber, construction, cement, glass, and copper industries was likely to be substantial.

The "sunrise" industries that many hope will replace these long-established building blocks of the American economy, at least in the value of their products and sales, are nearly all in the high-technology and services fields. In manufacturing, the leading candidate is of course the electronics industry: computers, communication equipment, medical instruments, fiberoptics, bioengineering, and robots. In each area, most of the inventions, ideas, and innovations that made the industry possible were American in origin. But much of the actual development and many current innovations are in fact Japanese accomplishments.

What once was a clear American edge is now much less certain. In a study conducted in 1982 by Japan's Society of Science, Technology and Economics, Japan was found to be inferior to the United States in 56 key technologies but superior in 51. Given the intensity of Japanese government organization, planning, and support for industrial improvement, this does not bode well for a continued American edge. The U. S. share of world trade in high-technology products, in fact, has already dropped from 30 percent in 1962 to 22 percent in 1982. Japan's rose from 4 percent to 12 percent in the same period.

The robotics industry neatly sums up the problems and prospects in the future high-technology world. The first patents were those of an American inventor, but only Japanese companies would undertake development. Currently, robots are in greater use in Japan and the Japanese robot-producing industry has a significant edge on American producers. Americans are, however, investing heavily to catch up and perhaps hold on to much of the growing U. S. market for robots. Currently only about 5000 robots are in use in the United States, mostly in the automobile industry. But anywhere from 100,000 to 150,000 are expected by 1990. When a variety of "smarter" robots are available, of course, hundreds of thousands of current jobs will be eliminated. Experts disagree sharply about the net employment effects of the imminent arrival of widespread robotization, but few expect the people displaced to become the makers or servicers of the complex electronic robots. At best, their children might.

The other major area of employment expansion expected as part of our economic transformation is in services, a category that includes more than half of all current workers. Some of this increase is expected to occur as the new technological capabilities are brought to bear on current practices, such as in reorganizing manufacturing procedures and much increased data processing. The "knowledge society" will in effect develop many new jobs by eliminating many more existing jobs. A good share of the increase in services, however, is expected to occur in fast-food restaurants, hospitals, and other areas of traditionally low pay and marginal job security.

Enthusiasm for an "information society" booming with new service occupations has been tempered by critics who argue that long-term prosperity depends on a healthy manufacturing sector. Acknowledging that the service sector has grown rapidly in recent years, they still insist that no economy can sustain itself simply by passing papers, talking, and eating. One such critic sardonically comments:

> Perhaps unemployed steel workers (or their children) can sit at computer terminals making American Express travel arrangements for a convention of paralegal personnel, systems analysts, and economists with a taste for McDonald's hamburgers.

Other second thoughts about the ongoing transformation have come from observers who argue that "smokestack" industries are needed to provide at least a residue of jobs and productive capacity in those fields. Giving up on such industries would place an unsustainable burden on other sectors and perhaps endanger national security. The romantic appeal of high-technology occupations far exceeds the real prospect of employment in those fields, and most people will have to hope for jobs in more familiar areas. The problem will be to find such jobs as industries' demand for labor declines. Not only are massive redistributions of job opportunities among middle- and lower-class workers involved, but also the question of maintaining an acceptable level of employment and some balance among industrial capabilities.

2 *The New Technologies and Economic Productivity*

OXFORD ANALYTICA

Western economies are beginning to feel a rising and long-term tide of applied science and technology advances. Modest assessment of its impact foresees significant effects in productivity and employment. More lyrical observers predict a cultural revolution, the groundwork for a renaissance of vitality in individual achievement and human fulfillment. Over the next decade the growing effects will be confined to new products, employment, productivity and technology-provoked optimism.

Lasers and fiber optics will have a continued economic impact by refining production processed in a very wide variety of applications and as components in telecommunications systems, information processing, and entertainment (videodisc and holography, including 3-D television). As capital goods for a relatively small market or as components of consumer goods, lasers are likely to continue their function of refining processes and improving products; thus, they will substitute for inferior goods rather than create many new products or new employment by the late 1980s. Fiber optics may generate considerable investment by enabling the creation of more efficient, larger-capacity communications networks.

Robotics and computer-controlled machines similarly hold potential for improved industrial productivity. Computer-aided design (CAD) and computer-aided manufacturing (CAM) will have a growing impact on production efficiency through the 1990s. Computer-controlled machines are substituting for conventional machine tools, and American technology superiority is well poised to increase U. S. share of world markets in these capital goods. The main impact, however, is likely to be on productivity and in displacing rather than creating jobs.

Likewise, the use of industrial robots will increase dramatically through the decade and will commensurately improve productivity and U. S. competitiveness. The rate of substitution will depend on alternative employment opportunities (some observers predict larger "replacement" industries in servicing and programming robots), the cost (unemployment compensation, social unrest) of additional unemployed workers, and union success in negating the gain in productivity from robots by demanding retraining or wage raises as work time diminishes. On the other hand, the use of robots by foreign industrial competitors, particularly Japan, may require that U. S. industry conform or seek protection with renewed urgency. The overall job-displacing effects of robots will not be dramatic in the next decade, though they will be significant in certain industries.

Finally, the use of computer-aided techniques of management (e.g., management information systems and computer-aided planning) will coordinate and enhance CAD and CAM to create CIM, *computer-integrated manufacturing*. These powerful tools will increase overall productivity, allow flexible production runs and greater control, and lower inventory levels. On the other hand, they will change the nature and organization of the workplace and influence levels of job satisfaction, stress, skills and productivity; CIM is not a panacea for problems in manufacturing. Applications are spreading, but aggregate use will be relatively limited through the 1980s. They will be first felt in metal-working industries in the northeast, central and mid-Atlantic states. The impact on the number and kinds of jobs will begin to be large in the 1990s, and the long-term potential impact is enormous.

Solar energy increasingly seems likely to be cost effective in the 1990s for the direct production of electricity by photo-voltaic conversion. With large-scale production of conversion equipment, such methods will be competitive unless oil prices falter. Commercial use of solar cells is fairly imminent as large private investment, mainly by oil companies, reduces the lead times. The most promising new process uses amorphous or non-crystalline (and thus cheaper) silicon printed onto thin plastic film. Mass production (with cost per kilowatt hour as cheap as oil-fired or nuclear electricity) for a wide variety of uses could occur by 1990. If so, this first real alternative to conventional energy sources holds great potential for transforming the economy and perhaps even society. However, the economic impact even in the early 1990s is likely to be conventional in reducing or containing rises in energy costs, as elements of total costs of production. The substantial economic effects would occur in the later 1990s.

Of much greater economic significance for the next decade than the above new technologies are of course *biotechnology and micro-electronics*. These industries are widely expected to form, in this decade, the industrial foundation for the 1990s and beyond, when joined by new energy sources. The economic impact will steadily affect productivity and jobs.

Biotechnology is likely to bring dramatic changes over the next fifteen years to a wide range of products and processes in a number of industries. The use of single-cell protein, enzymes, and various bacteria will continue to be more widespread in foods, agriculture, fuel, effluents treatment, natural resource recovery, pharmaceuticals, chemicals, and medical applications. For instance, enzyme technology could generate an entirely new type of chemical industry utilizing moderate conditions of pH, temperature, and pressure to produce complex chemical and renewable feed stock from wastes; it is likely to be cost effective before 1990. Moreover, the less conventional techniques of microbial genetics, such as protoplast fusion, gene amplification, and recombinant DNA technology, are likely to have significant commercial impact in the coming decade. The bacterial synthesis of human insulin, human growth hormones, and interferon is nearly commercially viable.

Genetic engineering in agriculture, however, is not so likely to have large impact on the national economy before 1995. Commercial achievement is expected in accelerated photosynthesis, acid-tolerant crops, herbicide resistance, and grains that fix nitrogen directly from the atmosphere, but this research will not be complete for at least ten years. Before that, recombinant DNA technology may lead to improved crops, new crop types, and techniques to manipulate plants. Many industrial laboratories are shifting emphasis and many others have been established in the past two years to exploit the large biotechnology opportunities in agriculture. In this and in industrial micro-biology generally it is clear that the surface of potential application and economic effect has been merely scratched. Minor plant improvements will occur in the 1980s, but economic effects of major breakthroughs will wait for the later 1990s.

Micro-electronics will continue its revolution as the world's fastest-growing industry as computer power continues to grow and as micro-processors spread to virtually every electrical product, commercial or consumer. While the trend holds profound implications for medicine, transport, education, and household life, *the area of greatest economic impact in the next decade is likely to be telematics, the information economy*. This is the development of a wide range of micro-chip-based systems of information processing combined with communications and control technologies. The technology is based on distributed processing to meet local, regional, or central decision-making, and therefore on distributed access.

High-performance computing is expected to generate growing excitement over the next decade but limited effect on the national economy. Industry specialists expect supercomputers to hold potential for profound social and economic change in the long run. New ideas, techniques and ways of building computers aim at more powerful, faster processors and memory chips, and machines that function in entirely new ways in order to think and learn. Systems that use "parallel processing"—allowing many operations simultaneously—would enable artificial intelligence applications: distinguishing between fragrances; vision; reading, hearing and speaking natural language; and the ability to reason and make judgments. "Fifth generation" or "artificial intelligence" computers would be simple for an untrained person to operate and could solve a range of problems that are not readily handled by existing numerical computers.

The applications are wide and important. An immediate use might be in expert systems, programs which mimic problem-solving abilities of human experts by encapsulating human experience, like medical diagnoses or financial advice. Machine vision would aid robotic assembly and military uses to guide land vehicles and aircraft. Supercomputers would be capable of being mass-produced and general-purpose enough for use in a very broad market by 2000. They are expected to enter economic sectors and social activities that have been little affected, thus far, by computers.

Certainly computers have more potential to affect a wide range of activities than do most other technologies, such as electronic chips, communications, biotechnology or even automobiles. Understandably, there is severe international competition for supremacy. The U. S. is still judged the world leader, with Japan threatening to overtake and Europe and Russia lagging behind.

Dozens of American universities and corporations pursue research on "fifth generation" computers. The U. S. government has boosted funding, mainly for military uses. American industry efforts involve mainly a cooperative R & D effort called Microelectronics and Computer Technology Corporation, involving about twenty large corporations. Current abundance of funds is allowing researchers to pick up the pace of research. There is a major U. S. initiative to retain primacy in the marketplace for computers, and it has new urgency.

Japanese projects to extend the limits of computing power aim to overcome U. S. dominance in the field. With a consortium of companies and government laboratories guided by the Ministry of Trade and Industry, Japan expects to build the fastest computer by far by 1990 and one with advanced artificial intelligence capabilities by 1992. Commercial models should be available in the 1990s. Already Japanese companies have high-performance computers that appear comparable to U. S. machines. If the U. S. falls behind in development and application of high-performance computers, it is likely to lose supremacy in many areas of scientific research and technology, and suffer in economic and national security. In addition, supercomputer research could help Japan gain leadership in the multi-billion-dollar business and personal computer field, as well as in development and promotion of many new products.

Supercomputers, artificial intelligence and the race with Japan are unlikely to have much effect on the U. S. economy by 1995. In the long term these topics will hold wide impact on jobs and markets, as well as U. S. image. However, even before the actual effects, news about these topics will affect

American life. They will introduce a new note of competitiveness, stress for high stakes in an area about which the U. S. has never had to worry. The National Academy of Sciences concluded that the U. S. is well positioned to retain its computing lead, that Japanese supercomputers are equal to but not greater than America's, and that there will be "a very closely run race in this area." The race for computer supremacy could determine which nation will dominate the world in economic, military and scientific fields in the next century. Certainly it threatens U. S. leadership position in computers, the field which, above all others, has come to symbolize the nation's technological pre-eminence.

TECHNOLOGY AND PRODUCTIVITY

There are *several general reasons for caution about the impact of technology on productivity in the next few years.*

First, there is a long time lag between innovations and their development into new products or processes. Even then, further development is necessary before these innovations can reach a mass market. Robots, for instance, are at present available to perform a few standard operations, and in practice have had to be designed for specific customers.

Second, even with innovations successfully embodied in available new equipment and accepted as more efficient than existing plant, it seldom pays immediately to scrap the latter. This is because the existing plant has generally already been paid for: thus, though its efficiency is lower, so too are its costs, compared to new equipment. Hence new types of equipment tend to get introduced over time, rather than as soon as they become available.

A third factor which will slow down the rate of adoption of innovative equipment will be its effect on the workforce. This is not only because, if it is labor-saving, it may displace existing employees. It is also because it is likely to require the development of new skills on the part of existing workers (e.g., reprogramming robots, using word processors), or to require workers with specific skills (e.g., computer programmers) who may be in short supply. Less specifically, but more important, there will be a quite long-run "learning process" as attitudes towards, coordination of, and facility with new methods gradually increase. An example would be management use of information systems. In the long run we can certainly imagine enormous savings in managerial resources as information processing and relatively standard decision-making become computerized. But the short and medium term will be a time of digestion; it is not impossible that there will be an *increase* in management to deal with the greatly increased amount of information available. Businesses will often be unclear about how to use this information in a sensible way, and as established patterns and subjects of communication break down, there will be hidden costs as relationships based on understandings, and procedures built up over long periods, become strained.

Fourth, the benefits of adoption of new methods in one area will in a number of cases depend upon the adoption of new methods elsewhere. For example, many gains in finance, real estate, and retail trade will depend upon interactive communications equipment with processing facilities in households. Related to this problem of "linkage" is that of standardization. This is most acute at the moment in computer software and peripherals. But it will become important in robots and some aspects of bio-engineering.

A final factor which will moderate the aggregate impact on productivity, at least initially, is statistical. Only certain parts of certain sectors are at the moment liable to major changes as a result of the new technologies. These include some manufacturing industries, both process and assembly line production, and some parts of the service sectors. Important industries (chemicals, autos, finance, retailing) are in this group, but together hardly represent more than 20 percent of GNP. So even if productivity growth doubled in those sectors, the aggregate impact would not be great.

In view of this, how much faith in technology's positive impact on the future is justified?

Broadly and immediately, there is a widespread notion that technology will offset or even solve the current structural problems of the U. S. economy; that is,

- that it will increase productivity, offset the disadvantage of relatively high U. S. wages, and so restore the U. S. competitive edge vis-à-vis foreign competitors;
- that it will open up new markets, both for capital and consumer goods, both in the U. S. and around the world, and so lead to a new boom;
- that it will utterly transform U. S. society, giving power for decisions to a new technocratic elite, which will be free from the errors of its predecessors and so will invest heavily in new technology of all kinds, creating a new "post-industrial," "communications" society, where there shall be no sorrow nor sighing, and the "former things shall pass away," like boring or dangerous jobs.

Are these observations accurate and their implications reliable? How positive, and how rapid, will the net effect of this wave of structural change be? It seems probable that the effect will be fairly rapid and on balance positive. But to say that is quite a different thing from some of the more excited predictions that have been current recently, not only in journalistic media but among industrial and business "futurologists" as well.

The three reasons summarized above for thinking that the advent of new technology will transform the U. S. economy roughly indicate bounds for economic effect.

First, the U. S. loss of competitive edge is not mainly due to higher U. S. wages. Some countries that have challenged specific U. S. industries most successfully (West Germany, Sweden) have higher average wages than the U. S. Moreover, some foreign competitors (notably Japan, but in specific sectors many others also) are introducing the new technology at least as fast as the U. S. There is no reason to think of the micro-processor generation of technology as peculiarly American, still less to imagine that it will help American business exclusively, to the disadvantage of competitors.

To the extent that American business actually does take advantage of the new technology, whether because of receptive attitudes, abundant sources of capital, or the availability of a large pool of computer-numerate labor, that technology does offer an opportunity to remedy some weaknesses that have developed in the competitive position of U. S. industry. But the remedy is not automatic, nor is it only on offer to American businesses. And it could even work against American businesses.

Second, the new technology will indeed open up new markets, both for consumer goods and for capital goods. It will not do so exclusively for U. S. manufacturers, even in the U. S. domestic market, let alone abroad. The new technology, it must be remembered, will at the same time be closing down old markets and making old products obsolete. How much larger the new markets will be than the ones they replace still needs to be established. On balance we think that substantial new markets will be opened up. But the experience of the past decade is not wholly reassuring. It has seen, for example, Japanese manufacturing overcoming an early American lead and penetrating the large American home market to take the lead in the new technology of video cameras and video recorders, as the Japanese had earlier demonstrated their ability to sell cameras, motorcycles, machine tools, television sets and automobiles in the U. S. market.

Third, employment effects over the next ten years are likely to be mixed. Overall there will not be massive unemployment. As the baby-boom generation ages, there will be fewer new entrants to the labor force, and some loss of jobs by automation will simply coincide with easing demand for new jobs. In addition, higher productivity, greater competitiveness and expanding markets for the new technology-induced products are expected to cause economic expansion and create new jobs, especially in services; employment will continue to shift from manufacturing to services. Finally, there will be upgrading and "skill-twist" among and within specific sectors. For instance, demand for clerical workers will decline while demand for middle and upper managers and sales personnel, jobs that require more education and training, will rise. Another example might be with robots: assembly-line jobs will be eliminated while (fewer) robot maintenance jobs will be created, but the latter will require considerably more technical expertise than the former.

Fourth, it is doubtful whether new technology will transform society as fast or as comprehensively as many prophets would have us believe. The history of innovation in the nineteenth and twentieth centuries suggests that while the market is swiftly saturated by useful and economically superior new products, not all new products turn out to be as competitive or economically superior as their sponsors at first hoped; and traditional products, processes, and methods of work organization will continue to overlap with new products and techniques for far longer than might be expected. The long run will see technology's revolution; to the mid-1990s the main social effect will be discussion of the coming changes. The social revolution is a long way off and is more likely to be a transformation anyway.

What lessons can we draw from the strictly economic effects of new technology?

First, to be realistic. Americans will need to be wary of the idea that today's "new technology" is utterly different in kind from previous generations of innovation, or that it marks a more fundamental break in the continuity of industrial and technical development than previous innovations such as, for example, the electric motor or the internal combustion engine.

Computers, like micro-processors, are machines, albeit extremely useful machines with an exceptionally wide range of applications. Their introduction, like that of any new generation of prime mover, will have complex and far-reaching economic and social effects. In economic terms, these effects will include (among many others) a sharp fall in certain types of employment, and an increase, numerically more modest, in other, higher-paid work (on average); an increase in productivity in many areas (for example, warehousing, accounting, design); a rapid increase in demand for certain new consumer goods; accelerated obsolescence of certain types of capital goods and their replacement by new types (robotics, computer-controlled machine tools, etc.); rapid advances in new techniques in communications, and changes in the pattern of office and plant work to take advantage of these.

In short, the new technology in micro-electronics, communications and control engineering offers the prospect of perhaps unusually rapid structural change—a process which has already begun, in part because of the progress the new technology has already made in the U. S. and elsewhere, especially in Japan, and in part for other reasons, such as the increasing dependence of the U. S. on imported minerals and fuels.

Second, micro-technology has certain tendencies and consequences which seem to run sharply counter to the traditional patterns, and to the traditional strengths, of the U. S. economy.

The U. S. economy, during its period of greatest success and supremacy, from around 1880 to 1960, was characterized by the following interconnected factors: wide dissemination of "medium" technology; mass production; relatively high wages; quick and universal distribution of new consumer products; low markup; high volume. A car in every garage, then two. A chicken in every icebox, then a dozen in every freezer.

Up to a point, computers can be treated as a new generation of consumer goods: microwave ovens, Atari games and home word processors replace stoves, Monopoly and typewriters. But there are comparatively limited markets for such substitutions. It is a very open question whether American-manufactured capital and consumer goods made possible by the new technology will really find markets in the U. S. or abroad that will compensate for lost markets for intermediate technology, let alone generate a new, high-technology boom.

In other words, as manufacturing in the intermediate technologies shifts to the newly industrialized countries (NICs), will high-technology production in the U. S. compensate? Even if it does not do so in volume, it still could help to improve competitiveness in manufacture of lower-technology products and thwart the shift to NICs. But in due time NICs will acquire high-technology processes and still out-perform the U. S. in the traditional or labor-intensive elements of production. The net effect is by no means favorable to the U. S.

Moreover, the U. S. will face stiff competition from Japan and possibly from Europe and the NICs in the high-technology sectors proper. Even if products which embody micro-technologies become widespread, as seems realistic, this does not assure a bright economic and social future for the United States. Japan could well beat the U. S. to dominate the new sectors at the upper end (semi-conductors, supercomputers, etc.), and the NICs beat it at the lower end. High technology can still improve U. S. productivity, but only in line with all other countries; but the new technology for industry in America and other countries might well tend to come from Japan. American complacency about its lead in technology will have evaporated by the 1990s.

3 *Corporate Restructuring*

GEORGE RUSSELL

It is known as downsizing, rationalizing, streamlining and, perhaps most commonly, restructuring. With a bow to the diet culture, some prefer to call it just plain slimming down. By whatever name it goes, a compulsion is sweeping through corporate America to bring about fundamental, long-lasting changes in the way it does business. U. S. corporations have always undergone periodic cutbacks in times of recession or strain, but this time the tone and scope of the effort are vastly different. Says Keith Stock, a partner in the Manhattan-based McKinsey & Co. management consulting firm: "What we're seeing is nothing less than a transformation of American industry."

Forced upon business by unprecedented global competition and financial turbulence, the change is so swift and powerful that it is churning across the business landscape with the force of an army of bulldozers. American companies have started the huge task of rebuilding themselves from the ground up, erecting a sleek new operating architecture to replace the unwieldly processes of the past. At corporate headquarters and on factory floors from New York City to Los Angeles, newly cost-conscious executives are on a relentless examination of the efficiency and effectiveness of everything they do. They are tearing up organization charts, selling off unsatisfactory product lines and closing down unprofitable plants at a rate never seen before. Their aim: to produce stream-

Source: George Russell, "Rebuilding to Survive," *Time* (February 16, 1987), pp. 44–45.

lined, combative concerns that can withstand the frenetic, competitive pace of the late '80s.

The task has the general aim of sharply cutting back on costs to make dramatic and durable improvements in long-term profitability and growth. Restructuring's theme is "back to basics." That means, among other things, an end to the corporate ethos of expansion for expansion's sake. It spells farewell to the notion, always more imagined than real, of the corporation as a kind of private-sector welfare state, with unlimited perks and unshakable job security. It also involves frequently deep retrenchment, as U. S. corporations cut back on marginal operations, strip away unnecessary layers of management and staff and refocus their attention on proven areas of profitability. Says James Brown, an executive director of the Conference Board, a business-sponsored research group: "Everybody is cracking down."

So far in the '80s, well over half of the names on the FORTUNE list of the 1,000 largest U. S. corporations have undergone some form of significant reorganization. Gulf & Western in the past four years has spun off some 65 diverse subsidiaries worth more than $4 billion. IBM has closed three domestic plants, cut back on employee overtime, and is reducing its U. S. work force 7%, to 225,000, through attrition and early retirement incentives. AT&T last year cut 32,000 out of a work force of 322,000, in an effort to save $1 billion annually. Among the jobs lost were 11,600 management positions. Many of those who left were coaxed along by payments of up to a year's salary.

United Airlines announced two weeks ago that it would cut about 1,000 employees, or more than a quarter of the Chicago headquarters staff, as part of a program to save $100 million in 1987. Battered USX, which lost $1.83 billion in 1986 and fended off the predations of Raider Carl Icahn, last week said it would shut down four steel plants and lay off about 4,000 of its 22,000 active steelworkers. Something of a storm was stirred up last week when the New York *Times* reported that CBS, which has already pruned some 1,200 of its 15,500 employees, would ask its news division to slice $50 million from its $300 million budget. That draconian figure was denied by Chief Executive Officer Laurence Tisch, but the company admitted that it was still looking for ways to improve efficiency. Hundreds of other large corporations are planning or already carrying out slimming-down programs, including Exxon, Union Carbide and Time Inc.

What is remarkable about the current cutbacks is that they come at a time when the gross national product is expanding at a respectable 2.5% annual rate. The goal of the slimming exercise, then, is not merely to compensate for hard times—though a quick fix for short-term profits is always welcome—but to have a permanent effect. Says Alfred Rappaport, an accounting professor at Northwestern University's Kellogg Graduate School of Management: "Restructuring will be a way of life for a long, long time." Reason: the forces that prompted the movement are still growing in power and momentum.

Chief among those forces is foreign competition. In addition to traditional rivals in Europe and Japan, American companies face an ever expanding roster of formidable competitors in developing countries from South Korea to Brazil. By late 1986, imports amounted to 14.5% of the GNP, up from 10.6% in 1982. One of the industries that has been hit hardest—and made the most radical adjustments—is autos. Struggling General Motors, where profits declined 26% to $2.9 billion last year, has laid off 6.5% of its 578,000 workers since 1981 and announced plans to close twelve major plants by 1989. At the same time, GM will reduce the number of managers and other salaried workers by 25% by the fall of 1989. Similar moves are under way even at Ford, which earned $3 billion in 1986 to overtake GM as the most profitable American automaker. Ford plans to cut its salaried payroll by about 20% by 1990.

Another force behind restructuring has been the avalanche of corporate mergers and acquisitions. More than 4,000 of those unions, worth a record $190 billion, took place last year. After most of the buyouts, the merged company eliminates staff duplications and unprofitable divisions. In the past six years, for example, General Electric spent $11.1 billion to buy 338 businesses, including RCA, a $6.3 billion acquisition. During the same period, GE shed 232 businesses worth $5.9 billion and closed 73 plants and offices.

The buyout spree has created yet another powerful incentive for restructuring: fear of takeover. In many cases, corporations have fought off raiders only by buying up huge amounts of their own stock, and along the way accumulating huge amounts of debt. Once the threat has passed, firms have been forced to restructure to regain profitability. In other cases, they have slashed costs and boosted profitability precisely to keep their stock prices above the level at which they would attract bargain-hunting takeover sharks, who are likely to chop far more brutally and indiscriminately than the present managements. No less a titan than ITT warily shook off a takeover bid by Raider Irwin Jacobs in 1985. That effort gave renewed impetus to a slimming exercise already begun by ITT Chairman Rand Araskog. Since 1980, Araskog has sold off more than 100 businesses, and last year he cut ITT's work force by 100,000, or 44%, and slashed headquarters staff from about 850 to 350. Says Araskog: "Corporate executives have to learn to do things for themselves. Pick up the telephone if it rings. Draft their own memos."

The corporate fitness trend is cresting at a time when some Government officials have taken pointed aim at businessmen for their inefficient ways. Last November, Deputy Treasury Secretary Richard Darman stirred controversy when he used the terms bloated and corpocracy to describe the U. S. business hierarchy. Darman's epithets rebutted executives who blamed federal tax and budget policies for problems with U. S. competitiveness. Both Darman and other officials, however, acknowledge that Big Business is changing its ways. Robert Ortner, chief economist for the Commerce Department, acclaims the present restructuring efforts of corporate America as "amazing."

Amazing, perhaps, but like any radical surgery, however necessary, inevitably painful. The new leanness of U. S. business means, above all, a crackdown on heavy payrolls. A large portion of the layoffs from restructuring have taken place in manufacturing. From 1979 to 1986, total U. S. manufacturing employment declined from some 21 million jobs to 19.1 million. But partly because of this slimming down, U. S. manufacturing productivity—hourly output—has risen by an average of 3.8% annually over the past five years, compared with 1.5% in the '70s. But no such productivity improvement is yet evident outside of manufacturing. Says Treasury's Darman: "We have to make ourselves more efficient in the service sector."

High corporate rank has provided no immunity from the restructuring effort that has taken place so far. "The efficiency problem," Darman points out, "is a white-collar problem even more than a blue-collar problem." Between 1983 and 1987, some 600,000 to 1.2 million middle- and upper-level executives with annual salaries of $40,000 or more lost their jobs. An additional 200,000 to 300,000 such executives are expected to receive pink slips over the next two years.

At many once paternalistic companies, the cost cutting has produced stunning changes in the corporate culture. Eastman Kodak, which has always prided itself on being a home away from home for its workers, has closed its employee bowling alley and billiard rooms, and no longer provides dinners with dance bands. Reluctantly abandoning its virtual guarantee of job security, the company trimmed away nearly 13,000 of its 129,000 employees last year as part of a program to save $500 million annually. Says Kodak Chairman Colby Chandler: "The principal object is to make the company more agile, more competitive and more flexible."

Even with those goals in mind, cutting even one job, says AT&T President Robert Allen, is "painful." Cost cutting can also hurt the companies if it is done sloppily or with too little thought to the future. Robert Reich, a Harvard professor of political economy and management, cautions against "slash-and-burn management" that sacrifices employee loyalty and teamwork with an eye only to short-term profits.

Executives and academics agree, though, that most companies have no choice but to shape up. Says General Electric Chairman John Welch: "The managers in the 1980s who hang onto losing business ventures for whatever reason won't be around in 1990." A less somber view is that the corporations that rid themselves of bureaucratic excess now stand to be among the healthiest entrants in the strenuous competition of the future.

4 The Contradiction Between Capital and Community

BARRY BLUESTONE AND BENNETT HARRISON

Deindustrialization does not just happen. Conscious decisions have to be made by corporate managers to move a factory from one location to another, to buy up a going concern or to dispose of one, or to shut down a facility altogether. These things never happen automatically nor are they simply a passive response to mysterious market forces. The planning behind such decisions is usually intricate, often costly, and extensive.

Deindustrialization is the outcome of a worldwide crisis in the economic system. The very successes of the long postwar expansion generated conditions that ultimately turned the normal, and often healthy, disinvestment process into a torrent of capital flight and wholesale deindustrialization. During the boom years, U. S. economic expansion abroad generated enormous short-run profits, but in the course of doing this it helped to establish excess (unused) productive capacity in one basic industry after another. Through their multinational subsidiaries and the profitable sale of patents and licenses to foreign enterprises, the leading American firms even helped to generate their own future competition. In the 1970s this competition came back to haunt them in virtually every major industry: steel, automobiles, shipbuilding, and electronics, to name a few.

With no rational way to divide up the international market, U. S. firms found themselves subject to intense world competition and as a consequence,

Source: Barry Bluestone and Bennett Harrison, *The Deindustrialization of America* (New York: Basic Books, 1982), pp. 15–21.

shrinking profits. One possible reaction to this situation would have been to try to meet the new competition in the old-fashioned way—an active search for new markets, increased research and development, and investments in more efficient technology. Some American firms took this route, but many more decided instead to abandon the competition altogether (as in electronics), to reduce their investments (as in steel), or to focus all their energies on reducing labor costs and circumventing public sector taxes and regulations. In a desperate attempt to restore, or preserve, the rates of profit to which they had become accustomed in the halcyon days of the 1950s and 1960s, American corporate managers in the 1970s went to extraordinary lengths to shift capital as rapidly as possible, from one activity, one region, and one nation to another. In the process, the industrial base of the American economy began to come apart at the seams.

To be sure, capital flight has always been a tactic that management wished to have at its disposal in order to "discipline" labor and to assure itself of a favorable business climate wherever it set up operations. But only in the last two decades has systematic disinvestment become, from management's perspective, a *necessary* strategy, and from a technological perspective, a *feasible* one.

It is crucial to view this development in the light of post World War II economic history. It was not only international competition that was threatening corporate profits. The postwar series of labor victories that successfully constrained the flexibility of management by regulating the workplace and forcing the corporate sector to underwrite part of the costs of the "social safety net" also contributed to the profit squeeze.

From the middle of the 1930s to the 1970s, organized labor in the United States won major concessions on a broad set of issues that ultimately limited capital's flexibility in its use of labor. One indicator of this loss in flexibility and of the subversion of unquestioned managerial discretion can be found in the sheer size of the contract documents negotiated between unions and employers. The initial agreement between the United Automobile Workers (UAW) and General Motors (GM), signed in 1937, covered less than one-and-a-half pages and contained only one provision: union recognition. By 1979 the UAW-GM contract, with its extensive array of provisions covering each of the company's 140 production units, contained literally thousands of pages printed in proverbial small type. In exacting detail the contract specifies hundreds of items from wage scales and a cornucopia of fringe benefits to limits on subcontracting and the pacing of each machine and assembly line; it even goes so far as to establish some rules governing the introduction of new technology.

Each of these rules and regulations was forced into place by labor for the explicit purpose of increasing job security and limiting the discretionary power of management. With the important—indeed absolutely critical—exception of limiting the right of management to reduce the aggregate size of its labor force, these incursions by organized labor were highly successful. As long as management *had* to deal with labor where workers were well organized, it was constrained to operate within the set of rules that unions had long struggled to secure. Moreover, using the power of the State, labor won important concessions from industry through the regulatory process. Minimum wages, fair labor standards, occupational health and safety provisions, equal employment opportunity, extended unemployment benefits, and improvements in workers' compensation constitute only a partial list of the gains made by labor during this period. Taken together, these victories limited management's ability to extract the last ounce of productivity from labor and thus the last ounce of profit.

During the heyday of American economic power, from 1945 to about 1971, industrialists were able to reap healthy profits while affording these concessions to organized labor. The so-called social contract between labor and management even proved advantageous to the corporate sector, for it assured some semblance of labor peace needed for continued economic expansion. Corporations did not complain as bitterly in the early 1960s when they were earning an average annual real rate of return of 15.5 percent on their investments. Their attitudes changed dramatically, however, when profits began to slip near the end of the decade. By the late 1960s, the profit rate for non-financial corporations had declined to 12.7 percent. It fell further as the result of increased international competition. By the early 1970s, the average rate had declined to 10.1 percent, and after 1975, it never rose above 10 percent again.

Management found that it could no longer afford the social contract and maintain its accustomed level of profit. Instead of accepting the new realities of the world marketplace, one firm after another began to contemplate fresh ways to circumvent union rules and to hold the line on wages. Of course, labor was not initially ready to concede its hard-won victories; therefore to accomplish its goal of reasserting its authority, management had to find some mechanism for disarming organized labor of its standard weapons, the grievance process, various job actions, and work stoppages. The solution was capital mobility. If labor was unwilling to moderate its demands, the prescription became "move"—or at least threaten to do so. For one enterprise, this entailed disinvestment. When entire industries adopted this strategy, the result was deindustrialization.

The capital mobility option had always been available to some extent. The early American canal systems and the coming of the steam locomotive allowed the transfer of production to new communities along the new transportation routes. What is different today is the distance and speed over which that transfer can take place. Satellite-linked telex communication and wide-bodied cargo jets provide a technical environment that has allowed production to become far more spatially "footloose" than ever before. The linking of communications systems to computers permits central management to coordinate worldwide operations at lightning speed, while jet aircraft permit the movement of physical commodities at near the velocity of sound—a far cry from the 3.2 mile per hour of barge traffic on the Erie Canal in the 1840s.

The capital mobility option provided by the new technology has shifted the fulcrum of bargaining power in favor of capital to an unprecedented degree.

It gives employers the ability to effectively insist upon smaller wage improvements or, as has been seen in an increasing number of core industries, actual wage rollbacks. In essence, the capital mobility option provides industry with the power to make "take it or leave it" propositions stick. There are some signs in the auto, steel, and rubber industries that the strategy is actually beginning to work. But the cost to workers and communities in the form of plant closings and labor displacement has been enormous.

The capital mobility strategy is not merely aimed at organized labor. The newly enhanced ability to move capital between regions within the same country provides corporate management with the necessary economic and political clout to insist upon reductions in local taxation, and therefore cuts in community services and the social safety net. The competition between local governments to retain existing capital or to attract new private investment is leading to an extraordinary retrenchment in social programs as we have known them. The most noteworthy victories of the corporate sector in this regard are found in Proposition 13 in California and Proposition 2½ in Massachusetts. These substantial tax cuts were actively supported by the banking and business communities of each state, which suggested that failure of passage would seriously undermine the "business climate" in their regions. The interregional rivalry to attract capital, in light of the high degree of threatened or real capital mobility, has led the business press to write that the nation is in the midst of a new civil war—a new "war between the states." Without the heightened ability to move capital, this could not have happened.

The election of Ronald Reagan elevated the "civil war" to a new level. To reindustrialize America, the federal government is insisting on creating a "good business climate" in the United States through extreme cuts in corporate taxes, drastic reductions in the government's guarantee of the social safety net, and the virtual deregulation of the private sector. Washington has joined the corporate sector in declaring a class war on workers and their communities.

Ironically, there were those in the business community who foresaw this a long time ago. In a special commentary in October 1974, *Business Week* predicted a slower growing economy in which

> some people will obviously have to do with less, or with substitutes, so that the economy as a whole can get the most mileage out of available capital. . . . Indeed, cities and states, the home mortgage market, small business, and the consumer, will all get less than they want because the basic health of the U. S. is based on the basic health of its corporations and banks: the biggest borrowers and the biggest lenders. Compromises, in terms of who gets and who does without, that would have been unthinkable only a few years ago, will be made in the coming years because the economic future not only of the U. S. but also of the whole world is on the line today.

The editors went on to observe that the idea that income and resources would have to be redistributed to big business would be a hard pill to swallow. To get the American people to swallow it, they predicted, was going to require a "selling job" beyond anything that any country had attempted in modern times.

At the root of all of this is a fundamental struggle between capital and community. In a brilliant paper, planning theorist John Friedmann of UCLA, has developed a particularly dramatic formulation of the contradiction between the imperatives of capital and people's need for community and economic security. It is so beautifully formulated that it is worth quoting at length:

> Two geographies together constitute a "unity of opposites." I shall call them *life space* and *economic space*. Although both are necessary for the sustenance of modern societies, they are inherently in conflict with each other. Over the last two centuries, economic space has been subverting, invading, and fragmenting the life spaces of individuals and communities.
>
> Life space is at once the theater of life, understood as a convivial life, and an expression of it. . . . Life spaces exist at different scales [and] are typically bounded, territorial spaces. . . . Places have names. They constitute political communities.
>
> In contrast, economic space is abstract and discontinuous, consisting primarily of locations (nodes) and linkages (flows of commodities, capital, labor, and information). As an abstract space, it undergoes continuous change and transformation.
>
> Economic space is open and unlimited; it can expand in all directions. Indeed, its continuous expansion is vital to the reproduction of capitalist relations as a whole. Expansion occurs ruthlessly. . . .
>
> We can see the result in the dissolution of life spaces and their progressive assimilation to economic space. The capitalist city has no reverence for life. It bulldozes over neighborhoods to make way for business. It abandons entire regions, because profits are greater somewhere else. Deprived of their life spaces, people's lives are reduced to a purely economic dimension as workers and consumers—so long, at least, as there is work.

What deindustrialization ignores is that "people want to improve their community, not abdicate from it." An unfettered investment policy destroys communities and personal assets while it creates an industrial "refugee" crisis of serious proportion—whole subcommunities without life space *or* work. Ultimately the process of creative destruction is unsustainable, as Joseph Schumpeter himself reluctantly admitted in sections of his classic work seldom cited by his more recent admirers:

> Can capitalism survive? No, I do not think it can . . . its very success undermines the social institutions which protect it, and inevitably creates conditions in which it will not be able to live and which strongly point to socialism as the heir apparent.

In the long run, people simply will not accept the degree of instability, insecurity, and other expressions of "pure rationality" that the process requires. As Schumpeter implied, capitalism and democracy are ultimately incompatible.

The Great Reindustrialization Debate is over precisely these issues. How do we build a stable, humane, equitable community and still have economic growth? And how can we go about the business of constructing a productive economy which produces livelihoods without destroying lives?

PART II
The Consequences of the Structural Transformation for Work and Workers

Deep economic, technological, and social changes are creating massive changes in the nature and organization of work. For some time now, global, national, and local economies have been redistributing jobs to meet the demands of capitalist development. These economic shifts are creating two major social problems—structural unemployment and the pauperization of work.

Traditional manufacturing before 1980 was characterized by strong unions, rising wages and other benefits, and work for the masses of semiskilled blue collar workers. Now, with plants relocating out of the country or in the sunbelt, manufacturing jobs being replaced by robots and other computerized forms, and the new jobs mainly in the information/service sector, the pattern has changed to nonunionized labor, low wages and benefits, and the dislocation of many of the unskilled and semiskilled. The result for the society is a greater proportion of the people locked in unemployment, poverty, and even homelessness. Or, if they work, a large number are in marginal jobs, working part time, or doing homework (that is, piecework—whether making products or doing word processing on a personal computer). As Harrington has put it: "So it is that part-time work, sweatshops, and 'homework'—three forms of extreme exploitation in the early period of industrialism—have returned in the age of the computer" (1986:83). Put another way, again by Harrington, the trends have "opened the way to the society of three speeds with a privileged elite, a downgraded middle, and a permanently precarious bottom" (1987:81).

In the new occupational structure, a distinction can be made between "good" jobs and "bad" jobs (Oxford Analytica, 1986:261–264), a bimodal view of employment sometimes described as that between primary and secondary labor markets. This redistribution of jobs and job opportunities is ignored by those who proclaim a triumphant economy with millions of newly created jobs. Several trends contradict the optimism of "the great American job machine." First, while the service economy can generate jobs in large numbers, many new forms of work push people to poverty. The overwhelming proportion of jobs are in the low-wage service sector that grew from sixty-five million workers in 1981 to seventy-five million in 1987. Second, contingent work arrangements (including part-time work, temporary work, employee leasing, and independent contracting) offer more jobs but leave workers and their families with low pay and without traditional benefits. The rapid growth of these forms of employment represents a fundamental structural change in the social organization of work. Third, many low-skill jobs are being replaced by jobs that demand more education and training. The computerization of the workplace is creating a mismatch of people and jobs. Many Americans lack the skills required to fill the new jobs in the information society.

Technology and the shift from manufacturing to services and information have produced serious contractions in industrial work. While the service economy creates a vast array of jobs, it eliminates standard, relatively secure, high-paid factory production jobs. Two million manufacturing jobs have disappeared permanently since 1981. The number of new jobs added to the economy does not compensate for the precipitous decline in manufacturing. Instead, the shift in job configurations is creating a new kind of worker: one who used to be part of the middle class and who may be employed but is now on the verge of poverty. The changing distribution of jobs is accompanied by a variety of costs to workers that are spread unevenly across class, race, and gender lines. This will be explored further in Section III.

Will high technology offset the erosion of traditional job sources? High-tech industries (those devoting a large percentage of investment capital to research and development) have, in fact, a low number of jobs. According to Bureau of Labor Statistics and Data Resources, Inc., high-tech industries in the United States will generate only between 750,000 and 1 million jobs in the whole country by the year 2000. This is less than half of the 2 million jobs lost from the U. S. manufacturing industry since 1981. And of these new jobs, fewer than one third will be for engineers and technicians. At the same time, the unskilled jobs in high tech such as those of assembly line operators are always under threat from low-cost foreign labor (Forester 1987:244). Global competition and capital flight will continue to remove lower-skilled jobs from the work force.

The readings in this section draw attention to the effects of technology as well as the effects caused by the structural reorganization of the labor force. These forces are closely related. For example, advances in technology have contributed to the rapid growth in temporary employment. Robots and other forms of computerized automation are redefining the demand for labor. New technologies and automation imposed by global competition beyond American control are changing working arrangements for millions. As Joan Smith has put it: "Dangerous jobs have been eliminated only to be replaced by new kinds of work with their own distinct hazards.

Scientific advances have meant that old skills are eroded to be replaced in many instances with a few jobs that require greater skills and many more requiring absolutely none" (Smith, 1987:6). The distinction between "good" jobs and "bad" jobs is not limited to reward structures of pay scales and promotion opportunities but has as much to do with the quality of work and with the skills and judgement required to complete the job. The displacement of work and workers is related to the reorganization of the labor process itself. This is a compelling issue in the new workplace. New technologies can change work by fragmentating it and reducing the skills required. Deskilling varies by occupation and industry, yet much work today lacks autonomy, variety, mental challenge, and judgment. It is indeed degraded and deskilled. Furthermore, technological advances can be used for managerial surveillance and control of workers. New managerial practices combined with information technologies now present a serious challenge to labor and to the unions.

The articles in this section examine the transformation of jobs and working conditions through new forms of work organization and new technologies. These changes are placed within the context of a global economy, increased capital mobility across national and regional boundaries, and the shift from a manufacturing–based to an information and service–based economy.

REFERENCES

FORESTER, TOM. 1987. *High Tech Society: The Story of the Information Technology Revolution.* Cambridge, Mass.: The MIT Press.

HARRINGTON, MICHAEL. 1987. *The Next Left: The History of a Future.* New York: Henry Holt.

OXFORD ANALYTICA. 1986. *America in Perspective.* Boston: Houghton Mifflin.

SMITH, JOAN. 1987. "Terminal Illnesses." *The Women's Review of Books IV,* 7 (April):6–7.

U. S. News and World Report. 1987. "The Great Jobs Mismatch." *U. S. News and World Report* (Sept. 7):42–43.

THE NEW SOCIAL ORGANIZATION OF WORK

5 *Brave New Workplace*

ROBERT HOWARD

The hush of an executive office is interrupted by the soft click of fingers on a VDT keyboard and the staccato sound of a high-speed printer spitting ink onto a page. On an automobile assembly line, spot-welding robots eerily dip and sway in a finely choreographed electronic ballet. White-coated workers load stacks of silicon wafers in the antiseptic atmosphere of a climate-controlled "clean room." In the dim light of an industrial design department, engineers peer at the multicolored images taking shape on their terminal screens.

These are just a few of the many familiar scenes signaling the comprehensive computerization of work. As American business spends $50 billion every year on new workplace technologies, this computerization is causing a major transformation of working life.

Word processing, electronic mail, advanced telecommunications, and other computer-based systems have inaugurated a new era of "office automation." According to SRI International, there are some 1.5 million electronic work stations currently in place in American offices; by 1990 they should multiply more than tenfold to 17.5 million. And the work of one out of every four white-collar workers will take place before their incandescent glow. In the factory, industrial robots, computer-controlled machine tools, even entire "computer-aided" design and manufacturing systems are the components of the most important reorganization of manufacturing work since the arrival of

Source: Robert Howard, *Brave New Workplace* (New York: Viking, 1985), excerpts from pp. 15–35.

the assembly line more than seventy years ago: "computer-integrated manufacturing."

But computerization is doing something more than merely bringing new tools to the workplace. It is reinventing American industry itself. . . .

In order to understand what the computerization of work means for workers requires penetrating behind technology's visible screen to the attitudes and assumptions, the ideals and "imprecise values" of the people responsible for managing new technology in the brave new workplace today.

Who are these people? They are manufacturing engineers and systems analysts, information systems experts and outside technology consultants. They plan new work systems, design or choose the technology to be used in them, and oversee its implementation in the workplace. While not exactly a clearly defined profession, they do constitute a unique occupational subculture with its own traditions, perspectives, and professional common sense. Call it the realm of the "technology managers," a growing new management class in the brave new workplace.

When we explore it, we are immediately confronted by a curious fact. While the technology of the computer and its arrival in the workplace are themselves relatively new, the ideas about work organization, human motivation, and social purpose currently shaping the use of technology at work are surprisingly old. The computerization of work may indeed constitute a "great hope," but for these technology managers it means something considerably more than new tools, new products and markets, or even newfound economic success. It is nothing less than the welcome realization of an ideal that has tempted industrial managers throughout much of the twentieth century—a captivating image of the workplace as a realm of perfect control. And armed with their chosen instrument, the computer, they are determined, if not to eliminate, then to render harmless the intractable imperfections of human work once and for all.

Young, bright, extremely ambitious; as Steve Taylor puts it, "I'm on a fast track." Twenty-seven years old, a graduate of the University of Michigan Business School, and a certified public accountant, Taylor is currently the executive assistant to the president of a major Houston bank holding company. However, we meet to talk about his previous job, when Taylor traveled the country as a systems analyst and office automation consultant for the Management Information Consulting Division of Arthur B. Andersen & Co., a prominent Chicago management consulting firm.

"There's a lot of room for computer consultants out there," Taylor begins. "It's a *big* business, let me tell you." As new work-place technologies proliferate, and as more and more companies see in them the solution to their productivity problems, corporate managers are turning to technology consultants like Taylor for organizational models, implementation plans, and advice. Such expertise does not come cheap—"no reputable consulting firm will do a feasibility study for less than $30,000 to $50,000," says Taylor, and supervising the actual implementation of a new system can cost twice as much. Nevertheless, technology consultants are in high demand. For what they have to sell is as important as technology itself—an idea and an approach for its use at work. Steve Taylor puts it this way: "Computer systems are just a way to expand management control."

Taylor's comment brings to mind another corporate manager with the same last name—Frederick Winslow Taylor, the turn-of-the-century mechanical engineer, management consultant, and founder of American "scientific management." Like his contemporary namesake, he was living during a period of enormous change at work. The small workshops and family-owned enterprises that had typified nineteenth-century industrialism were giving way to larger corporations. New immigrants were pouring into America's industrial workplaces. As factories grew larger and as the personal bonds of loyalty and dependence that had often linked the worker to the boss in the small firm were erased by the impersonality of the big corporation, the question of how managers were to retain control over work became urgent. Taylor's solution was to replace the personal authority (whether charismatic or coercive) of the traditional boss with an objective "science" of management. Taylorism, as it came to be known, was based on the radical separation of the execution and performance of work from its design. The organization of work was entrusted to a new school of men trained in the techniques of scientific management. They divided jobs into their smallest possible components, easy-to-learn "specialities" to be mastered by the unthinking worker. They performed time and motion studies on those fragments of work to determine "objective" work standards and quotas. Finally, they strived to routinize and standardize work in order to realize that uniform "one best way" of working that would simultaneously guarantee economic efficiency and management control. In his magnum opus, *The Principles of Scientific Management,* first published in 1915, Taylor espoused the principle that would become the unquestioned orthodoxy of American management for more than a generation. "In the past," he wrote, "man has been first; in the future, the system must be first."

Seventy years later, the principles of Taylorism are alive and well in the assumptions of many technology managers. Indeed, one might even say that they are giving Taylor's idea of a dominant "system" new substance by building it into the hardware and software of computer-based work systems. Whereas Taylor wanted to subordinate man to this system, his successors today see computer technology, first and foremost, as a means to eliminate, or at least minimize, the human element in work. Just as Taylor called for the thorough standardization and routinization of tasks, they see computerization as a mechanism to spread that systematic standardization into areas of the workplace and industries where it has rarely been seen before. They also share Taylor's fascination with measurement, and they are using the computer's vast information-processing capabilities to measure and monitor work more intricately and more closely. Finally, by arrogating the design and implementation of new technology to themselves, they are taking the separation of execution and design inherent in Taylorism to a new level.

Of course, since the dawn of the Industrial Revolution, businesses have strived to replace human labor with machines. It should be no surprise that, at its simplest level, using the computer to expand managerial control means getting rid of the human element in work. Using technology as a tool for job elimination in order to cut labor costs is a prominent theme in the way technology managers talk about their jobs. They call it "cleaning up the work process" or "getting rid of excess heads." Steve Taylor says that "our job was eliminating people. In fact, that's the way we had to justify our fee. We always had to be thinking, 'How many people can we get rid of? How can we get the paperwork through faster?' The whole thrust of office automation is to reduce the workforce"—Taylor estimates by about 20 to 30 percent—"in one fell swoop."

A "human factors" supervisor at Bell Labs in Holmdel, New Jersey, Judith Reitman, also reflects this major goal. "There are a lot of places in the Bell System where things cost money," she says, "and the biggest cost is people." Reitman says that of the many ways to cut those costs, the most important is "replacing a piece of the person with something automatic." For illustration, she describes a recent technological innovation that has begun to appear in Bell operating companies around the country known as the Audio Response System. It is a computerized voice synthesizer that provides a caller with assistance in obtaining the number he or she is seeking. When the caller gives an individual's name and address to the directory assistance operator, the operator types the information into the computer, pushes a "green key," then disconnects and lets the computer take over from there. The computer locates the name in its data banks and provides the proper number to the customer (twice, just in case it's missed the first time around).

"There are some 40,000 directory assistance operators out there," explains Reitman. "To shave even one second off the average call can save $24 million a year." The computer-generated voice can eliminate an average of five seconds per call. The faster an operator can complete each call, the sooner she is available for the next. The more calls she can squeeze into her shift, the less operators you need. "You just don't need as many people that way," says Reitman.

Examples like these of "replacing a piece of the person with something automatic" are easy to understand. The computer voice replaces a few seconds of the operator's time. An industrial robot replaces the human movements of a worker on an assembly line. What is harder to grasp is the capacity of the computer to eliminate the need for mental as well as manual labor, to capture not only the actions of work but its internal logic as well. But it is precisely this quality that distinguishes it from previous workplace technologies and allows it to eliminate the human element in work at higher and higher levels of the occupational hierarchy. . . .

What is to ensure that the jobs created by new technology will make up for those it eliminates? And even if there is no overall employment crisis, what are the institutional mechanisms to guarantee that those workers most affected by technological displacement—middle-aged manufacturing workers, say, or the women who by a vast majority occupy the lower rungs of the occupational ladder in the service sector—will have the skills and the training to shift to other jobs? And what, if anything, should we do about the social paradox best expressed by a May 1983 declaration of the Business Council (a prestigious business lobbying group made up of some of the nation's top corporate executives), which announced that advanced automation was a key corporate strategy for economic recovery but because of the labor-saving advantages of this technology, many of their employees laid off during the recession would not be called back to work?

Among economists and public policy-makers, these are questions of enormous controversy. Among technology managers themselves, they are rarely considered. One manager who has thought about them is Tony Friscia, a consultant on the "factory of the future" at the Yankee Group, a Boston management consulting firm. "It's only a gut feeling," says Friscia, "and I don't really have any statistics to back it up. But there's got to be a five-to-ten-year period, probably around 1985 to 1995, when there are going to be a lot of victims. Because there are certain people whom you're just not going to retrain. What is the guy who's fifty years old and spent the last twenty years in some factory going to do? You're not going to train him to be a robotics technician; for the most part, you want young guys for that."

Friscia has no answer to his question. "Ultimately, those who survive, survive, and those who don't, don't. I guess that's the capitalist way of looking at things. In theory, it makes a lot of sense. Maybe even in practice, when you look at things over the long term, it makes sense. But if I were a labor negotiator, the fundamental issue for me would be 'How are you going to save the people I'm responsible for?' Or, if I were a legislator, I'd want to be damn sure that there weren't going to be so many people uncared for that I was setting myself up for some kind of social upheaval. A society just can't survive when a substantial portion of the population is out of work."

But Tony Friscia is something of an exception. Few technology managers are so reflective about the social implications of their work. The simple reason is that they don't have to be. While the corporation enjoys the benefits of labor-saving technology, it can avoid most of the costs by "externalizing" them onto society as a whole. Far more typical are the comments of a Citibank executive whom I interviewed.

"There's a question for society in all this," he told me. "What happens to those people who can't make it in this superefficient world that we're creating? It's true; over the long run, fewer people will work at Citibank. Who knows? Maybe we'll all work a twenty-hour week. But that's not our problem; that's up to society to decide."

Another goal of scientific management was to standardize work, on the theory that the more routine and predictable a particular production process, the easier it would be to control. As with the elimination of manual and mental labor, computerization takes standardization to a new level. . . .

Standardized work is controllable work. Perhaps the most effective instrument in the Taylorist system of control is the systematic measurement of work by time and motion studies. But whereas the measurement of work under traditional Taylorism was occasional, limited to the visits of the time-study specialists, the computer's vast information-processing capacity allows technology managers to extend the monitoring of production over time, making it constant. They talk about increasing the "visibility" of work—as if computer screens were windows peering into the heart of the brave new workplace.

Since graduating from MIT in 1979, Joe Kracunas has been designing and installing a direct numerical control (DNC) machining system at GE's Aircraft Engine Group in Lynn. The DNC system links sixty computer-controlled machine tools to a central computer. Programs to run the tools are loaded in the host computer in a bright white air-conditioned control room, then electronically transferred to each individual machine out on the shop floor. DNC is a major part of the computerization of factory work, because it ties each individual machine into a factory-wide system. What's more, it allows management to track machine performance with a comprehensiveness never before achieved. "With this new system," explains the twenty-seven-year-old Kracunas, "you can monitor every machine out there in the shop. The real payback is in factory management and control."

In effect, DNC creates an information network on the factory floor. Just as information is sent to the machines and operators on the shop floor—for example, part programs and machining instructions—it is also extracted from them. Currently the DNC system at Lynn collects data on machine utilization: how often the machine is cutting parts and at what feed-rate settings. When a particular machine has been turned off for more than fifteen minutes, the computer prompts its operator for an explanation by means of a message displayed on the terminal perched on the top of each machine's control panel. The operator's answer is entered into the machine's utilization record. Planned enhancements include attaching special electronic probes to each machine in order to monitor oil pressure, power supply voltages, servomotor outputs, and other indicators of machine performance. "We want to attach sensors to different parts of the tool," says Kracunas, "sort of to monitor its health." At some point in the not too distant future, the DNC system at Lynn will be so sophisticated that every time someone comes to fix a malfunctioning machine, he will arrive with a computer printout detailing the machine's current status and entire performance history.

Early in 1984, Kracunas installed a feature he first saw at another GE facility. At the front of the control room, hanging from the ceiling, a large video screen displays a map of all the machines on the shop floor. Every machine is one of three colors: those running smoothly are bright green; those momentarily turned off are yellow; and those "down" for maintenance are a flashing red.

Another example of these electronic windows looking into the brave new workplace is a computer system called manufacturing resource planning, or MRP. MRP concerns itself with factory scheduling and inventory control, some of the most costly problems in industry. Nearly 85 percent of the time that a part is in a factory, it is simply waiting for something to happen to it—waiting to begin a production run, waiting to move from one machine to the next, or simply waiting to be shipped out to the customer once it is finished and packed. The average machine tool spends only 6 percent of its available time actually cutting metal, for example. The cost of inventory buildup is enormous. Half the space in the average factory serves as a warehouse—parts sitting around. One of the thorniest challenges for most manufacturers is how to set up a system for the continuous flow of parts through the workplace.

MRP systems are complex computer programs which do just that. They work backward from the delivery date of a factory's products, in order to orchestrate an elaborate schedule of raw materials, machine time, tools, labor, and other resources. They define the precise interactions between each component from the beginning to the end of a nonstop production process. Ideally, raw materials are delivered on the very day they are to be used, then travel from one machining process to another in an uninterrupted flow across the factory floor. More than a hundred such MRP systems have been developed and put to use in more than ten thousand manufacturing sites. Some of the companies that use them have been able to cut back their inventories by as much as 40–50 percent.

It would be hard to overestimate how excited this can make technology managers. "MRP is not just a computer system; it's a whole new management philosophy," one told me. "At our plant, we treat MRP like a religion," says another. Tony Friscia, from the Yankee Group, describes the implications: "If something breaks down, the office *immediately* knows about it," he says. "If there is an inventory need, the office is *immediately* notified. When you have access to all that information, then your ability to manage improves significantly. You can eliminate all the paper passing back and forth from hand to hand. You don't have to wait until tomorrow or next Monday morning for an update on your inventories or your production runs. You've got it *now.* That's what they mean when they talk about improving management control."

Such intricate control of work is admirable when applied to machines or products moving through a shop or factory. But what about when it is applied to people? It does not take long to realize that when the managers of the brave new workplace talk about expanding managerial control, they mean control over workers as well as over products or machines. Here, the legacy of Taylorism reaches its most dangerous expression.

At one point during a long conversation about the computerization of work in the banking industry, a vice-president at a Citicorp subsidiary passes a large computer printout across the expanse of his desk. It is an MIS Daily

Operating Report for a clerk in the firm's Chicago office. MIS stands for Management Information System, and it is key software in any office automation package. With it, managers can correlate large quantities of information about workflows and performance—even in offices hundreds of miles away. This particular clerk's printout indicates the percentage of his (or, perhaps, her) day spent on the telephone and that spent processing records of the financial services the firm provides. It compares the time it took the clerk to complete each item with a standard time allocated by Citicorp industrial engineers. The manager explains that a good productivity rating would be in the neighborhood of 91 percent. On this particular day, the clerk's is a feeble 38.8. "If I were the manager there, I would want to know why," he says, in passing.

The MIS report also lists the clerk's "utilization" rate—the portion of the day spent actually processing forms. Multiplying the productivity and utilization ratios, says the manager, gives an overall "operating index," an "all-inclusive measure" of performance. Finally, the MIS form reports this clerk's "backlog," item by item (those processing jobs in the computer yet to be completed). Each day the system automatically "ages" the items, so managers can pinpoint workers with an insufficiently rapid "turnaround."

"We monitor productivity on a daily basis," explains the Citicorp vice-president. "The system lists every clerk, every day. It tells us the work he got in and the work he got out." He pauses, lifting the printout off the desk for emphasis. "And we can get one of these for every work station in the business."

The use of technology as an instrument of control not only over work processes but over the people who perform them is nowhere more clear than in the way technology managers talk about their own jobs managing the design and implementation of computer systems in the brave new workplace. Where Frederick Taylor divided the organization of work (reserved to the initiates of scientific management) from its execution (left to workers), they are applying this principle to technology itself. How these designers of the brave new workplace treat the "users" of new technology is a case study in the expanding corporate control of work. At its worst, it is little more than an exercise in manipulation.

The consulting job that Steve Taylor remembers most vividly is the computerization of an office at Texas International Airlines in Houston, now a part of Continental Airlines. He was sent by the Chicago office of Arthur B. Andersen to design and implement what he calls a "work measurement and performance evaluation system" in the company's revenue accounting department. Some 130 clerks, mostly women, performed various processing jobs in the office: they priced tickets, tabulated the amounts on passenger coupons from Texas International flights, and did some fifteen other clerical tasks. Taylor's job was to create a computer system that would automatically monitor their work, record their output, and calculate their productivity by means of a mathematical index.

According to Taylor, his work at Texas International was a classic example of the technology consulting process. The first step was to meet with company executives. "Any computer consultant works for top management," he explains. "You come into the corporation, sit down with upper management, and help them decide what they want. It's a top-down approach. Top management gives us the mantle, and we truck on down to the workplace and put their directives into effect."

At Texas International, top management expressed the usual desires for increased productivity, for people to work, in Taylor's phrase, "faster, smarter, and better." But corporate executives also expressed another concern. Clerks in the office had recently voted to join the Teamsters union. The "performance evaluation" system was management's way of showing them that it was still in charge.

Receiving the go-ahead from top management, however, was only the first step in a long, carefully planned procedure. Next, Taylor had to move to the workplace itself. "If you're going to do a study, you've got to have information, right?" Before he could rationalize the revenue accounting office, he needed to know precisely how the office clerks did their work. And the best source for that information was the clerks themselves.

At this stage, says Taylor, a key role is played by middle management. "Middle management's place is to decide how to break it to the employees," he says. Taylor held meetings with office supervisors to discuss this delicate problem. Finally, someone suggested, "Let's tell them this is 'methods improvement.'" The announcement was made that an outside consultant was "going to help you do a better job."

Thus, the accounting clerks at Texas International were neither surprised nor concerned about the appearance of the young stranger in their office, asking questions day after day about their jobs. As Steve Taylor says, with equal parts of amazement and pride, "They didn't know what was going on." Step by step, he began to understand how work was done in the revenue office. He sat down with workers for long intensive interviews probing them on how they did their jobs. Unbeknownst to the clerks, he was also analyzing their descriptions and redesigning the work process from scratch. "You determine what's good, what they should be doing, and what can be automated." Having formulated a plan to streamline office work, he then began slowly to introduce it to the workforce. He suggested new work procedures and started timing workers as they did their jobs. "Slowly but surely," he says, "we're going to teach them something that outlives us."

Of course, the rationalization of the revenue accounting office did not always proceed smoothly. For one thing, the fiction of "methods improvement" had to be maintained at all costs, and at times Taylor felt as if he were on a psychological and emotional roller coaster. "After a while," he says, "an awful lot of people kind of suspected that something was different" about the way they were doing their jobs, different in a way they didn't particularly like. But for Taylor, this only added to the excitement and challenge of his own. "I loved it when I worked on that job," he recalls fondly. "It was always so crazy because it was so political." The wrong word to the wrong person and "you could really get burned. Because then, the whole scheme would be off."

Fortunately for Taylor, he succeeded in maneuvering his way through the different groups in the office without blowing his cover. Now he was ready for the final step of the work rationalization process—the creation of the actual computer system. The computer integrated productivity data from the various jobs the clerks performed. It could be programmed to produce reports on individuals or on the office as a whole on a monthly, weekly, or even daily basis. The reports listed how many items in each category a particular clerk had processed, and tabulated a productivity index rating her overall efficiency. The printout included what Taylor calls "a nifty little trick." It itemized each worker's absences as well. "If they were gone, it would show up on the printout," he says. "You know—'Wednesday, no activity.' "

As soon as the system was up and running, Taylor and the office managers at Texas International noticed a curious phenomenon. Office productivity would dip, like clockwork, at the same two points of every day—around nine in the morning and three in the afternoon. Taylor traced this distressing decline in performance to what he calls "abuses" in the office's flextime system. The clerks had the option of arriving at work anytime between six and nine in the morning. Depending on their hour of arrival, they would leave between three and six in the afternoon. When the last of the office workers arrived at nine, those who had been working since six would stop to say hello. Productivity, says Taylor, would plummet. And when those who had been working since six left at three, the same thing would happen again. "So," says Steve Taylor matter-of-factly, "management decided to cut back on their flextime."

There are moments when the work of technology managers like Steve Taylor sounds almost like a crusade: to eliminate inefficient practices, to pare wasted actions, to squeeze out every second of "dead time" like so many drops from a too-absorbent sponge. Because these changes they are planning for work are so far-reaching (and have such enormous implications for those doing the work), the managers of the brave new workplace also come armed with a promise. It is, of course, a promise of efficiency—that the new workplace technology will ensure competitiveness and guarantee our economic and social prosperity—but it is also a promise about the quality of working life. Put simply, they claim that their computerization of work is going to make for better jobs.

One facet of this claim is that technology will eliminate "dog work," the boring, dirty, dangerous jobs that no one wants to do anyway. Speaking about the computer-generated recording that "replaces a piece of the person with something automatic," Judith Reitman adds that "this kind of replacement gets the workers very happy. You may be reducing the workforce, but it's for a very tedious job. There are fewer workers but the jobs are much better."

Technology managers also argue that the changes brought about by computerization ultimately have a positive effect on skilled jobs. Thus, after describing the automation of the senior designer's "steam path," John Eskesen argues, "If the logic of design and drafting rules can be programmed into a computer, then by definition, it's unskilled labor. What the computer has done is to take on the repetitive detailing work at many levels of sophistication." It is like a chemical solution that leaches out "mundane" tasks, leaving behind a pure precipitate of ever more challenging and more innovative "conceptual work."

Indeed, many see the capacity of computerization to introduce a kind of "flexible specialization" as the occasion for a total reinvention of work, a reinvention that will result in more integrated and less fragmented jobs and will allow workers to perform more creative, useful, and interesting tasks. "Given the training and the opportunity," says a top technology manager in the auto industry, "most people would like very much to expand the horizons of their jobs." For him, computerization "will broaden the spectrum of worker involvement and allow the individual to input more of his personality into his job."

All these possibilities are potentially true. But when the technology of the computer is conceived primarily as a means for managerial control, the claims of the technology managers are often proved wrong on both counts. The assumption of improved work quality is regularly contradicted in the brave new workplace today. And even the promise of increased efficiency turns out to be far more difficult to achieve than most managers tend to think.

THE NEW SOCIAL ORGANIZATION OF WORK

6 Job Content: Job Fragmentation and the Deskilling Debate

HEIDI I. HARTMANN, ROBERT E. KRAUT, AND LOUISE A. TILLY

[Harry] Braverman's seminal work (1974) has shaped the terms of the debate about the effects of information technology on the content of jobs. He argued that the introduction of technology into the office is a mechanism to rationalize and fragment office work. As in a factory, office work is broken into many subtasks, each performed by a "detail" worker, who loses the integrative contact with the total product and who loses variety in the job. The consequences from the employer's perspective are to reduce the skill requirements of office work—that is, the average employment experience and education needed to adequately perform the job—and therefore to reduce labor costs. The consequences from the employee's perspective are to reduce the quality of the job by reducing the variety of tasks performed, the mental challenge, and job autonomy and responsibility.

Studies have documented many episodes in which the introduction of new information technology coincided with a fragmentation of jobs and a decrease in the skill levels required to do them. For example, Murphree (1984) describes the fragmentation of the job of legal secretary in the large firm she studied. The fragmentation was not necessarily caused by the new technology, but it facilitated an ongoing process of routinizing work. The variety of tasks that made the job of legal secretary both interesting and highly paid were assigned to

Source: Heidi I. Hartmann, Robert E. Kraut, and Louise A. Tilly (eds.), *Computer Chips and Paper Clips: Technology and Women's Employment* (Washington, D. C.: National Academy Press, 1987), pp. 136–143.

other workers: paralegal aides composed first drafts of simple contracts, specialized legal librarians searched for legal citations, and messenger services hand-delivered urgent documents. With word-processing equipment, even the task of producing a manuscript was divided into an initial data-entry stage and a later proofreading and editing one, with different personnel performing each function. Relieved of many of their duties, the legal secretaries found themselves doubled up, working for more than one lawyer, primarily in a gatekeeping role. In this case, new technology was a concomitant of decreased variety, less challenge, and less responsibility.

Researchers have also identified cases in which information technology incorporated substantive knowledge of a job and its procedures, an establishment, or an industry, leaving less for a worker to know. In the insurance industry, the skilled work of assigning risks or assessing claims has increasingly been codified into computer software, so that less skilled, less experienced, and less educated clerks can perform the work once performed by skilled clerks and professionals (Baran, 1985). In social science, sophisticated statistical analyses that once were done only by professionals are now performed by undergraduates or research assistants using a statistical analysis computer program. In supermarkets, knowledge of brands, in-store promotions, and arithmetic is less necessary to a checkout clerk using a bar-code reader and an intelligent cash register: for example, when produce is coded, the clerk need not know the difference between root parsley and horseradish. In these cases the uses of information technology have led to the creation of more routine jobs.

In contrast to these cases in which technology is a tool to restructure work and fragment jobs, other researchers have argued that new office technologies can reintegrate jobs that had been previously fragmented for other reasons and require more skill and responsibility of workers. At the clerical level, for example, by using centralized data bases, customer service representatives can handle all of the transactions associated with a client's account, taking orders, entering data, making adjustments, and answering inquiries. At a more professional level, using new workstations managers can have greater control over more stages of their work through computer-based tools for searching corporate data bases, calculating the consequences of investment strategies, creating illustrations, outlining, writing, checking spelling, formatting, and sending a finished report to a mailing list of recipients.

Moreover, starting from the assumption that the most routinized jobs in an industry are the first to be automated, some analysts have argued that automation increases skill requirements for the remaining jobs, especially generalized cognitive and problem-solving skills; increases worker responsibility; and increases coworker cooperation. For example, if most routine banking operations are performed by customers themselves using automatic teller machines, the banking tasks remaining to human tellers are nonroutine, e.g., handling problem inquiries that may require considerable knowledge of banking procedures, problem-solving skills, and skill at dealing with people. Similarly,

Baran (1985) notes that, as some work by insurance professionals is being deskilled and redistributed to clerks and machines, the remaining professionals function as exception handlers and do more skilled work. Attewell (1985) also notes that computerization is eliminating the routine work of insurance examiners, such as calculation of deductibles or identification of potential duplicate payment, while at the same time leaving examiners more time to make decisions about dubious claims. Similarly, social scientists using modern statistical software are relieved of the tedium of calculation and can undertake more challenging intellectual tasks.

From these studies it is clear that both job fragmentation and deskilling and job reintegration and upgrading are occurring, but it is not clear which trend is predominant. Aggregate data, although flawed, show little evidence of widescale deskilling either within particular industries or in the labor market as a whole and, indeed, show some evidence of increased skill requirements. Like most of the research reviewed in this chapter, the data are suspect, and the conclusions based on them should be viewed as tentative. The wide variety of skills needed in the workplace and their uneven distribution across jobs and industries and over time makes comparison of skill levels exceedingly difficult. To assess the skill requirements of jobs, researchers have often resorted to the use of one or two proxy measures, such as years of experience or years of education of workers. These measures lose detailed qualitative and quantitative information about skill differences, and they may reflect changing tastes of employers or demographic shifts in the population unrelated to skill requirements of jobs.

Researchers using assessments of the skill requirement of jobs from the *Dictionary of Occupational Titles* (DOT) and similar sources have found little evidence for the deskilling of white-collar work, either within particular occupations or, more broadly, across occupations. For example, Attewell, using the Bureau of Labor Statistics' skill-level categorizations for 13 occupations within the insurance industry, found that between 1966 and 1980 (a period of intense automation) 4 showed statistically significant downgrading, 6 showed significant upgrading, and 3 showed no trends. He also found a substantial increase overall in the proportion of the insurance work force in higher-level white-collar occupations (from 38 to 60 percent). Using DOT data, researchers have found little change in the skill requirements of a sample of industrial and service occupations from 1949 to 1965 or more broadly from 1965 to 1977 and either no changes in the aggregate skill of the nonfarm sector from 1900 to 1970 or a small aggregate upgrading of skills from 1960 to 1976.

To the extent that some automated jobs indeed require less skill than nonautomated ones, one still needs to ask about the role of technology in these changes. Scholars are now debating whether new technology encourages job routinization and fragmentation, is a beneficiary of it, mitigates it, or is independent of it. Certainly, scientific management of office work was occurring well before the current introduction of microelectronics. Some have argued that job fragmentation and routinization result from the inherent conflicts of interest between capitalists and workers, and others have argued that it is a bureaucratic response to increases in office size and paperwork resulting from the increased size of firms, their movement into national markets, or government record-keeping requirements. Finally, others argue that fragmentation and routinization are necessary preconditions for the application of the simplistic computer technologies that are available at the current stage in the automation of office work but are not necessarily consequences of the automation. According to this view, work that has already been simplified is a prime candidate for automation.

STAGES OF TECHNOLOGY

What causes the contradictory impressions conveyed by the literature on job fragmentation and deskilling? In part the disagreements are the result of comparisons of experiences at different stages in the evolution of the technology, variations in the uses of the equipment, and differing social and economic circumstances under which new technology is introduced. Given rapid changes in the technology itself, in its uses, and in the cumulation of its effects, conclusions from one wave of technology use may not generalize to later waves.

An examination of the social organization of word processing is an instructive case. Large pools of workers are much more likely to be the modal organization of personnel if the word-processing equipment is old (pre-1978) rather than new (post-1981) and if the firm introducing the technology is large (Benoit et al., 1984). Women who work in large pools have more homogeneous and boring work than do other word-processing workers. They spend more than 90 percent of their time doing typing tasks, while those who serve several clients or who work as private secretaries for a single principal have a much greater variety of tasks.

It is not clear whether these differences in the social organization of word processing reflect the differing technical capabilities of old rather than new equipment (i.e., minicomputer-based rather than microcomputer-based word processors) or changes in management philosophy resulting from experiences with previous implementations (i.e., a "learning curve") or other social influences. But the point remains that researchers trying to understand the role of technology in job fragmentation would have come to different conclusions had they studied large or small firms using old or new equipment.

LEVELS OF ANALYSIS

A second cause of the disagreement in the research literature results from ignoring important level-of-analysis distinctions among the units of work that are being skilled or deskilled. The skills and skill requirements of tasks, individuals, jobs, occupations, firms, industries, and the labor force as a whole can change with the introduction of technology, but a change at one level has no necessary implication for changes at other levels. A particular task may require less skill to perform when it is done on a computer, but depending on the variety of tasks comprising the job and the knowledge of the worker in that job, the job itself may become more satisfying and challenging.[1] This paradox may be compounded as tasks migrate between jobs and occupations and as new workers fill the transformed jobs.

For example, consider the use of a word-processing system. A standard word-processing system today allows text editing of manuscripts, formatting (e.g., centering, underlining, justification, pagination, and setting fonts and point sizes), sorting, and syntax and spelling checks. As a direct result of the word-processing system, the skill and educational requirements needed to perform some specified tasks have decreased: to sort the entries in a bibliography, a secretary needs less facility with the alphabet; to proofread a manuscript, less knowledge of spelling and grammar. In addition, some of the procedures needed to achieve a certain result have changed (e.g., centering or underlining text), although it cannot be determined in advance whether the change in procedure represents an increase or decrease in skill requirements. Finally, some tasks in creating a manuscript electronically have no counterpart in a typewritten manuscript: for example, setting the size and style of fonts is more similar to traditional production editing jobs than to typewriting jobs. It is not clear in the abstract what the net effect of these changes in tasks will be on the skill required to produce a manuscript. Even if, overall, less training, skill, and education are needed to create a manuscript using a word-processing system than using a typewriter, there is no necessary connection between this decreased skill requirement and the skill exercised by particular typists or secretaries, the skill requirements of the job of secretary within a particular establishment, or the skill or educational requirements of the secretarial occupation as defined in private and public occupational classifications.

The skill requirement of jobs depends on the way their tasks are organized. Some jobs may be composed only of tasks that have been clearly deskilled when new technology was introduced: for example, the typing, proofreading, and correcting functions have been separated, and a skilled job is replaced by several new ones, some of which clearly require less skill. In contrast, a secretarial job can be organized, as it is in many universities, so that secretaries perform many services for the principals for whom they work. In this case, the increased speed at performing some of the tasks may lead to more time and

[1] This discussion proceeds as if the skill and education needed to perform a job were solely functions of the tasks comprising the job. Clearly, however, the skill requirements of a job have political and social as well as technical foundations. For example, the increased educational requirements for secretaries found in employment advertisements over the last 30 years may reflect the supply of college-educated women, the association of education with other "desirable" class and racial characteristics, or the quality of contemporary education more than they reflect changes in the secretarial occupation.

energy available for other tasks, including, in the case of university secretaries, advising students, monitoring grants and contracts, and maintaining records. This kind of change is likely to lead to skill enhancement for the job as a whole.

Again, the amount of skilling or deskilling of tasks and jobs has no necessary implication for the skills that a particular job incumbent can use. Clearly, as new technology is introduced, work procedures will change, and some of a worker's knowledge will become obsolete while new knowledge will have to be acquired. But knowledge change is not equivalent to deskilling. In the case of an individual, whether a deskilled job is one that is not challenging because it does not require the individual's mental or physical capabilities is a function of both the skill level required by the job and the skill level brought to it by the job holder. When the same people hold jobs before and after the introduction of technology that lowers the skill requirements of those jobs, they are unable to use their knowledge, skills, and experience, and their jobs become less challenging and satisfying. But the introduction of technology that lowers a job's skill requirements often affects new labor force entrants, rather than incumbent workers, who may retire or be promoted, transferred, or laid off. The new entrants—often with less employment experience and education—may well be challenged by the level of work. Indeed, the deskilling of some white-collar jobs may be the vehicle by which less advantaged social groups gain white-collar work. And for the same reason, less advantaged workers may face greater job insecurity because those jobs may be at risk of further deskilling or elimination through new technology.

In addition to changing the skill level required to perform certain tasks, the introduction of new technology is often the occasion for the reorganization of existing tasks across job categories. In the law firm that Murphree (1984) studied, for example, the shifting of duties to specialized personnel (e.g., legal librarians using data-base computers) was one element in the deskilling of secretaries' jobs. In universities many professors do much of their own typing on personal computers, taking on data-entry tasks from secretaries, while at the same time achieving more control over the form and content of their writing. In the insurance industry, some of the less skilled tasks that examiners once performed have been transferred to data-processing clerks, upgrading the jobs of both the examiners and the clerks. When tasks are transferred between occupations, shifts in the skill requirements of one job may be associated with complementary shifts in the skill requirements of another.

Occupations are the aggregation of similar jobs across the economy. Determining whether an occupation has been deskilled entails problems similar to those in determining whether a job (rather than a task) has been deskilled. In addition, one must contend with shifts in the definition of the occupations. In particular the Census Bureau and the Bureau of Labor Statistics periodically adopt new occupational classifications. These changes have made all comparisons across time, for both the numbers of workers and quality of work, extremely difficult.

Finally, in asking whether the skills of the labor force as a whole have been affected by technology, one must remember that occupations and industries that require many or few skills can grow or shrink independently of changes in skill requirements for particular occupations. Thus, as Simon (1977) predicted, the skill requirements of the labor force as a whole could rise if low-skill jobs die out. Of course, the reverse could also happen. In this regard it is interesting to note that the greatest new job growth is expected to be in fairly low-skilled occupations, such as food handler and janitor.

CONCLUSION

Examples abound of both positive and negative effects on skills and job quality when new information technologies are introduced. Using the research literature to ascertain which tendency prevails is difficult because of the different, coexisting stages of technology, the variety of uses of technology, and the differing social and economic circumstances of the workplaces studied. Even the same technology is used in different ways in different workplaces. No study of technological change, the quality of employment, or the effects on skills is based on a representative sample of workplaces and technologies in use. Moreover, information technology and its use are currently changing rapidly; even a well-designed study might miss important new effects on skills.

The available research studies also differ in their levels of analysis. Changes in tasks that occur with new technological capabilities do not necessarily translate into similar changes at the level of the job (or the labor force as a whole). As noted, easier word entry could make a typing job more boring and repetitive or it could make it more challenging if time is freed for other tasks. Thus, the available research literature and the methodological difficulties do not allow the panel to conclude whether there has been an increase or a decrease in the skills required or the quality of jobs with the use of new information technologies. And whatever the overall net effect technology has on skill requirements, there will be some jobs that require less skill and experience after new technology is introduced while other jobs will require more. Consequently, the changing skill requirements associated with technology, regardless of the direction of change, are bound to produce gaps between the skills that job incumbents and entrants have and the skills that jobs require. These gaps are a problem that require attention on their own.

REFERENCES

Attewell, Paul. 1985. "The Automated Office: A Case Study," unpublished manuscript. Department of Sociology, State University of New York, Stony Brook.

Baran, Barbara. 1985. "The Technological Transformation of White-Collar Work: A Case Study of the Insurance Industry," paper prepared for the Panel on Technology and Women's Employment,

Committee on Women's Employment and Related Social Issues (Washington, D. C.: National Research Council.

Benoit, Carmella, Alfred Cossette, and Prisco Cardillo. 1984. *L'incidence de la Machine á Traitement de Textes sur l'Emploi et le Travail.* Montreal, Quebec: Ministere du Travail.

Braverman, Harry. 1974. *Labor and Monopoly Capital: The Degradation of Work in the Twentieth Century.* New York: Monthly Review Press.

Murphree, Mary C. 1984. "Brave New Office: The Changing World of the Legal Secretary." Pp. 140–159 in Karen Sacks and Dorothy Remy (eds.), *My Troubles are Going to Have Trouble with Me: Everyday Trials and Triumphs of Women Workers.* New Brunswick, N. J.: Rutgers University Press.

Simon, Herbert A. 1977. "What Computers Mean for Man and Society," *Science* 195:1186–1191.

Ind. that require many or few tasks can grow or shrink independently of changes in skill requirements

THE NEW SOCIAL ORGANIZATION OF WORK

7 Automation and Labor in the Computer Age

HARLEY SHAIKEN

The flexibility of microelectronics extends computer technology to every corner of manufacturing. On one automobile assembly line, for example, 98 percent of over 3,000 welds on the car body are performed by swiftly moving robots, computer-controlled mechanical arms. In an aerospace machine shop, the complex contours of a jet-engine support are carved out by the whirring cutter of a numerically controlled machine tool, itself directed by a computer that is simultaneously supervising over a hundred other machines. In a diesel engine factory, bearings are removed from one of over 4,000 identical bins by a cart that glides silently down 100 yards of darkened aisle and lifts its forks three stories into the air. In an engineering office, the structural properties of a part are tested by an engineer sitting at a cathode-ray tube—a TV-like screen—in Cologne, Germany, and then the finished design is instantaneously transferred to a computer in Dearborn, Michigan.

Each of these operations represents an extraordinary change from conventional methods and an impressive technical achievement. The most far-reaching development of all, however, is the linking together of these separate processes into a highly integrated manufacturing system, the computerized factory. The transformation is often called computer-integrated manufacturing (CIM), or sometimes by the longer title computer-aided design

Source: Harley Shaiken, *Work Transformed: Automation and Labor in the Computer Age* (New York: Holt, Rinehart and Winston, 1984), pp. 1–11.

and computer-aided manufacturing (CAD/CAM). Whatever it's called, the result is a revolution in the way things are made and the way work is organized. The technical basis for this revolution rests on the information processing capabilities of computers and the ability of microelectronics to bring computer power directly to the point of production whether that happens to be a machine tool or a word processor.

In the last several years, the capabilities of computers and microelectronics have soared while prices have plummeted. During the 1970s, for example, the cost of computer power declined an average of 50 percent every two and a half years. This trend has contributed to the explosive growth of industries that manufacture new highly flexible forms of automation based on these technologies. Look at the production of robots. The revenues of robot producers have soared 50 percent annually between 1978 and 1981. Or, consider the manufacture of computer-aided design equipment, a marriage of computer technology to the design process. The leading firms in this area have been growing at over 30 percent a year. The recession in the early 1980s, of course, has dampened the market for automated equipment but this was often a case of deferred rather than cancelled purchases. (In fact, the 19 percent growth in robot sales in 1981 would have been highly desirable for most industries.) Overall, General Electric, which has moved heavily into this area, predicts that by 1991 the market for all forms of factory automation could exceed $29 billion a year.

The workplaces already using these new technologies span the gamut of manufacturing: industries from automobiles to machine tools, production volumes from millions of parts to one-of-a-kind prototype work, and factory types from five-million-square-foot transmission plants to storefront machine shops. General Electric, for example, as if to make its prediction on automation growth a self-fulfilling prophecy, spent over $2 billion to modernize its own factories in 1980 alone. And, despite its battered condition after over four years of weak sales, the auto industry plans to spend $80 billion in a five-year program to bring out a new generation of small cars. By 1985, 255 domestic auto plants will be rebuilt utilizing the latest in computer technology, and billions more will be spent to convert thousands of supplier factories.

This far-reaching technological changeover raises some pivotal questions that affect the entire society. Since any advanced industrial economy is heavily dependent on manufacturing, changes of this significance in the technology of production are bound to have repercussions far beyond the factory walls. For example, will these new technologies spur economic growth and the creation of new jobs or will they be introduced in a way that decimates employment? Will automation free workers from soul-destroying work or will it lay the basis for subordinating people to machines in a new and degrading way? Will computerization be used to rebuild the nation's industrial base or will it facilitate the ability of multinational corporations to operate independently of any given country?

I argue that the answers to these vital questions are not predetermined and depend on more than the technical capabilities of computers and microelectronics. The technology provides choices; what is selected depends on who does the choosing and with what purpose in mind. The following chapters explore the nature of the technological choices now being made, the ways in which they affect life on the job, and the changes they are spurring in the structure of the workplace. Looking only at the impact of technology, however, is too narrow and static a focus. It implies that the nature of the changes taking place are fixed and the consequences unavoidable. Instead, I want to uncover the social and economic forces that mold automation itself, and probe the other possibilities for structuring work that might be available.

Computers and microelectronics present a dazzling array of alternatives in the workplace: more worker autonomy or greater managerial authority, more skills or fewer skills, increased hierarchy or more democratic decision-making. In a market economy, the selection and assemblage of these electronic building blocks into a production system is largely the prerogative of management. Even in industries where there are strong unions, a clause inevitably exists in the contract that cedes the right to run the business to management. Take paragraph 8 of the UAW-GM agreement. It states:

> . . . the products to be manufactured, the location of plants, the schedules of production, the methods, processes and means of manufacturing are solely and exclusively the responsibility of the corporation.

Management's central criterion for production is maximizing profitability. This theoretically leads to a search for production methods capable of producing the most output with a minimum of inputs, whether the input is materials, energy, or labor. In meeting this goal, some of the design choices are largely technical in nature. The scheduling necessary, for example, to assure that a load of steel arrives at the same time that a part is due to be machined is basically a question of coordination. Other choices are social. They affect what people do in the workplace. Workers, however, may have very different ideas than managers about what should take place in production, and the way these conflicts are resolved ultimately involves power.

For managers, power can be extended through the design and deployment of technology. A central theme of this book is that, driven by considerations of profitability, management's development of new machines and systems reflects the desire to increase control over production, over the activities of workers as well as over the movement of materials. This propels the design of automation in a direction that seeks to minimize the disruptive potential of workers. There are two ways to do this. One is to automate the job entirely; this is not always technically possible or economically feasible. The other is to reduce the input of the worker on the job. This calls for designs that reduce skill requirements, transfer decision-making off the shop floor, and exert tighter control over the workers who remain. The result can be more boring and stressful work in a more tightly controlled work environment.

Developing automation in this direction carries with it some high long-term costs for society. In fact, there is a fundamental contradiction between the potential of computerization to enrich working life and increase productivity and the development of the technology in the pursuit of authoritarian social goals. The moral cost is that people's lives become diminished through degrading their work. The productivity loss stems from the inability of systems that reduce the input of workers to fully utilize the skill, talent, experience, and creativity that only human beings can provide. Moreover, in seeking to bypass human input at almost any price, new systems often achieve a breathtaking complexity that is prone to breakdown and consequently requires even greater human input. Ironically, the drive to eliminate any dependence on workers who operate machines can result in a new and greater dependence on those who repair the equipment.

The productivity loss is frequently hidden because it is difficult to quantify. The output of new machines is compared to the production of the conventional machines they replace, which often results in spectacular improvements. Part of the improvement, however, stems from the fact that the new machines are more capital-intensive. A more meaningful comparison is the actual output of the new technology compared to what its potential output might be with design changes that more fully utilize the human capabilities available. Looked at this way, the drive for control winds up as a brake rather than a spur to productivity—a truly ironic twist since the design of the technology is justified by its ability to increase efficiency.

In most ways, automation based on computers and microelectronics forms a continuum with previous technologies. The latest form of computer chips, for example, is based on development work that goes back decades and on a body of scientific knowledge that has evolved over centuries. In other ways, however, this technology is qualitatively different from what has come before. Given its extraordinary capabilities, it could aptly be termed superautomation. There are three key differences in this technology: its flexibility, scope of application, and rate of technical change.

Consider first the flexibility of computer-based automation. When the term automation was first coined in the late 1940s, reportedly by a Ford Motor Company executive, it referred to transfer devices such as those that shuttle engine blocks between machines without human aid in automobile engine plants. Skill was built into the equipment in the form of steel cams to guide machine movements and fixtures to hold parts. In order to modify a part, however, the entire production process had to be rebuilt, calling forth welding torches, the remachining of parts, and plenty of skilled workers. Often, the price of the modification approached the original cost of the equipment.

In contrast, computer automation is characterized by its flexibility.

Programmed instructions adapt general-purpose machines to specific applications. If there is to be a change in the product, then new instructions to the machine are virtually all that is necessary. A robot, for example, might be used in the welding of the body of a large car one year and in the production of a subcompact the following year.

This flexibility leads to the second difference, the scope of application. The "hardness" of previous forms of automation limited their use to mass production where the high volume could justify the expense of the equipment. Computer automation, on the other hand, is economically viable for a few pieces as well as for several million. Moreover, the sophistication of the control systems allows for the automation of many types of jobs that previously could only be done by humans. The result is a technology that is simultaneously revolutionizing the office, the assembly line, the supermarket, the engineering department, the hospital, and the loading dock.

The third difference is the rate of technical change. In the past, the design of automation evolved slowly over years if not decades. For example, an engine line brought into an auto plant in 1964 would probably only be marginally different from one introduced in 1954. The development of computer technology, however, has been proceeding at an exponential rate. New levels of capability are often achieved in a matter of months rather than years.

Taken together, the properties of computer technology lay the basis for an unprecedented restructuring of the organization of production. The technology is used to integrate the various parts of the factory, speed the transfer of information, and remove the limits of time and space from the direction of a corporation. Managers gain the ability to mold production to their social and technical purposes whether on the shop floor of a single plant or throughout the operations of a multinational company.

Computers and microelectronics play the role of the central nervous system in the factory. Information pours out of computer terminals located on time clocks, machine tools, inspection stations, design terminals, and warehouses. There is a vertical integration from the designer in the engineering office to the machine tool on the shop floor. The designer sketches a part on a cathode-ray tube, and when the design is finalized, a set of instructions is automatically produced to guide the machine. In other words, the designer "talks" directly to the machine. Further, the technology makes possible a horizontal integration on the shop floor, linking together a wide range of production activities. Computer systems not only control the operation of machine tools, but track raw materials coming into the shop, inventory completed parts, monitor robots on the assembly line, and schedule production. And these different systems are designed to be able to interact: Machines "talk" to machines.

The information-gathering potential of the technology is impressive. Imagine the logistical problems of building a jet aircraft. Thousands of complex and expensive parts have to be machined in the right order and at the right time to be put into hundreds of intricate subassemblies, which in turn become part of the finished plane. To accomplish this at the sprawling McDonnell Douglas plant in St. Louis, 439 terminals in 41 buildings pour 100,000 transactions a day into a central computer. In manufacturing alone, 25,000 different work orders are processed at any one time.

In the latest systems, the collection and transmission of data occurs as the event is taking place, literally at the speed of light. In the past, reports of shop floor activities were transmitted by verbal command and written memo, and except for the most pressing emergencies, communication routinely took place after the fact. Information more or less drifted up to the desired level. Now production problems are relayed instantaneously to the designated point in the production hierarchy, and routine information is compiled on a much more frequent basis. If the parts produced on a machine begin to drift out of preset limits, for example, a computer instructs the machine to make the proper corrections or, if the problem is more serious, the maintenance office is signaled. When managers need information, they can retrieve it instantly from the computer rather than wait for a weekly or even monthly report.

The same information-gathering ability that expands the horizons of the manager, however, can also be used to monitor more closely what workers are doing, weaving a net of control around their activities. The key issue is not the ability of computers to provide greater access to information, but *who* defines the data to be collected, who has access to it, and for *what* purposes it is used.

Telecommunications combined with computers remove the geographic limits from management. A transnational company is able to maintain the advantages of centralized control over managerial decisions in far-flung subsidiaries without sacrificing the benefits of decentralized operations. Sitting at computer terminals, engineers or managers can view the identical data at the same time from different areas of the plant, the country, or even the world. Basic policy decisions can be made at world headquarters and then refined as engineers throughout the company's operations interact with the same body of information. When sitting at computer terminals, these engineers are, in effect, in the same room for the purposes of planning and communication even if one happens to be in Paris, another in Detroit, and perhaps a third in São Paulo. Walking away from the terminal, the engineer is at the point where production takes place and therefore is immediately able to receive the kind of feedback that only comes from being on the scene.

Some of the more potent capabilities of superautomation are based on the ability to tie separate computerized machines and systems into larger networks. The integrated nature of these networks is mirrored by the corporate activities of the firms that are manufacturing automated equipment. General Electric is an excellent example. In April 1981, GE unveiled its master plan to strengthen its position in every area of factory automation. The plan involved heavy research and development expenditures, large capital commitments to internal product lines such as computerized machine controls, and a series of outside acquisitions. The corporation announced a $58 million expansion of its R&D center in Schenectady, New York; the construction of a $55 million microelectronics center in Raleigh, North Carolina; and the formation of a new industrial electronics business group. Then the company unveiled some of its outside acquisitions: $235 million to acquire Intersil, one of the largest producers of integrated circuits, a pivotal building block for automated systems; $170 million to buy Calma, the third largest producer of computer-aided design equipment; and an undisclosed amount to acquire Automation Systems, a robot organization to be used as the nucleus for a major thrust in this field. The emphasis in this entire endeavor is on fitting these various systems together, and all of this corporate activity represents only a beginning.

In pursuing this analysis of superautomation, it is important to separate the technical possibilities of computerization from the social purposes that now influence its design and deployment. There may be an enormous difference between what the technology is capable of and the particular way it is currently being used. A computer's ability to collect data while an event is taking place, for example, does not have to lead to surveillance of workers. Unfortunately, technical possibilities and social purposes do not come in neatly labeled packages but are often intertwined in a complex and rapidly changing environment. To unravel them and see their relation, the analysis must be anchored in the realities of the workplace: the real world of whirring cutters, electronic controls, managerial directives, and worker activities. It might be simpler to theorize about the potential of automation without this taxing technical detail, but the analysis would fail to take account of the interaction of production systems with people that constitutes the work environment. Moreover, the development of technology is a dynamic process that changes in response to its environment. What happens on the shop floor today, both the cutting of metal and the activities of workers, influences the ways in which technology will be designed in the future.

THE NEW SOCIAL ORGANIZATION OF WORK

8 *Women and Contingent Work*

KATHLEEN CHRISTENSEN

For the sake of flexibility, growing numbers of Americans, most notably women, are forfeiting the basic rights of full-time employment such as health benefits, pensions, and advancement opportunities. These workers comprise what is increasingly known as the contingent workforce—those workers who are not in conventional full-time employed positions. Although the purpose of this article is to examine the causes and consequences of contingent work for women, the issue of language must first be addressed.

The term contingent work is conceptually problematic. In its current usage it covers a multitude of arrangements: employees who maintain employee status but alter the *amount of time* on the job, working part-time schedules; workers who alter their *contractual agreements* with their employers and are hired on as independent contractors, temporary workers, or leased employees; and workers who alter their *place of work,* typically working at home. As currently used, the term contingent work mixes and matches these dimensions of work.

Although these conceptual problems exist, it is important to note that any use of the term contingent work be cognizant of exactly which dimensions are being addressed. The inclusion of home-based workers within the contingent workforce is conceptually the most troubling because the only point in common these workers have involves their place of work. Home-based workers

Source: Kathleen Christensen, "Women and Contingent Work," *Social Policy* 17 (Spring 1987), pp. 15–18.

can be self-employed contractors or business owners, part- or full-time employees, or temporary workers. In effect, the home becomes the locational manifestation of the employment status of the worker. As a result, I would argue that home-based work be eliminated as a category of contingent work and focus instead on the workers' employment status.

For purposes of this article, contingent work covers three types of employment arrangements: part-time work, temporary work, and independent contracting. The former implies a change in work time, while the latter two involve changes in the contractual agreement between the worker and the company. In order, however, to discuss women as contingent workers, this trend must be placed within the broader picture of women's labor-force attachment.

WOMEN'S LABOR-FORCE ATTACHMENT

Women have entered the labor force in record numbers in the United States. Nearly two-thirds (65 percent) of all women between 18 and 64 years of age are now gainfully employed, representing 44 percent of the total labor force. All indications are that their percentage share of the labor force will continue to increase, reaching 46 percent by 1995.

The most dramatic change in workstyle for American women can be seen among mothers. In 1987, an American mother is more likely to be gainfully employed than not to be. This represents a dramatic departure from the patterns and norms of 25 years ago. In 1960, less than a third (30.4 percent) of all American mothers were gainfully employed, but by 1986, almost two-thirds (62 percent) worked. The absolute numbers are as impressive as the percentages. While only 6.6 million mothers worked in 1960, 20.0 million worked in 1985.

The Women's Bureau of the U.S. Department of Labor estimates that the majority of American women work because of economic need. Approximately two-thirds of working women are widowed, divorced, separated, have never been married, or have husbands whose annual incomes are less than $15,000 a year.

One of the net results of these changes in workforce participation is that women now spend more years working. An American woman in the 1980s can expect to spend nearly 30 years of her life gainfully employed. Again, we see a pattern different from that in 1960 when a woman could expect to spend 20.1 years in the workforce, compared to 41.1 years for a man. By 1970, a woman's worklife expectancy rose slightly to 22.3 years, while a man's dropped slightly to 37.8. By 1980, the most recent year for figures, a woman could expect to work nearly three-fourths the time a man would: 29.4 years for a woman compared to 38.8 years for a man. In other words, over the last 25 years, a woman's worklife expectancy has significantly increased, while a man's has slightly decreased.

Perhaps to be expected, women can now be found in virtually every occupation and at every level of the occupational pyramid in America—they are moving up and out through the American workforce. Although, the executive, managerial, and administrative positions have historically been the domain of men, we seen evidence that is changing. In 1975, women held slightly over one-fifth (22 percent) of these positions; by 1985, women accounted for over one-third (36 percent) of these jobs. In fact, the percent of women holding managerial positions (9 percent in 1985) is beginning to approximate the percent of men (13 percent) who hold comparable positions.

Despite these advances, however, the majority of women in America remain concentrated in female-dominated professions, which tend to be lower paying and often offer limited opportunities for advancement. For example, 29 percent of all working women are in administrative-support positions, which includes clerical jobs, in contrast to the 9 percent who are managers.

The difference in women's occupational affiliations shows up in their earnings. Those who worked full-time, year-round in 1985 had median earnings of $15,600, while men working full-time, year-round had median earnings of $24,200.

During the last quarter of a century, in a word, America has witnessed dramatic changes regarding women's participation in the labor force. Yet, this is just the beginning of the story. There is further evidence that several macrolevel changes to the American economy may result in more changes in women's labor-force attachment, giving added impetus to the rise of the contingent workforce.

STRUCTURAL CHANGES TO THE DOMESTIC ECONOMY

In the United States, as elsewhere in the developed world, several forces are converging to affect the structure, design, and distribution of jobs. First, the internationalization of the economy has forced Americans to recognize that competition from abroad is exerting as much of an influence on jobs as is competition from within the American economy. Employers are trying to find ways to cut labor costs, maintain quality, and remain competitive in a world economy that frequently can compete more favorably than ours.

Technology constitutes a second driving force affecting both the design and distribution of jobs. Office work increasingly is being automated and estimates are that by 1995, 15 million computer terminals will be used in America. Although a recent National Science Foundation report by Vassily Leontief and Faye Duchin (*The Future Impact of Automation on Workers*) found that technological change would result in absolute gains in jobs in most occupations, they projected an absolute decline in the number of clerical jobs, as well as slow growth for those in management. Technology cannot be divorced from the first force, the internationalization of the economy, either. Advanced telecommunication equipment is making it possible to transport data-entry work offshore.

The third factor affecting the design and distribution of jobs is perhaps the most profound: the transformation of the American economy from an industrial to a service economy. Service-related jobs in retail and business have grown at a much faster rate than industrial jobs over the last decade. These new service jobs vary in their skill requirements: some necessitating higher educational and skill levels, others calling on minimal training. Moreover, service-sector jobs have a time and space independence that industrial jobs do not. Not only can the jobs be moved to different regions of the country to save on labor costs or to provide amenable work environments, they can also be done off-site from a centralized office, contributing to the rise of white-collar home employment.

The combination of foreign competition, technological change, and the growth of the service sector is profoundly affecting the design, the structure, and the nature of jobs in the United States.

THE CONTINGENT WORKFORCE

Most contemporary discussion of workers is based on the notion of a full-time salaried employee. Strategies and plans are premised on this notion. While most Americans continue to be full-time salaried workers, increasing numbers are hired on contingent bases—as part-time, temporary, or independent contracted workers.

Although data are limited at this point, there is evidence that women constitute a significant share of these three types of contingent workers. Nearly three-quarters of all part-time jobs in the United States are held by women. Temporary agencies, particularly those who hire clerical workers, more often fill their jobs with women than men. For some of these women, contingent work represents a work accommodation that serves their lifestyle and economic needs, for others the accommodation may not be in their best interests—particularly over the long run.

Part-Time Jobs

In America, part-time work is defined as work that involves fewer than 35 hours per week. In 1985, an average of 18 million Americans, constituting nearly one out of six workers, held part-time jobs with nearly three-quarters held by women. Wives living with their husbands constitute the major share of these part-time working women, representing about 60 percent in 1985.

Part-time workers typically are not given discretionary company benefits such as health insurance, paid sick leave, maternity leave, or vacation time, nor pension coverage. However, their employers must contribute to the worker's federal benefit coverage, including their social security account, as well as to their unemployment and workman's compensation coverage.

Temporary Agencies

According to a recent report by the Washington-based Bureau of National Affairs, the number of people employed by temporary help firms grew by more than 400 percent from 184,000 in 1970 to 760,000 in 1986. Today temporary work accounts for one out of every 13 new jobs. Estimates are that this figure will increase to 1,060,000 by 1995.

Until recently temporary work usually involved and has been associated with clerical workers. A trend is emerging, however, in which such professional and technical workers as doctors, dentists, lawyers, and computer specialists are seeking and being hired on temporary bases. Since figures on temporary workers are far from adequate, it is impossible to judge precisely how many are doing this work voluntarily or what their gender composition is.

Independent Contracting

Independent or sub-contracting involves transferring work from an employer's organization to another firm or individual. Sub-contracted work is particularly prevalent in production processes in real-estate construction, automobile and steel manufacturing. It often is a controversial practice because employers can contract work out as a way of avoiding unionizing their workers or having to deal with unions.

What is of interest to us is what appears to be an increasing use of independent contractors: individuals hired on a freelance basis to do work that would otherwise have to be done in house by employees. Both the Internal Revenue Service and the Department of Labor's Fair Labor Standard's Act define these contractors as self-employed.

The common-law test for independent contracting status is the extent to which workers have *control* over the nature or pacing of the work, as well as the *opportunity for incurring profit or loss* from their efforts. If the worker exercises minimal control and has no real opportunity to gain or lose, that worker would likely be an employee and eligible for all benefits given employees, rather than a self-employed independent contractor.

Self-employed independent contractors are paid by a contract that is negotiated on an hourly, piece-rate, or service-rendered basis. They are given no discretionary health benefits by the company. They are entirely out of the mainstream of the company and typically not eligible for skill upgrading or advancement opportunities.

Since they are self-employed, the "employer" makes no contributions to the worker's social security, workman's compensation, nor unemployment insurance. The extent to which employers may be fraudulently hiring workers as independent contractors as a ruse to avoid paying them either discretionary company benefits or contributing to their federal benefits has become somewhat controversial.

EMPLOYERS' ADVANTAGES FROM CONTINGENT WORK

Employers cite several reasons for finding contingent work attractive. The first being *cost containment*. Employers pay only for the work done. Furthermore, they rarely pay any discretionary benefits such as health insurance, paid leave, job-retraining opportunities, or pensions.

The second reason is *flexible staffing*. Employers want to avoid the situation in which they found themselves in the early 1980s recession when they had to lay off massive numbers of workers. They now want a workforce that has built-in flexibility; it can be contracted, expanded, or redeployed according to demand. The typical model for this flexible workstaff is one in which there is a stable, secure core of salaried employees, and part-time, temporary, and contracted workers constitute peripheral rings around the core that can be altered according to the shifting needs of the company. Their "flexibility" can also provide a protective buffer to the salaried employee in the core.

A third reason driving employers' consideration of a contingent workstaff is the desire to attract a *highly skilled workforce*, in which some workers want higher degrees of flexibility and autonomy than would be available to them on a salaried-employee basis. This particularly can be seen among professional and technical independent contractors and temporary workers.

A fourth reason applies only to independent contracting. Some employers use this practice to *avoid the unions*. By contracting out work to either firms or individuals, the hiring organization can avoid unionization, which they could not do if they had the work done in-house.

Although these forces are fueling the growth of contingent work from the perspective of the employers, some of them are expressing concern that a contingent workforce will result in decreased loyalty to the company and potentially to decreased productivity and quality of work. Although a contingent workstaff may be profitable in some senses, it can be costly in others.

WHY WOMEN FIND CONTINGENT WORK ATTRACTIVE

There are several factors that make contingent work appealing to some American working women, particularly those with families. The first factor is the desire for a flexible work schedule. With over 60 percent of mothers working, there is a strong desire to find a way to accommodate the demands of both work and family. As noted before, two-thirds of all part-time jobs are held by women, with 60 percent of those being held by women living with husbands.

Temporary agencies reflect women's desire for work that corresponds to family schedules. They report that they experience a drastic reduction in the number of available working women during the summer months, which correspond to school vacations, and an upswing in availability beginning with the school year in September. Temporary work, and to a more limited extent part-time and contracted work, can prove more responsive to the seasons of parenting than can the conventional full-time 9-to-5 jobs.

The second factor is the desire for a supplemental income, when a full income is not necessary. For the one-third of women who do not work strictly out of economic need, contingent work can provide extra money, without exerting extreme demands on the family woman who has full responsibility for the children and home. It must be noted, however, that by accepting part-time work, women trade off money for flexibility, since part-time work pays disproportionately less per hour than does full-time employment for exactly the same work. Part-time workers earn an average of $4.50 an hour compared to $7.80 an hour for full-time work. While some women have the luxury of trading off the extra income for the flexibility of part-time work, others don't and are penalized by this lack of parity.

The third factor is that many women can't find full-time jobs, so they take these contingent arrangements as a way simply to survive. Evidence of this can be found in the rise of women holding two or more jobs. Over the last 15 years, the percent of female dual-job holders doubled from 2.2 percent in 1970 to 4.7 percent in 1985. The rate for men stayed constant for the last 25 years at approximately 6 percent.

Of these dual-job holders, 40 percent of them hold multiple part-time jobs, while over half (57 percent) hold a part-time job in addition to a full-time one. If they are single, divorced, separated, or widowed, they take on multiple jobs for financial reasons. If they are married, they are more apt to do it out of choice as a way to build a business, save money, or accommodate the demands of a family. Sometimes it is easier to split a day with two part-time jobs, rather than one full-time job, in order to take care of their family.

PROBLEMS AND PLUSSES OF CONTINGENT WORK

Whether women pursue contingent work out of choice, or by default, there are some potential problems that cannot be ignored. The most obvious problem is the lack of fringe benefits. For married women who are covered by their husbands' plans, this may not be a problem in the short run, but for other women, particularly single heads of households, it can be serious. Perhaps the most troubling aspect of these contingent forms of work will be the lack of pensions, an area in which women are already particularly vulnerable.

Retirement income in the United States primarily comes from employment. According to the Women's Bureau, women are far less likely than men to be covered by private pension plans—41 percent of women versus 51 percent of men in 1979—or by social security paid for by their own work efforts. This pattern will only be exacerbated by hiring practices such as those found in contingent work, which will pay little to nothing toward either a private pension fund or the federal social security account of the woman worker.

Another potential problem lies in the lack of advancement opportunities for contingent workers. As currently designed, part-time workers are rarely considered for promotions with the same seriousness with which full-time workers are. There is no evidence that either temporary or contracted workers would be considered at all since they are entirely out of the employee pool.

The third potential problem has to do with whether workers hired on contingent bases will be able to get back into the core salaried-employee pool. If, in effect, companies are restructuring jobs and converting them from employee lines to temporary, contracted, or part-time jobs, there may be in absolute terms fewer full-time jobs available when women would be ready to consider them. Obviously this is a speculative concern at this point, but one that must be addressed, if we are to avoid a de facto segregation of the workforce in which women would increasingly find themselves on the periphery of corportions with limited access back into the core.

A further potential problem could be the steady increase in the number of women dual-job holders. If women cannot find full-time work, they increasingly may be forced to take on two or more jobs. Since part-time work pays disproportionately less than full-time, then their earnings will be consistently cut by taking multiple part-time jobs. In this scenario, women could find themselves working more and earning less. As it is, with median earnings of $15,600, women would have to take one full-time and one half-time job just to equal men's median earnings from $24,200.

Just as there are potential pitfalls, there are also potential gains. Part-time, temporary, and contracted work can provide women with opportunities to stay in the workforce during times in which they would not or could not work full-time because of family responsibilities. Several national surveys in the United States reveal that mothers prefer part-time over full-time work when their children are young.

Furthermore, these types of contingent work allow women to maintain their skill levels and supplement their incomes when full-time work is neither necessary nor desirable.

These types of work embody a flexibility currently not found in most jobs in the United States. The challenge that we will have to face is to ensure that these types of hiring practices enhance, rather than hurt, women's earning power, both in the short and long terms.

CHANGING OCCUPATIONAL AND REWARD STRUCTURES

9 The Rise and Fall of the Blue-Collar Worker

PETER F. DRUCKER

Whether high-paying jobs are growing or declining in the American economy is being hotly debated. But as important as the numbers is the fact that the new high-paying jobs are not where the old ones used to be.

For 30 years, from the end of World War II to the mid-1970s, high-paying jobs in all developed countries were concentrated in unskilled blue-collar work. Now a majority of the new high-paying jobs are in knowledge work: technicians, professionals, specialists of all kinds, managers. The qualification for the high-paying jobs of 20 years ago was a union card. Now it is formal schooling. The long and steep rise of the "working man"—in numbers, in social standing, in income—has turned overnight into fast decline.

There is no parallel in history to the rise of the working man in the developed countries during this century. Eighty years ago American blue-collar workers, toiling 60 hours a week, made $250 a year at most, or one-third the price of that "low-priced miracle," Henry Ford's Model T. And they had no "fringes," no seniority, no unemployment insurance, no Social Security, no paid holidays, no overtime, no pension—nothing but a cash wage of less than one dollar a day. Today's employed blue-collar worker in a unionized mass-production industry (steel, automotive, electrical machinery, paper, rubber, petroleum) working 40 hours a week earns about $50,000 a year—half in cash

Source: Peter F. Drucker, "The Rise and Fall of the Blue-Collar Worker," *The Wall Street Journal* (April 22, 1987), p. 36.

wages, half in benefits. Even after taxes, this equals seven or eight new small cars, such as the South Korean Excel, or 25 times the worker's 1907 real income (if food were used as the yardstick, the increase would be even larger). And the rise in social standing, and especially in political power has been greater still.

SOCIETY'S STEPCHILDREN

And now it is suddenly all over. There also is no parallel in history to the abrupt decline of the blue-collar worker during the past 15 years. As a proportion of the working population, blue-collar workers in manufacturing have already decreased to less than a fifth of the American labor force from more than a third. By the year 2010—less than 25 years away—they will constitute no larger a proportion of the labor force of every developed country than farmers do today—that is, a 20th of the total. The decline will be greatest precisely where the highest-paid jobs are. Blue-collar automobile employment in the U.S., 15 or 20 years hence, will hardly be more than half of what it now is, even if there are no imports at all—and automobile blue-collar employment is already down 40% from its peak, less than 10 years ago. No wonder the unions do not regard the fast growth of high-paying knowledge jobs as a compensation for the steady decline in the numbers, power, prestige and income of their constituents. Yesterday's blue-collar workers in manufacturing were society's darlings; they are fast becoming stepchildren.

This transformation was not caused by a decline in production. U.S. manufacturing output is steadily expanding, growing as fast as gross national product or a little faster. The decline of the blue-collar worker is not a matter of "competitiveness," of "government policies," of the "business cycle," or even of "imports." It is structural and irreversible.

There are two major causes. First is the steady shift from labor-intensive to knowledge-intensive industries—e.g., a drop in pouring steel and a steady rise in making pharmaceuticals. All the growth in U.S. manufacturing output in the past two decades—and it has about doubled—has been in knowledge-intensive industries. Equally important is the world-wide spread in the past 40 years of two American inventions (or discoveries), "training" and "management." In a complete reversal of all that economic history and theory had taught, these two methods enable a country with the labor costs of an "underdeveloped" economy to attain, within a very short period, the productivity of a fully "developed" one.

The first to understand this were the Japanese after World War II. By now everybody does—the South Koreans, for instance, or the Brazilians. The most telling example are the "maquiladoras," the plants on the Mexican side of the U.S. Mexican border, where unskilled and often illiterate people produce labor-intensive parts and goods for the U.S. market. It takes three years at most for a maquiladora to attain the labor productivity of a well-run American or

Japanese plant even in turning out highly sophisticated products—and it pays workers less than $2 an hour.

This means that manufacturing industry in developed countries can survive only if it shifts from being labor-intensive to being knowledge-intensive. Machine operators getting high wages for doing unskilled, repetitive work are being replaced by knowledge-workers getting high wages for designing, controlling and servicing process and product, or for managing information. This shift also fits in with demographics. In every developed country more and more young people, and especially young males, stay in school beyond the secondary level and are no longer available for blue-collar jobs, even for well-paying ones.

These are changes so sharp and so sudden as, for once, to deserve being called "revolutionary." Yet their impact is different from what everyone expected, and different also from what economic and political theory had taught.

This applies particularly to U.S. unemployment. In Britain and Western Europe the decline in blue-collar jobs in manufacturing has indeed, as unions predicted, resulted in stubborn unemployment. But in the U.S. the decline has had marginal effects at most. Even the massive job losses in the steel and automotive industries have barely left a trace in national unemployment rates. To be sure, the current 6½% unemployment rate for both adult men and adult women is probably somewhat above the rate of "natural unemployment" (the rate needed for normal job changes)—but not by much, considering the age structure of the working population. And the 4½% unemployment rate for married men is, if anything, below the natural rate and constitutes virtual "full employment." "Hidden unemployment"—that is, people who have given up looking for a job—is very big in union propaganda but probably quite scarce outside of it. A larger proportion of American adults than ever before in peacetime history—almost two-thirds—is in the labor force and working. One explanation for the low unemployment rates is surely that American workers are singularly adaptable and mobile—far more so than anyone would have thought possible. But, equally significant, blue-collar labor in manufacturing may also have already shrunk to a point where it only marginally affects total employment and unemployment rates, consumer spending, purchasing power and the economy as a whole. This would mean that we should stop looking at manufacturing employment as the economy's bellwether and should look at manufacturing output instead; as long as its volume continues to rise, the industrial economy is healthy almost regardless of employment.

Equally novel is the behavior of wage costs in the U.S. That unions give priority to the maintenance of nominal wages rather than accept lower wage rates to gain higher employment has been one of the axioms of modern economics. It still holds in Europe. But America's unions have shown an amazing willingness to make sizable concessions on wages—and even on work rules—to prevent plant closings and massive layoffs. In the U.S., at least, the principal cost-rigidity inhibiting the "self-correction" of a market economy surely no longer lies in wage costs (as economics has assumed since Keynes) but in the cost of government.

Every labor economist and every labor leader would have expected the decline of the blue-collar worker to lead to "labor militancy" on a grand scale. Some politicians still expect it—for instance, the Rev. Jesse Jackson in the U.S., the "Militants" in the British Labor Party and the "Radicals" among the German Social Democrats. But so far there has been labor militancy in only one developed country: Canada. Elsewhere there is much bitterness among the rank and file. But it is the bitterness of resignation, of impotence rather than of rebellion. In a way, the blue-collar worker has conceded defeat.

And this may underlie the most startling, and least expected, development: the political one. It is almost an axiom of politics that a major interest group actually increases its political clout for a long time after it has begun to lose numbers or income. Its members join ranks, learn to hang together lest they hang separately, and increasingly act and vote in concert. The way in which farmers in every developed country have maintained political power and increased their subsidies despite their rapid decline in numbers since World War II is a good example.

POLITICAL STRENGTH ERODED

But though it is only 10 or 15 years since the decline of the blue-collar workers first began, their political strength has already been greatly eroded. In the midst of World War II, John L. Lewis of the United Mine Workers defied the country's most popular president—and won. Thirty years later, another coal miners' leader—this time in Britain—forced a prime minister to resign. But in 1981 President Reagan broke the powerful and deeply entrenched air traffic controllers union; and a few years later British Prime Minister Margaret Thatcher broke the union that had driven her predecessor into political exile. And both President Reagan and Prime Minister Thatcher had overwhelming popular support. The labor vote may still be needed for a "progressive" candidate to be nominated. But then, in the election, labor's endorsement has become a near-guarantee of defeat, as shown by Walter Mondale's debacle in the U.S. presidential election of 1986, by the German election this January and by numerous British by-elections.

In little more than a decade before World War I, the blue-collar worker rose from impotence to become a dominant economic and social power in Western Europe, and his party the largest single political factor. The U.S. followed suit 10 years later. This transformed the economy, the society and the politics of every developed country, transcending even two world wars and tyrannies beyond precedent. What then will the decline of the blue-collar worker—and its counterpoint, the rise of the knowledge-worker—mean for the rest of this century and the next one?

CHANGING OCCUPATIONAL AND REWARD STRUCTURES

10 *The Changing Occupational Structure*

ROBERT KUTTNER

There is a good deal of evidence that job opportunities in the United States are polarizing, and that, as a result, the country's future as a middle-class society is in jeopardy. What the decline of the middle class would mean to the country can only be guessed at, but it presumably would be unwelcome to the millions of parents who hope that their children can move up the economic ladder; to American business, which needs a middle class to consume its products; and to everyone who is concerned about fairness and social harmony.

As the economy shifts away from its traditional manufacturing base to high-technology and service industries, the share of jobs providing a middle-class standard of living is shrinking. An industrial economy employs large numbers of relatively well-paid production workers. A service economy, however, employs legions of keypunchers, salesclerks, waiters, secretaries, and cashiers, and the wages for these jobs tend to be comparatively low.

This increasing concentration of jobs at the extremes of the country's range of earnings is an issue quite distinct from high unemployment—the most visible threat facing the work force today. It is also an issue distinct from the country's poor rate of economic growth. Restoring growth may make "everyone" better off, statistically, but growth by itself has little bearing on the distribution of wages. Moreover, the government's existing income-support programs simply

Source: Robert Kuttner, "The Declining Middle," *The Atlantic Monthly* 252 (July 1983), pp. 60–72.

are not designed to address this new problem. The fact is that statistically, most dollars of government transfer income go to the elderly, through Social Security and Medicare. Thus, even counting all the welfare checks and food stamps for the poor, and all the dividend coupons, interest payments, and capital gains for the rich, the working-age population still depends on jobs for fully 85 percent of its income.

During the three decades after World War II, real income grew steadily, along with the number of good jobs. Between 1958 and 1968, manufacturing added 4 million jobs; state and local government added another 3.5 million. The child of a working-class family who graduated from high school in the 1950s and 1960s could choose among a number of relatively well-paid jobs—precisely the jobs that are disappearing now. Manufacturing, which pays on average almost three times the minimum wage, added fewer than a million jobs between 1968 and 1978, and has lost nearly 3 million jobs over the past four years. Construction and production work, taken together, account for about one job in eight today, compared with one in four in 1950. According to Barry Bluestone, a professor of economics at Boston College and co-author of the book *The Deindustrialization of America,* "The pattern of wages in the old, mill-based economy looked just like a normal bell curve. It had a few highly paid jobs at the top, a few low-wage jobs at the bottom, and plenty of jobs in the middle. But in the new services economy, the middle is missing."

The erosion of the middle of the labor market is easy to misinterpret, because its roots are multiple. During the 1970s, the entry into the work force of an unprecedented number of women and of young adults born during the baby boom resulted in too many workers for the jobs available, and depressed wages. The decline of the middle also has something to do with the explosive growth in world trade since 1960. As manufacturing technologies have become more mobile, and multinational firms more footloose, production jobs have migrated from the U.S. to countries where wages are low. In addition, technology itself has helped to provoke the shifts in the job market. For example, fewer American workers would have been needed to make steel in 1980 than in 1960 even if the pressures of global competition had not been a factor, because new machines have made many of their tasks redundant. Finally, the high rate of unemployment caused by these trends has tended to drive wages down further, especially at the low end, since it forces unskilled workers to compete for their jobs with unemployed people who are willing to do the work for less.

Although demographic shifts, stepped-up world trade, unemployment, and especially the advance of technology all have had an effect on the shape of the job market, middle-level jobs have been disappearing ultimately as a result of the ways in which technological gains are being distributed. When a machine replaces a production worker, both the firm and consumers as a group benefit. The loss falls mainly on the worker who is displaced. If that loss is generalized to millions of high-paid workers, they suffer as a group, and the economy as a whole suffers a loss of worker purchasing power. Thus the lack of a mechanism to distribute some of the financial gains from technology to the work force comes back to haunt the entire economy.

As it happens, production work is becoming heavily automated at a time when the most powerful agents of distribution—the trade-union movement and the public sector—are in decline. There is nothing intrinsic in assembling cars, mining coal, or pouring molten steel that requires high wages. These jobs *do* pay well mainly because of the efforts of strong unions. Just as high wages are not a necessary characteristic of factory work, low wages are not a necessary characteristic of a service economy. But because unions have organized few fast-food workers, typists, bank clerks, or even computer assemblers, most of the new service jobs that are replacing factory work pay far lower wages. And the chief realm of a service economy in which wages *are* good—public service (teaching, social work, nursing, and so forth)—can no longer be relied on to take up the slack. Although the number of jobs funded by state and local government grew substantially throughout the 1960s and 1970s, it peaked in 1980, and had fallen by 200,000 at the end of 1981—the most recent year for which figures are available. Whether or not one favors strong unions or the pervasive presence of government in society, unions and government undeniably have operated as instruments of a relatively egalitarian distribution of wages. As their influence declines, so does their impact on the national pattern of earnings.

Labor statistics show that wage and salary income is becoming more unequally distributed in the economy as a whole, and in service industries in particular. It also turns out that the fastest-growing industries are those with the most extreme wage spread.

The Bureau of Labor Statistics ranks earnings in ascending order by percentile. Thus, a worker at the eightieth percentile makes more money than 79.9 percent of the rest of the population. A study conducted in 1980 by two economists at the Department of Labor, Peter Henle and Paul Ryscavage, found that between 1958 and 1977, the earnings of men at the twentieth percentile increased by 130.6 percent (with no adjustment for inflation), but the earnings of men at the eightieth percentile increased by 206.7 percent. The book *Services: The New Economy,* by Thomas M. Stanback, Jr., and three colleagues, published in 1981, concludes, "For services as a whole, the most important observation is that there tends to be a heavy concentration of employment in better-than-average and in poorer-than-average jobs. In contrast, in manufacturing and construction, the distributions are more heavily weighted toward medium and above-average jobs."

Statisticians at the Department of Labor have been comparing the salaries paid in the twenty fastest-growing occupations with those in the twenty that are in the sharpest decline. Although their results will not be released in full until later this year, Ronald Kutscher, an assistant commissioner of the Bureau of Labor Statistics, says that workers in the fastest-growing industries are earning, on average, $5,000 a year *less* than workers in industries that are declining or that are growing most slowly.

There is, of course, independent evidence of widening inequality in America, apart from changes in the distribution of jobs. The number of Americans living in poverty, which dropped to about 23 million in 1974, is greater than 30 million today. The 20 percent of households in the country's highest tax brackets take 41.9 percent of all income paid in the U.S.—the highest share in two decades. Paradoxically, the entry of more women into the labor force has exaggerated the inequality of family incomes. On average, the working wife of a man who makes $50,000 earns $5,000 more than the working wife of a man who makes $10,000. Homeownership, for years the badge of membership in the middle class, has become a nearly impossible feat for those who do not have a house already. A new house of average price is beyond the budgets of more than 75 percent of all households.

One might fault the Reagan Administration for cutting taxes at the top and social programs at the bottom, but the widening polarization of the labor force remains a phenomenon distinct from these policies and largely ignored in the national debate. Long after the Reagan Administration ends, this problem will plague the American economy.

In 1982, the Department of Labor ranked the number of actual job openings in 1980 by category. These openings reflect turnover as well as net job growth. The top fifteen job categories, with a single exception, are ones that middle-class parents hope their children will avoid:

Job	*Openings, 1980*
Retail salesclerks	757,750
Managers and administrators (not elsewhere classified)	711,793
Cashiers	617,973
Secretaries (not elsewhere classified)	599,216
Waiters and waitresses	465,628
Cooks (except private household)	437,341
Stockhandlers	358,393
Janitors and sextons	333,309
Bookkeepers	304,789
Miscellaneous clerical workers	299,940
Nursing aides and orderlies	284,332
Child-care workers, private household	277,525
Building-interior cleaners	259,528
Typists	250,276
Truck drivers	245,377

Of the top fifteen, only one, the grab-bag category "Managers and administrators (not elsewhere classified)," suggests a well-paid line of work. The first profession to appear on the department's roster is job number thirty-six, accountants, with 140,108 openings. Far down on the list come computer programmers (30,953) and electrical engineers (23,089). According to the most recent comprehensive forecast of the Bureau of Labor Statistics, released in

1982, the economy will generate some 19 million new jobs between 1980 and 1990, about 3.5 million of which will be "professional and technical." Low-wage service and clerical work will account for almost 7 million new jobs.

Within the manufacturing sector, the job structure of the one major growth industry—microelectronics—is highly stratified. Computer scientists and engineers typically have more than a college degree and earn $30,000 a year and more. The workers who assemble computers, however, earn about $12,000 a year. The number of skilled technicians whose wages fall between these two extremes is surprisingly small. The rate of productivity of the electronics industry is accelerating so fast, in fact, that electronics is not expected to be an important source of new production jobs, good or bad. The more significant effect of the computer on work is being realized not in the computer's manufacture but in its application. The Bureau of Labor Statistics projects that the U.S. will require about 120,000 more computer programmers and 125,000 more electrical engineers during the next decade, but it will require some 3 million more secretaries and office clerks.

By 1990, 600,000 more janitors, 500,000 more salesclerks, and 400,000 more fast-food workers will also be needed. To be sure, the economy has always included low-wage service jobs, but in the past it also included a larger middle range of relatively well-paid production jobs. These were the jobs that created the postwar, blue-collar middle class. Today, a welding job in an automobile plant is performed more cheaply by a robot or a foreigner; that job once paid somebody $12 an hour, and there are few openings at $12 an hour in restaurants, hotels, or computer-assembly plants.

Given the 10 percent unemployment rate, it is tempting to conclude that lots of low-wage, "entry-level" service jobs are just what the economy needs. Unfortunately, although these jobs do provide an entry into the work force, few offer opportunities for advancement. The fast-food industry, for example, employs a small number of executives, and hundreds of thousands of cashiers and kitchen help who make about $3.50 an hour. With some variation, keypunchers, chambermaids, and retail sales personnel confront the same short job ladder.

Statistics do not do justice to the story. A look at actual offices, stores, and factories is even more revealing. The Boston area, with its heavy concentration of financial services (banking, credit reporting, real estate, and so forth) and high-technology firms, can serve as a model of where the economy as a whole is headed. To get a practical glimpse of the emerging labor market, I interviewed personnel executives at several companies there.

The work force of financial-service industries grew 32 percent, to 5.4 million, between 1973 and 1982—a period when durable manufacturing lost workers. At Blue Cross Blue Shield, which falls within this sector of the economy, the work force grew 46 percent, to 88,000, between 1970 and 1982. As in most insurance companies, the main activity of Blue Cross Blue Shield is collecting premiums and paying claims. Computerization and remote data transmission have made these tasks more routine and have allowed them to be performed almost anywhere.

The company's offices in Massachusetts employ about 4,000 people, of whom some 400 are rated managers, another 1,000 are rated professionals (attorneys, actuaries, programmers, doctors, nurses), and some 2,600 are rated clericals. According to Theron R. Bradley, who oversees employee training for Blue Cross Blue Shield of Massachusetts, computerization has not had the same effect on everyone who works for the company. For professionals like him, it has provided new possibilities. Bradley says, "As the manager of training, If I wonder who in the company has taken which courses, I can have the computer research that at the touch of a button. If I need to analyze the results of testing, I can have the computer produce research that I could not have imagined allocating the resources for a short time ago. Our accountants can now spend their time doing what they were trained to do—setting up accounting systems and interpreting the results. In the past, a lot of their work was glorified bookkeeping." At the same time, however, the computer has had a different effect on Blue Cross Blue Shield's large clerical force. Bradley says, "In the past, the lowest six levels of our clerical structure looked like an inverted pyramid. At the very bottom were jobs like filing, sorting, mail delivery, and basic data-entry jobs. You'd almost have to try *not* to be promoted in order to avoid rising through the clerical ranks. Most people did get promoted. With good oral and written communication skills, you could move fairly easily into a supervisory or junior management job. Those senior clerical jobs are disappearing, because the computer is taking over so much of the decision-making. We have a lot more routine jobs at the bottom, and a few more complex jobs at the top, where you need someone to analyze hundreds of pages of data."

Blue Cross Blue Shield's largest job category is that of claims examiner. The company recently opened "outreach" offices in the towns of Plymouth and Rockland, outside Boston, to house the newly organized Blue Shield claims department. The work is part-time, in order to attract mothers of young children, students, and others who will accept reduced hours. As it turns out, all of the claims examiners on the day shift at Plymouth are women. The day shift runs from eight until two. "You can drop Junior off at school, and drop down to our office at the local shopping mall," Bradley explains. Each clerk sits at a computer screen with a pile of claims, and types in the subscriber's name or number, the code number of the medical procedure claimed, the fee, and the identification number of the doctor. The computer verifies that the subscriber is covered. A different department handles more complicated claims, and still another mails out reimbursement checks.

A claim takes about a minute to process, and a good clerk can dispatch 300 a day; the computer keeps track of productivity. On average, the clerks are paid between $4.50 and $5.75 an hour, but wages are adjusted every two months to reflect each clerk's output. Blue Shield provides a fifteen-minute paid lunch break and a week of vacation after one year. It provides no health insurance.

When the company announced its plans to move to Plymouth, a Blue Shield Day was held at a local vocational high school to recruit applicants. Bradley says, "We locate these offices in areas of high labor demand so that we don't have to outbid everybody else's wages. We have no trouble finding people. Every time we open one, we are swamped with applications."

According to one of the clerks at Plymouth, who requested anonymity, few of the women who began work with her at the office when it opened, in March of 1982, are there today. The most demoralizing thing about the job, she says, is the isolation. Virtually all communication is between the clerk and the machine. Since pay is based on output and the job itself requires no human interaction, there is every incentive to keep social conversation to a minimum. "Last Christmas," my informant says, "they organized a little office party. They had a 'Christmas Grab,' where everybody picks a name out of a hat. You bring in a little gift for somebody and they bring one in for you. We realized that nobody knew anyone else's name."

The new system has indeed improved the productivity of the company's claims department, but more intensive use of technology and higher output have not translated into better jobs. Despite the space-age technology, each clerk can be trained in only four weeks. The work is more routine, and the company's labor costs are lower.

It is intriguing to consider whether the loss of middle-income jobs reflects technological imperatives or merely management preferences. Much has been written on the tendency of technology to degrade work, but there are enough instances to the contrary to suggest that managers have substantial latitude in how they choose to reorganize work.

Banks provide examples of technology both making work more tedious and making work more challenging. Like armies, banks divide their troops into officers and enlisted personnel. Behind every officer is a cadre of backroom clerks. Although in theory the lowliest teller can aspire to be an officer, in practice there is little opportunity to move up from the clerical force into the officer ranks. For example, First National Bank of Boston, the nation's seventeenth largest bank, hires about a hundred people a year for its loan-officer-development program. As many as ten may be promoted from within. Most new loan officers, however, are recent graduates of colleges or business schools. The competition is keen, and the bank will pay more than $30,000 to attract a promising MBA holder. Clerical workers at First National outnumber officers by more than two to one. The 621 tellers earn an average salary of $6.59 an hour, or about $263 a week, and below that rank there are thousands of other, lower-paid clerical jobs.

One job that has become more routine as a result of technology is that of money-transfer clerk. Like most banks, First National has automated its money-transfer operation, plugging into a computer-connected network. When a customer requests that money be transferred from one bank to another, the clerk punches up a standard format on the computer terminal, checks whether funds are available, enters the number of the bank to receive the funds, and electronically consummates the transaction. According to Steven Bavaria, the bank's vice president in charge of personnel, "We are able to use more junior people for transfers since that was automated. We even use some part-time high school students. Because we build more safeguards into the system, the people operating it can concentrate all their attention on five or six variables: the account number, the amount of money being transferred, the name of the bank, and so on, rather than having to type and proofread an entire Telex message."

In contrast, the computerization of First National's letter-of-credit department has upgraded the status of several dozen employees. A letter of credit is a document from the bank to a foreign supplier's bank, guaranteeing payment for goods that are about to be shipped. Formerly, processing a letter of credit involved assembling pieces of paper, with a chain of clerks verifying shipping and credit documents and typing forms. Now the bank's letter-of-credit department is mostly what Bavaria terms "junior professionals," each with a computer terminal. "The same person can review the documents, create the file, send the letter, and, when we have verification that the order has been shipped, authorize payment," Bavaria says. Most letter-of-credit employees earn between $300 and $400 a week, and the position is sometimes a way station in the officer-training program.

Gross statistics on the earnings of clerical workers nationwide suggest that this kind of upgrading is unusual. The vast majority of clerks, men and women, earn less than $300 a week. For full-time female clerical workers, the average weekly salary in 1980 was about $220. Of the total 17 million female clerical workers, fewer than 500,000 earned as much as $400 a week. Fewer than a million earned more than $335. The study by Henle and Ryscavage indicates that clerical wages are becoming increasingly skewed to the low end of pay scales.

Retail sales is another growth industry in which wages are clustering more and more at the extremes. The number of people working in the Department of Labor's "wholesale and retail trade" category has grown from 8.7 million in 1950 to more than 18 million today. In 1981, nonsupervisory workers in trade and retailing earned an average of about $150 a week. The 2.4 million salesclerks earned an average of $178 a week, and the 410,000 manufacturers' sales representatives earned an average of $434.

In theory, the shift to an economy based on services and information ought to be compatible with the equitable distribution of income and jobs. Substituting technology for human labor makes the country as a whole richer. Some economists argue that as the economy grows more productive, technology will gradually eliminate many routine, dead-end jobs, replacing them with high-skill positions and more leisure time for all. The market does not necessarily convert rising productivity into an acceptable distribution of income and leisure, however. The pattern of development in much of the Third World demonstrates that it is quite possible for an economy to attain a high rate of growth based on rapid technological progress, and still function as a society in which one half of the work force waits on the other half, and makes a good deal less. Low wages tend to foster more low wages. If you have to pay a welder $12 an hour, it may make economic sense to substitute capital for labor and build an automated welding machine. But if fast-food workers cost only $3.50 an hour, automation may be a long time coming. It is often the high-paying jobs that are replaced by machines, not the low-paying ones.

Few economists expect the booming high-technology field to be an important source of new jobs: its productivity is too high. According to projections by Data Resources, Inc., a corporate consulting firm, the entire high-tech industry, broadly defined, will provide fewer than a million new jobs over the next decade. The American Electronics Association projects that the number of jobs for skilled electronics technicians will grow by only about 15,000 a year. Moreover, the relatively low-skill assembly jobs in microelectronics plants, which do not pay well to begin with, are rapidly being automated or shifted to countries in which wage scales are even lower.

The Hewlett-Packard Company—a California-based manufacturer of computers, computer software, scientific instruments, and testing devices, with assets totaling about $4.3 billion—offers a good example of technology's effect on the high-tech labor force. The company prides itself on "the Hewlett-Packard way"—an informal style of management with few barriers between the executive suite and the shop floor.

The medical-products group, one of five major corporate divisions, makes an array of sophisticated electronic monitoring devices used by hospitals to analyze blood chemistry, to monitor fetal functions, and to check heartbeat, blood pressure, and oxygen intake. It employs some 2,700 people in two plants near Boston. During the past two years, the number of engineers and computer scientists at the company has grown by about 1,250 nationwide, but few technicians or assembly workers have been hired. Despite the group's steady growth in sales, averaging between 15 and 20 percent a year, productivity has also been rising so rapidly that the company has barely avoided layoffs.

Hewlett-Packard's large plant in the suburban town of Waltham occupies a long, low brick building. The shop floor is quiet, clean, and well lit. There is no assembly line. In the main section of the floor, employees sit at benches attaching the final few components to printed circuit boards that have been produced mainly by machines elsewhere in the plant. James Phelps, the group personnel manager, says, "For years, loading circuit boards was like building the pyramids. You needed armies of human labor. All the components were hand-inserted. Today you can design a circuit board by computer, and the same computer program will load the machine that produces the board, put the components into the right sequence, and produce most of the board."

At the end of the long shop floor is the sequencing machine—about fifteen feet long and five feet high. In bins above the machine are supplies of dozens of different components. Following a computer tape, the machine loads the components in proper sequence onto what look like ammunition belts for a Gatling gun. The belts then feed the circuit-board-inserting machine, and out come nearly finished circuit boards, with capacitors, resistors, and diodes all in place. Only a few parts remain to be inserted by hand. One worker tends the sequencing machine. She was trained for the job in a few days. A second semi-skilled worker tends the inserter.

Automation is evident, and equally impressive, elsewhere in the plant. A printed circuit board begins with a layer of inert fiberglass sandwiched between two sheets of copper plate. Holes are punched into the blank board to receive the components. In the "printing" of circuits onto the board, an ink screening process is used to indicate the tracks that will eventually be etched into the board. In the past few years, machines have been introduced that plate the copper, punch the holes, and even do the screening. In every case, a few machine tenders now do the work of several production workers, and they need not be highly skilled technicians. Workers were trained to use the automatic screening machine in a single day; there was no upgrading of status, job description, or pay.

According to James Phelps, "We've peaked in terms of our need for the classical technician. The machinery is more and more self-diagnosing. There's less of a judgment factor. The computer spits out an instruction to change such-and-such a component, and you do it. We can upgrade a $5-an-hour assembler to a $6- or $7-an-hour junior technician with little advanced training. A few years ago, we were recruiting skilled technicians as aggressively as we're recruiting engineers today. We're not doing it now. Organizations that train technicians should realize that they're preparing people for a smaller and smaller marketplace."

A few miles away from Hewlett-Packard, the GenRad Corporation makes electronic testing instruments, which are used in quality-control operations throughout the electronics industry. According to Richard Cambria, GenRad's vice president for personnel, the instruments are having the same effect on GenRad's production force that they are having on GenRad's customers. There is a reduced need for the traditional job of "test technician," because most testing has become integral to production. Cambria says, "It used to be that the assemblers would assemble, and the inspectors would occasionally throw the work back at them. With the quality control built right into the manufacturing process, we've been able to get rid of an entire layer of inspectors."

In principle, the proliferation of computers will require skilled repair people to maintain them. The Bureau of Labor Statistics projects about 130,000 new repair jobs for all types of office machinery over the course of this decade. But according to Michael Odom, a human-resources executive with Digital Equipment Corporation, the nation's second-largest computer maker, computer repair is likely to be a field of short-lived growth. Many of the new minicomputers on the market are self-diagnosing. If a circuit malfunctions, the screen will literally tell the user which panel to pull out. The circuit boards slide out like trays in a baker's rack. Digital gives its customers a supply of yellow mailing envelopes, with instructions to send the defective board back to the company for a replacement.

"Things that break are usually things that move," Odom says. "The circuit board is being condensed into a chip. You don't repair a chip. The old personal computers had 100 chips; the new ones have nine. Pretty soon, computers will have no moving parts other than an on-off button. When the whole operation is on a few chips, the greatest likelihood of malfunction is probably random failure caused when a circuit is struck by an extraterrestrial gamma ray.

"When the systems were more complex, we needed more sophisticated technicians to maintain them," Odom adds. "As the computer becomes a mass-market commodity—like a stereo—sales, distribution, and repair are all becoming routinized and simplified."

Odom believes passionately in educating people to work with computers. He is currently on loan from Digital to the Boston city school system to help develop a computer curriculum. But he scoffs at those who expect high technology to fill the gaps in the middle of the labor market. "It's just wrong to hold out hope for high-tech as a mass provider of jobs," he says. "Increasingly, it's an elite industry of professionals. The job growth is in the industries that use our technology, and I worry about what kind of jobs those will be and who will get them."

Within manufacturing generally, automation will affect the middle of the labor market in ways that are not always easy to predict. One important innovation just beginning to be applied is "computer aided design/computer aided manufacturing," or CAD/CAM. Using a computer to make and modify designs, based on an engineer's specifications, increases the productivity of a draftsman many times. The same computer that produces the blueprint can instruct the machinery to build the part. A number of junior colleges have introduced CAD/CAM programs to retrain and upgrade a whole generation of draftsmen. But, according to Lon Bonczek, a corporate vice president of Computervision, a leading CAD/CAM manufacturer, just over the horizon is something called CIM, or "computer-integrated manufacturing." The job of CAD/CAM technician, which is regarded as hot new opportunity, will diminish, because the engineer himself will sit at the computer terminal and transform his specifications directly into a product.

The economy, of course, will still require technicians to perform traditional jobs. During the 1980s, about 200,000 more auto mechanics will be needed. But contrary to expectations of a smaller but more highly skilled blue-collar work force serving high technology, there is little evidence on the shop floors or in federal statistics of a boom in demand for skilled technicians.

The Bureau of Labor Statistics projects that by 1990 about 1.5 million new technical jobs will be created, accounting for 8 percent of the total job growth. But these "new technical jobs" include dental hygienists and surveyors as well as electronics technicians.

If the changing shape of the nation's economy indeed signals a change for the worse in the distribution of jobs, there has been surprisingly little fruitful discussion of what is to be done. Public-policy debate focuses primarily on the gross unemployment rate, and on the need for government to create jobs or to subsidize job training. The quality of work, aside from industrial health-and-safety hazards, and the distribution of job opportunities are not visible public issues.

Some economists believe that demographic shifts are responsible for the polarization of the work force, and that demographic shifts will eventually restore equilibrium. In this view, the explanation for both the depression of average wages and the widening gap between high and low wages is the disproportionate number of young people and of women who entered the work force during the 1970s. Young workers characteristically earn less than older ones, because of their inexperience, and women earn less than men, because of discrimination, and because of the poor salary histories that past discrimination has engendered. So it is often argued that as the work force ages and discrimination recedes, wages will fall closer to the center. Also, some economists say that as the baby-boom generation is absorbed into the work force, labor supply and demand will again balance out and employers will have to pay more to attract workers.

The trouble with this theory is that it fails to acknowledge the changing range of occupations, as well as the opportunity to rise within them. The demand for workers is increasingly skewed to the extremes as a result of technology—a pressure independent of supply. And as the youthful work force ages and more women seek to rise from clerical status, the competition will become even more intense for the scarce good jobs. Demographic shifts are likely to aggravate the competition for at least a generation. In 1975, about 16.7 million workers were between thirty-five and forty-five years old—the prime earning period. In 1995, almost 34 million people will fall into this age group. Twice as many middle-aged people will be competing in 1995 for the limited supply of breadwinning jobs. A recent study by the economists Martin Dooley and Peter Gottschalk, who teach at McMaster University and Bowdoin College, respectively, considered whether or not demographics alone could account for the inequality in wages during the 1970s. Even after correcting for the age, sex, size, and education of the work force, Dooley and Gottschalk found that wages remained unaccountably skewed.

Others argue that as the productivity of the economy continues to rise, wages must logically rise as well. Isabel Sawhill, the former director of the National Commission for Employment Policy, says, "In the long run, the cure for low wages is increasing productivity."

This cure does not treat the distributional malady, however. It is entirely possible for the economy as a whole to grow more productive while wages and occupations continue to concentrate at the extremes. At the GenRad Corporation, company revenues per worker rose from $59,000 in 1981 to $72,000 in 1982 to an estimated $83,000 in 1983. Nevertheless, the wages of assembly workers have not grown nearly that fast, and the company is having no trouble filling those jobs.

The fragmentation of the economy into dissimilar realms has been debated for years by economists. John Stuart Mill, in his work *Principles of Political Economy,* published in 1848, noted the fact that a common laborer could not reasonably aspire to be an artisan, and analyzed the barriers to free competition among workers for available jobs. Several generations of American students of trade unionism, beginning with the institutional economist John R. Commons in the 1920s, have pursued this line of inquiry. In the early 1970s, the labor economists Peter Doeringer and Michael Piore began to write about dual labor markets. The "primary" labor market, they found, is typically union-organized or professional, with relatively good wages and benefits, fairly predictable careers, and job security. The "secondary" labor market, in contrast, is made up of less desirable, low-wage jobs. Its defining characteristic is instability, with high turnover and a large proportion of part-time or seasonal work. Piore, Deoringer, and other labor-market theorists also point out the predominance of white males in "primary" jobs, and argue that the dual system both reflects and reinforces discrimination. A surprisingly large number of primary jobs, they note, do not have high formal entry requirements; skills are learned instead on the job. Through informal referral networks, white men are steered into the better jobs while blacks and women end up with the inferior ones.

Economists of the rival "human-capital" school have responded to dual-labor-market theorists by arguing that if such a schism exists at all, it reflects mainly differences of skill and education. They say that undoubtedly some racial and sexual discrimination persists, and is to be deplored, but if workers are only marginally literate, with little training and sloppy work habits, it is hardly surprising that they should find only marginal jobs. Indeed, they say, it is fortunate that society provides jobs for such workers at all. According to the human-capital theorists, the obvious way to upgrade jobs is to upgrade the qualifications of the work force. This approach has informed most federal job-training efforts, including the Manpower Development and Training Act of 1962, CETA, and, more recently, President Reagan's retraining initiative.

A favorite counter-example by the dual-labor school is the mobilization of factories during World War II. All kinds of people who had never seen the inside of a plant learned a variety of skills on the job. But jobs were available then. If the demand for workers today is increasingly shifting toward professional jobs at one extreme, and low-wage, routine jobs at the other, it is no solution to upgrade the qualifications of workers. Although nobody should oppose the improvement of basic education for its own sake, education will have little impact on the kinds of jobs available.

Ronald Kutscher, of the Bureau of Labor Statistics, reports that true shortages of qualified workers exist only in a few engineering specialties, and in the highest-skilled blue-collar occupations, such as journeyman tool-and-die maker. Since the real problem is the supply of good jobs rather than the supply of good workers, an emphasis on education and training will make the work force even more frustrated than it is now. According to Isabel Sawhill, during the 1970s one college graduate in five was forced to take a job that did not require a college degree.

If retraining and better basic education are not by themselves solutions to the polarization of the supply of jobs, society is not entirely without remedies. It can intervene in several ways to influence the mix of jobs, to maximize career possibilities within the workplace, and, if nothing else, to assure that routine jobs at least pay decently. Most such measures, however, are either unpopular or relatively untested.

The simplest way to promote relatively equal wage scales is to encourage trade unionism. According to James Medoff, a professor of economics at Harvard who specializes in labor markets, unionism by itself tends to increase wages by between 10 and 20 percent. In retail trade, for example, workers in unionized stores earn an average of $179 a week, according to the Bureau of Labor Statistics. Non-union retail workers earn only $111 a week. Unionism is a force for wage equality for several distinct reasons. First, unions understandably seek to organize workers from the bottom up, both for reasons of solidarity and to ensure that workers whose wages are low do not underbid union members. Second, unions tend to negotiate standard wage rates across firms and regions. Unions are also the principal lobby for a minimum wage, which tends to improve the distribution of wages generally. It is interesting to note that the minimum wage, which was once 54 percent of the average wage, has fallen to a record low of 39 percent—another reflection of widening inequality, and of the declining power of unions.

When unions represent most workers, they undoubtedly improve the overall distribution of wages. When they represent only about one worker in five—as in the United States today, where union membership is at a forty-seven-year low—they tend to reinforce the trend toward a dual labor market, because their members enjoy a degree of market power to bargain for cost-of-living and productivity adjustments that other workers lack. In countries with strong unions, the labor movement acts as a prime force for what is sometimes called a solidarity wage policy—the deliberate narrowing of the range in wage and salary income. In Sweden, where both the trade-union federation and the Social Democratic Party have recognized the limits of taxation and income transfer as instruments of redistribution, wage equalization is seen as a more efficient version of the welfare state. If wages are more nearly equal to begin with, there is less need for social programs and high taxes.

Unions can be a force for egalitarian labor markets insofar as they are the principal constituency for clear career ladders, apprenticeship programs, and on-the-job retraining. Although some unions, notably those for the building trades, have acted to restrict opportunities for apprenticeship, others have sought to expand them. In the tool-and-die industry, for example, where it takes as long as four years to train a skilled journeyman, many small shops are reluctant to invest that much effort in a worker who is free to leave for a better-paid job. The International Association of Machinists has long lobbied—unsuccessfully—for publicly subsidized apprenticeship training. The United Auto Workers recently bargained for a joint program with Ford, which allows both active and laid-off workers to train for new jobs, partly at company expense. In health care, which is fragmenting into a number of paraprofessional specialties, the Service Employees International Union has fought a rearguard action to allow orderlies and practical nurses to train on the job for better positions, rather than letting hospitals recruit all their technicians from technical schools and their registered nurses from four-year colleges.

Unions can also influence the way technology is applied. Officials of the United Food and Commercial Workers (UFCW) report that the new optical scanners are popular among unionized check-out clerks. A clerk has more time to chat with customers, while the scanner rings up the groceries. The scanners obviously help management improve inventory control. In bargaining over the introduction of the new technology, however, the union made sure that it would not be used to tie wages or promotions to output. This contrasts sharply with the policy of Massachusetts Blue Cross Blue Shield, whose claims examiners are not represented by a union and whose automated equipment accomplishes precisely what the UFCW sought to avoid.

Although American unions are in decline, and therefore lack a pervasive influence on how management makes use of technology, management's own interests are sometimes served by the upgrading of jobs as a result of technology. While many Marxist economists, notably the late Harry Braverman, have contended that the "de-skilling" of the labor force is an inevitable consequence of advanced capitalism, the contrast between the status of the Blue Cross Blue Shield claims clerks and that of First National Bank's letter-of-credit clerks suggests that management has some latitude in the way it applies technology to the tasks at hand. William Spring, a White House adviser to President Jimmy Carter on employment policy, believes that few managers pay close attention to internal labor markets. While most corporations are nominally committed to a policy of promotion from within, maintaining a broad middle sector of the work force is rarely a corporate goal for its own sake. And the wish to cut short-run costs seems to encourage just the sort of de-skilling that the Marxists denounce.

The other unfashionable remedy for the polarization of work is the public sector itself. Tom Wolfe, in his essay "Mau-Mauing the Flak-Catchers," which described an anti-poverty office in San Francisco, noted a truth about public-service agencies that poor people themselves had perceived immediately.

The real value of the war on poverty lay not in the programs offered *by* the poverty agencies but in the opportunity of holding jobs *in* the poverty agencies. The public sector has been the main avenue of upward mobility for blacks and women. In 1970, according to the economist Martin Carnoy, of Stanford University, 57 percent of all black male college graduates, 79 percent of all black female college graduates, and 62 percent of all white female college graduates held public-sector jobs.

As more and more of society's material goods are produced by fewer and fewer workers, the kind of society that results is ultimately a political choice. We can permit the owners of wealth to spend most of their earnings privately—and the service jobs will include salesclerks, waiters, and open-heart surgeons—or we can tax away some of those earnings and let the service jobs include more teachers and social workers. The choice exists within a realm of public policy that is the least charted of all: the means for redistributing the fruits of productivity and automation.

Harry Bridges, who was the president of the International Longshoremen's and Warehousemen's Union from 1937 to 1977, was among the first labor leaders to bargain explicitly with management on the consequences of automation. Beginning in 1960, as the West Coast docks were automated, Bridges negotiated the following arrangement: The shipping companies would be free to automate, but his members would not be laid off. The wages of longshoremen holding "A" union cards would rise with productivity, and the number of longshoremen would gradually be reduced by attrition. Bridges liked to say, "At this rate, by the year 2000 there will be one longshoreman left on the docks. But he's going to be the best-paid son of a bitch in the United States." As Bridges recognized, the deal was fine for union members holding "A" cards, but not so good for the work force as a whole. To generalize from Bridges's quip to an only slightly more absurd extreme, what would happen if all physical goods were produced by robots? The one worker who flipped the switch would boast astronomically high productivity. What should he be paid? And what would everyone else do for a living?

Twenty years ago, many futurists predicted that society would soon become so productive that the main problem would be what to do with leisure time. It hasn't quite worked out that way. The productivity of manufacturing is rising, but since 1973, average household income, adjusted for inflation, has declined about 11 percent, and average wages have dropped 16 percent. In more than 50 percent of families with children, both parents work. Almost half of all mothers of children under six years of age work. Far from the ideal of part-time jobs, shared child-rearing, and increased leisure, most parents in two-income families are instead working full-time, juggling baby-sitters, and wondering why they have no time even to watch television.

Among the various possible remedies, the Luddite answer is surely the wrong one. It makes no sense to preserve manual jobs doing hot, dirty work, when machines could liberate workers to do something else. But, still, the question is what? Other countries, notably in Western Europe, have made an effort to redistribute productivity by lowering the retirement age, shortening the work week, increasing vacation time, and promoting part-time work. Although none of these experiments has yet had impressive results—the European economy continues to generate more products with fewer workers, and to create more low- than high-wage jobs—the obstacles to a radical redistribution of technology's benefits seem to be as much political as they are economic.

The industrial West as a whole is coming under heavy pressure, both from technology and from trade, to lower its wages. If machines are available to replace most human production jobs, and if unemployed women in the Orient are eager to take the remaining jobs producing for Western markets at ninety cents an hour, then the postwar social contract that developed in Western Europe and North America is in severe jeopardy. That social contract included high wages, job security, and a costly welfare state. It enabled the West to defy Karl Marx's prediction that under capitalism the labor force could only become worse off. Now, paradoxically, one hears that in order to maintain our standard of living we must first lower it: to remain competitive, the West must jettison the very social baggage that has assured a comfortable passage. This conclusion somehow defies common sense. From the viewpoint of the entrepreneur, wages are an unfortunate cost of production. From the perspective of the entire economy, however, they are the stuff of social stability.

A distinguished company of essentially conservative political economists—Charles E. Lindblom, Karl Polanyi, Joseph Schumpeter, and even Adam Smith—have recognized what Lindblom was eventually to call "the limited competence of markets." Free markets are superb at energizing human creativity and innovation. They incur painful costs of transition, however, which are often imposed unfairly. Through no fault of their own, textile workers in North Carolina bear the costs of a new technology invented in Tokyo. The market is a marvelous engine of growth, but it is not very good at providing a socially defensible distribution of income. Over the course of this century, governments big and small have intervened to smooth out business cycles and distribute income more equitably; in this country, the devices have included minimum-wage laws, legal recognition of trade unions, Social Security, and a progressive income tax, among others. Perhaps we should add to the list of things that the market does imperfectly the distribution of work opportunity, and find ways to intervene here, as well.

Wassily Leontief, the Nobel laureate in economics, argues that the loss of purchasing power from the decline in wages brought about by automation could doom the economy to a paradoxical state of rising productivity and rising destitution. Leontief proposes an ever-shorter work week, in order to assure a fair distribution of jobs and leisure, and a greater reliance on government transfer payments to distribute income and maintain consumer purchasing power. Although millions of workers are unemployed, millions more are working longer at their jobs than they might wish: the two obvious categories are people near retirement age and working parents. Following Leontief's approach, the government might supplement the pay of workers who voluntarily cut back to a three-day work week at the age of sixty. It could also supplement the income of working parents so that they could work half-time for, say, five years at a full-time rate of pay. Millions of new jobs would thereby be opened up, and at the same time, those who chose to work fewer hours for a while would suffer no loss of purchasing power. This system would be expensive in tax dollars over the short run, but as the unemployment rate dropped, it would also generate economic growth, and hence more tax revenue. Unions could then welcome automation without fear that society as a whole would benefit at the expense of the livelihood of production workers.

Automation may be the ultimate solution to the inroads of foreign competition on American markets, and to the loss of American plants to poorer countries. As more efficient machines replace human workers, smarter capital equipment matters more than cheaper labor. But for automation truly to serve as a remedy of this country's economic ailments, society must become far more explicit about how the benefits of automation are distributed. Leontief's idea—automation combined with redistribution—would require an even greater role for trade unions and for the state than they now play. Although the government as the redistributor of income and trade unions as the defenders of wages have been declared all but extinct lately, they may be the best instruments we have to resolve an explosive dilemma. The debate about the distribution of wages and the content of work has scarcely started, but it is certain to become more urgent.

CHANGING OCCUPATIONAL AND REWARD STRUCTURES

11 *The Great American Job Machine: The Proliferation of Low Wage Employment in the U.S. Economy*

BARRY BLUESTONE AND BENNETT HARRISON

For more than a decade the United States has been in the enviable position of producing more new jobs than most of the rest of the industrialized nations combined. Indeed, Europe had virtually zero employment growth between 1973 and 1984, while the U.S. added nearly twenty million new jobs during the same period.

As a result, America has been justifiably termed "The Great Jobs Machine." While unemployment rates are clearly much higher than in previous decades, the economy has generated employment opportunity for millions in the "baby boom" generation and for an unprecedented number of women who have entered and remained in the labor force. Leaders of the Common Market and the Organization for Economic Cooperation and Development (OECD) have been so impressed by U.S. employment gains that they have sent delegations across the Atlantic to learn what they can from this apparent success.

Those charged with responsibility for current economic policy have been particularly pleased with recent employment gains. Since 1981, the total number of jobs in the U.S. has expanded by nearly 10 million, despite the severe 1981–82 recession, and the grand total has grown nearly every month since the end of 1982. The tax cuts initiated under the Reagan Administration and the

Source: Barry Bluestone and Bennett Harrison, "The Great American Job Machine: The Proliferation of Low Wage Employment in the U.S. Economy," a study prepared for the Joint Economic Committee of Congress (1986), pp. 1–7; 43–45.

expansion in the nation's money supply beginning in 1983 have indeed brought about a substantial economic recovery.

Yet for all of the jubilation surrounding this accomplishment, the recent employment record is not quite as good as the raw numbers seem to suggest. For one thing, the civilian labor force—the number of Americans who are working or who *want* to work—grew at an even faster pace than the rate of job growth. This has left the overall unemployment rate essentially unchanged while the absolute number of unemployed workers has increased by nearly one-half million. Moreover, the recent record of accomplishment is actually weaker than that of the period that immediately preceded it. In the four years 1976 to 1980, total employment rose by nearly 7.3 million, or 1.8 million net new jobs per year. Job creation since 1981 has been proceeding at a rate that provides 200,000 fewer new jobs a year. The slowdown in the annual rate of new job development has contributed to the much higher average unemployment of the 1980s. Indeed, if the earlier annual rate had been maintained into the mid-1980s, more than one million additional persons would be at work today and the civilian unemployment rate would be 6.3 percent, if not lower.

What is most important, however, is that none of the aggregate numbers reveal anything about the types of jobs created during this period, or how much they pay. As for the first question, the record reveals a continuation into the 1980s of a strong trend toward employment growth in the service sector, with literally no expansion whatsoever in employment in goods production. In fact, since 1981, the number of people employed in construction and manufacturing has declined by more than 500,000, while private sector service employment (including transportation and public utilities; wholesale and retail trade; finance, insurance, and real estate; and business and personal services) has been responsible for all of the total net growth in the number of civilian jobs. Clearly, the radical sectoral restructuring of the American economy continues apace.

How this shift in the sectoral composition of employment affects the distribution of real wages and salaries—our second question—is obviously important if one is concerned with the quality as well as the quantity of job creation in the U.S. Both the level and distribution of wages are likely to be affected by these sectoral shifts, and (perhaps even more strongly) by changes in the distribution of jobs within each sector. The redesign of full-time into part-time jobs, the disproportionate growth of part-time or part-year work, and the spread of wage freezes and concessions from one industry to another all suggest a real decline in annual earnings. In addition, demographic factors—including the continued rise in female labor force participation and the entry of the baby boom generation into the workforce—could also be affecting the wage distribution.

Our objective in this paper is not so grand as to statistically parcel out the impact of each of these diverse factors, but rather more modestly to provide some summary measures of the trends in the distribution of real labor income to which the various factors are contributing. Within that general framework, the

Net Increase in New Jobs (1979-1984) by Annual Wage Grouping

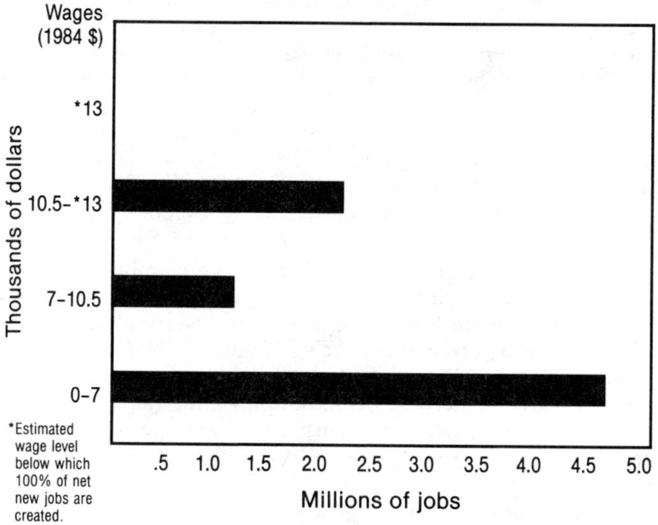

particular issue we wish to address in the present paper is straightforward: underneath the appearance of substantial overall job creation since 1973, is America proliferating low-wage jobs and perhaps even shifting toward an increasingly polarized labor market structure?

To answer these more specific questions, we investigate changes in the level and distribution of real annual wages and salaries earned by workers during the period over which the Great American Jobs Machine was apparently so productive—1973 through 1984. Using *Current Population Survey* (CPS) data from the U.S. Census, we tabulate the net new employment generated over this period according to whether the additional workers earned "high", "middle", or "low" real annual wages. These wage distributions are then recalculated by industrial sector, region, age, race, gender, education, and on the basis of whether the workers in question were employed full-time and year-round. Finally, we compare changes in the proportions of workers falling into these three earnings categories between two subperiods: 1973–1979 and 1979–1984. The comparison allows us to ascertain just how "good" the most recent job gains have been, at least in terms of income generation.

What do these statistics reveal?

- Compared to the first subperiod (1973–1979), the net new employment created between 1979 and 1984 has occurred disproportionately at the low extreme of the wage and salary distribution (i.e. below $7,000 in 1984 dollars). Specifically, during the 1970s about one out of every five net additional wage-earners found a job (or jobs) paying as little as $7,000. Since 1979, that fraction has risen to nearly six in ten.

- During the second subperiod (1979–1984), the number of workers with earnings as high or higher than the 1973 median ($14,024 in 1984 dollars) actually declined by 1.8 million, while workers with earnings less than the 1973 real median increased by some 9.9 million. While there have been some high-wage jobs created during this period, on a net basis, all of the employment increase experienced since 1979 has been generated by the creation of jobs which paid less than the median wage in 1973. Thus while the "middle wage" earnings category shows some growth during this subperiod, this growth was concentrated at the bottom end of the category (between 50% and 100% of the 1973 median).
- Of the major demographic groups in the U.S., white men have been the biggest losers, suggesting the increasingly universal impact of these changes in the wage structure. No longer are the "working poor" restricted to women and those who are members of minority groups. Between 1973 and 1979, fewer than one out of every four new jobs taken by white men fell into the low wage category. Since 1979, however, nearly 97 percent of net employment gains among white men have been in the low wage stratum. During the same period, white men have experienced a net loss of one million jobs paying $28,000 or more in 1984 dollars.
- Between the 1970s and the 1980s, white women continued to show small gains in their access to higher wage jobs, but unlike the case in the earlier period, the proportion of such workers in low wage employment also increased. Thus employment for white women has shown a slight tendency toward polarization—a growth in the low and high wage extremes of the distribution at the expense of the middle.
- Employment for both minority men and minority women shows a renewed trend toward low wages. Notable in this regard is the apparent reversal of the trend toward higher wage job opportunities for black men which was observed during much of the 1970s.
- The disproportionate expansion of the low-wage sector is found to be especially prevalent among younger workers (age 16–34). However, it is not restricted to them; it is also evident among older workers, those 35 years of age and older. Hence, the shift toward low-wage employment cannot be dismissed simply as a consequence of baby-boomers entering the workforce in the 1970s.
- The trend toward low-wage employment is found in all regions of the country, but it is substantially more pronounced in the Midwest. In the hard-hit industrial and agricultural Midwestern states, middle- and high-wage employment declined between 1979 and 1984 by more than a million jobs, while the number of low-wage jobs increased by more than 900,000. In no other part of the country have the income effects of deindustrialization—enormous downward wage mobility—been so pronounced.
- The tendency toward low wages holds for year-round, full-time workers as well as for those who do not work as often during the year. This implies that it is not merely the increase in part-time work that is responsible for low annual wages. The underlying distribution of hourly or weekly wage rates appears to be shifting toward the low-wage extreme as well, although we do not measure weeks or hours directly in this paper.
- In sum, the economic restructuring of the 1980s—including the loss of jobs in the manufacturing sector, the continued growth of the service economy, and the

reorganization of work toward more part-time schedules—has left in its wake a proliferation of low-wage jobs. If this pattern of development continues, the standard of living of a growing proportion of the American workforce will be significantly jeopardized.

CONCLUSION

The continuing decline in high wage manufacturing, combined with the expansion in the low-wage retail trade and service sectors, have led to the popular perception that America may be on the verge of losing its middle class. Writers often equate "middle class" with "recipients of middle-level incomes". In those terms, our results confirm an unmistakable trend in this direction for individual working people.

At the margin, the net additions to employment being generated in the U.S. since the late 1970s have been disproportionately and increasingly concentrated at the low-wage end of the spectrum. That 58 percent of all net new employment between 1979 and 1984 paid annual wages of less than $7,000 clearly supports this conclusion. Indeed, if the ratio of net new low-wage employment to net new total employment for the earlier period (1973–79) had prevailed for the period after 1979, the total number of new low wage positions would have been only 1.6 million, rather than the 4.7 million that we actually experienced. Hence, the strong and apparently enviable record of job creation since 1979 masked the fact that more than 3.1 million of the 8 million net new jobs represented additional low-wage employment over and above the number of low-wage jobs that would have been created under earlier conditions. That there was an absolute decline of more than 450,000 jobs paying high wages confirms the recent shift toward low-wage work.

Of all the demographic groups in the U.S., younger white men have been the biggest losers in the sense that *all* of the net new jobs held by this group after 1979 paid very low wages. In contrast, the share of all new jobs held by white women which pay high wages grew between 1973–79 and 1979–84, while the incidence of very low-paying work increased only slightly. The earnings of workers of color of both genders have not continued to improve as in the earlier period. Notable in this regard is the apparent halt in the previous trend toward a growing high wage share for non-white men. In the period 1973–1979, 42 percent of the net employment growth among minority men was in the high wage end of the distribution. In the years since 1979, that proportion has dropped to 6.4 percent.

The tendency toward the expansion of the low-wage end of the earnings distribution appears to be concentrated especially in the Midwest, but no region of the country is completely immune from it. Finally, the tendency toward low wages is particularly pronounced within the manufacturing sector (which started the period with the smallest low-wage stratum), but it appears even in high tech and in the broadly defined service sector.

A growing volume of research by Danziger and Gottschalk, Thurow, Bell and Freeman, Levy and Michel, and others all points in the same general direction—a tendency toward low wages in particular and growing income inequality in general in the United States. It would seem that a serious political debate over how this unsettling development might be reversed should be on the agenda of the next Congress.

THE HAZARDS OF HIGH TECH WORK

12 High Tech Control of Workers

ROBERT HOWARD

In December 1979, some 500 local union officials and staff members of the Communications Workers of America (CWA) met in Dearborn, Michigan, to hold the first conference on workplace technology ever sponsored by an American labor union. Not far from the former home of Henry Ford, patron saint of the assembly line, conference participants examined a very different kind of workplace, one shaped by the widespread computerization of work in the telecommunications industry.

The terminology they used was new, but the experiences that telephone workers described would have been familiar to the early mechanics and craftsmen in the American automobile industry: traditional skills made obsolete by new technology; jobs fragmented and downgraded to lower pay; work reorganized and rigidly centralized; workers subject to automatic pacing and monitoring, oversupervision, and job-induced psychological stress. Like the shift from craft to mass production in the auto industry, new technology in telecommunications has eroded workers' sense of control over their work lives. "Members feel that they have lost the challenge of the job, the authority to make decisions," said a local union president at the conference. "Also, they have lost their sense of accomplishment, freedom, and overall prestige."

The CWA technology conference reflects growing concern in the labor

Source: Robert Howard, "Brave New Workplace," *Working Papers for a New Society* 7 (November/December 1980), pp. 21–31.

movement about the implications of new workplace technology. "The widespread application of computers, silicon chips, and industrial robots, and the speed with which these and similar technological advances are being developed, portend vast changes in all segments of the workplace," writes Dennis Chamot of the AFL-CIO's Professional Employees Department. "In essence, our society is involved in a rapid and massive redesign of work."

For industry and government, this redesign promises to boost lagging productivity and stimulate the revitalization of American industry. Robert Noyce, of the Intel Corporation, calls high technology sectors like telecommunications, computers, and semiconductors the "jewels of American innovation." They are to play the same role in the American economy of the future that the automobile industry and its assembly lines played in the past. The only obstacle to this scenario—"the potential fly in the ointment," according to *Business Week* magazine—is opposition from organized labor.

Whatever the fears of the business community, most unions have addressed the technology issue in the narrowest of terms—whether technology will reduce jobs. What the CWA conference suggests, however, is that the new technology also threatens to radically transform those jobs that remain. In fact, like the assembly line before it, the computerization of work is ushering in a whole new economic era, one that will disrupt past customs and compromises established over decades of labor relations—to challenge labor's hard-won gains of the past fifty years.

To consider new technology, therefore, is to consider not only jobs, but power. "If labor does not find a way to control technology," writes Harley Shaiken, a technology consultant to the United Auto Workers, "then management will use technology to control labor." This struggle for control of workplace technology promises to be one of the most important arenas of conflict between corporations and unions in the decades ahead. And no industry offers a clearer example of what is at stake for American workers than that jewel of American innovation, the telecommunications industry.

Technological change is certainly no stranger to telecommunications. But the current wave of new technology is revolutionizing both the character of telephone service and the economic structure of the industry itself. Rapid advances in computer technology during the past ten to twenty years are blurring the boundaries between heretofore distinct industries. Telecommunications, data processing, and business machines are all beginning to merge into an undifferentiated mass of information processing services and equipment, supplied interchangeably by telephone or computer firms. And under the impact of this change, the commercial structure of telecommunications is being transformed.

In 1968, the Federal Communications Commission inaugurated the long range deregulation of key sectors in the telephone industry. The FCC's "Carter-fone" decision gave outside firms the right to manufacture and supply telecommunications equipment for connection to Bell System lines, and to provide "specialized communications services" to businesses. Over the last ten years, a score of corporations including IBM, TRW, MCI, and Rockwell International have entered the so-called "interconnect" market, which in 1979 totalled $730 million. Last April, the deregulation pendulum swung in Bell's direction. The FCC's "Computer Inquiry Two" opened the lucrative "enhanced services" market—computer and data transmission equipment at each end of the telephone line—to AT&T. The corporation recently announced what the *New York Times* termed "the most sweeping reorganization in its 103-year history," establishing an independent subsidiary as required by the FCC ruling in order to compete in the unregulated computer business.

In effect, telecommunications is rapidly turning into data communications. "What is at hand," according to *Bell Telephone Magazine,* in-house publication of AT&T, is nothing less than "the realization, at long last, of something that has been imagined for decades, if not centuries: The Information Society. And telecommunications will form much of the technical basis of that 'age of the mind.'" What the new age will also bring is a competitive battle of corporate giants with AT&T and IBM taking the lead in a fight for mastery of what has been estimated as a $400 billion integrated information industry.

The effects of such a struggle on work are predictable. AT&T has redoubled its efforts to reduce labor costs, increase productivity, and tighten its control over the Bell System workforce. The computerization of work makes these goals easier to achieve. A memorandum from an AT&T vice-president to Bell System managers describing a new computer system called MLT gives a taste of prevailing corporate policy. "A word of caution," the executive warned. "The savings available from MLT implementation will only occur if you plan for them to occur and then actually get the people off the payroll. Too often, we are intrigued with the capabilities of new mechanized systems and forget the real reasons for their existence. . . ."

According to AT&T, the changes in telecommunications will *not* reduce the total number of Bell System jobs. Overall employment is expected to grow during the next decade much as it has over the last twenty years, when the rapid expansion of telephone business more than made up for the 100,000 operators' jobs lost to new technology. But unlike earlier technological changes, the widespread computerization of work will entirely transform the occupational structure of the Bell System—with serious implications for the Communications Workers' union.

"Because of the growing need for 'knowledge' workers and the lessening need for 'tool' workers," says *Bell Telephone Magazine,* most of the growth in employment will be confined to the ranks of management. Bell's 260,000 managers could expand by as much as one-third—a rate of increase three times that estimated for nonmanagement personnel. Not surprisingly, the majority of these new jobs will be in two areas: mid-level management categories associated with the increasing computerization of work—electronic data processors, data systems designers, "the people who draw on the skills of the computer, who define and set up mechanized systems," in the words of one AT&T official—and sales and marketing personnel as AT&T gears up for the new era of competition ahead.

The few new nonmanagement jobs (Bell's nonmanagement workforce of 780,000 is expected to grow by one-tenth) will be limited to unskilled, primarily clerical, work. And this growth in unskilled jobs will be matched by a precipitous decline in the skilled crafts. From 1974 to 1978, craft employment dropped by almost 22,000 jobs, a trend that is certain to continue.

These developments prefigure a serious weakening of the CWA's power base. Unlike computers or semiconductors, telecommunications in the one high technology industry with a unionized workforce. The skilled crafts have traditionally been the backbone of the union's bargaining power. But current trends in industry employment not only shift most of the new jobs out of the union's bargaining unit; they effectively cut off unionized workers from the considerable new skills that computerization of work brings in its wake.

"Certainly, technological wonders . . . may exact some cost in human attitudes," says *Bell Telephone Magazine*: "the way employees feel about themselves, their work and their business." But, "the key to helping human and machine get along in the 1980s would seem to be to get them to know each other." This peculiar version of computer dating is already well under way in the plants and offices of the Bell System. Though they never use it themselves, the phrase that best describes what telephone workers express is Harry Braverman's "the degradation of work." What follows, then, is a close look at what workers in one industry have come to know about the new technology. It is a tour of the degradation of work in the Information Society.

The din of a cross-bar switching system sounds like a roomful of teletype machines amplified over a loudspeaker. Thousands of metal switches, stacked in long library-like rows, click and clatter in irregular rhythms as they make and unmake the connections that route telephone signals from one phone to another. When traffic is heavy, the noise is intense, a mad chorus of chattering metal. You have to raise your voice to be heard.

In contrast, the only sound produced by the new electronic switching systems (ESS) is a faint, hypnotic hum. As its name implies, ESS transfers calls electronically. It has no moving parts; it operates by stored-program computer control. ESS can handle 550,000 calls in an hour, four times the amount of its electromechanical predecessor. The "1A" central processor, the most advanced computer brain of the electronic switching system, can direct traffic, silently and efficiently, at the rate of 150 calls per second.

For about fifty years, electromechanical cross-bar switching has been the work-horse of the American telephone system. But since ESS was first introduced in 1965, it has grown to serve over 30 percent of the telephone central offices in the nation. By 1985, 50 percent of central offices will carry electronic switching. By the year 2000, the din of the cross-bar will be heard no more.

Electronic switching systems are at the heart of the current wave of technological change in telecommunications. According to the final report of the CWA technology conference, "the impact of the introduction of ESS has apparently affected employees in all job categories. . . . " But the job that has borne the weight of this change is that of the switching technician.

Switching technicians are responsible for the maintenance of central office equipment. They locate malfunctions, figure out what is wrong, and make the necessary repairs or direct other craft personnel to do so. Traditionally, the job has required a high degree of skill, and switchmen have enjoyed substantial autonomy. In a typical central office, there might be anywhere from twenty to thirty switching technicians and other skilled craft workers under the direction of a single foreman who rarely interferes in their work.

Electronic switching, however, radically changes this work environment. First, ESS cuts down on the need for switching technicians in some fairly simple and obvious ways. For example, the system monitors its own performance and diagnoses circuitry problems by itself; computers do much of the analysis and trouble-shooting that technicians used to do. Also, the system has no moving parts, thus eliminating the need for mechanical repair skills. Technicians simply replace one malfunctioning circuit-pack with another. The installation of ESS in the typical central office would probably cut the workforce in half.

But the effects of ESS on switchmen do not stop there. Despite all the rhetoric about computers enabling decentralization, the Bell System has used computer control of the switching process to centralize the monitoring and repair functions of the switching technician. "Automatic Switching Control Centers" now observe a number of central offices at once by remote telemetry. With the Control Center, the number of technicians at the central switching office itself can be reduced to one or two; some switching offices have none at all.

It is this move to the Control Center that has made the most difference in the way the switching technician does his job. Centralization replaces the autonomy of the central office with rows of video display terminals (VDTs) where anywhere from forty to sixty technicians keep watch on the system's performance. The foreman of the central office gives way to an extended management team. Dave Newman, a switchman in New York City, describes the Control Center as "a large room like a mass-production office, a clerical type of thing. Right away, that's culture shock, especially for the guys. You are just about literally chained to a desk."

Centralization also fragments the switching technician's job into isolated segments. For example, one traditional task of the switchman has been to identify and repair malfunctioning trunk lines. In the central switching office, a trunk line "trouble report" would arrive on a teletype; the technician would locate the trunk and put it out of service. Later, when he had the time to analyze the problem and devise a way to repair it, he would return.

Now that trunk line trouble reports are fed into the Control Center, however, they can all be assigned to two or three technicians. They spend their day in front of their VDT screens keying messages to the computer to put faulty trunks on busy. They do not fix the malfunctioning lines themselves; other technicians are sent out from the Control Center to do the necessary repairs on site.

"People *hate* the Control Center," says Ilene Winkler, shop steward at the mid-town Manhattan Control Center. "No one wants to work there. You sit there like an idiot in front of your terminal. We can't even figure out what our jobs are supposed to be anymore." According to Winkler and other technicians, the most momentous change associated with the shift from central office to Control Center is the enormous expansion of management supervision. Not only are there a lot more managers at the Control Center than at the old central office; increased supervision also means tighter control over the assignment and direction of work. Gone is the rarely interfering foreman.

Expanding management control of work has even been designed into the computer itself. By a new procedure known as "pricing and loading," clerks type each switching technician's schedule for the day into the computer. The schedule lists a series of jobs along with the estimated amount of time each should take. Technicians receive their assignments by accessing the computer at their own VDT. As each assigned task is completed, they also enter the amount of time it took into the computer. Pricing and loading takes away the last remnants of autonomy from a job that, according to Ilene Winkler, "just a few years ago was totally craft-controlled."

In theory, there is nothing inherent in the new technology of electronic switching that requires the systematic deterioration of the switching technician's job. True, ESS has made certain electromechanical repair skills obsolete, but it demands far more complicated technical skills and educational training—for example, knowledge of electronics-based systems and programming concepts. The problem is that access to these skills is restricted to Bell System management.

First, new maintenance tasks associated with electronic switching—the editing of computer programs, for example—are the sole responsibility of management engineers. According to Ilene Winkler, serious breakdowns at the mid-town Manhattan Control Center—what are known as "central processing troubles"—are immediately "escalated" up to management (unless they happen in the middle of the night; then, technicians are allowed to work on them). "There is no opportunity to learn on the technology," adds Dave Newman. "They don't train you to do that. They don't want you to do that."

Management also expands its own control at the expense of craft workers by a procedure best described as "phony upgrading." As switching control functions have been computerized, tasks that used to belong to technicians have been redefined as management responsibility. The steady encroachment of management on switching technicians' work has led to an ongoing conflict between union and company over who has the right to operate certain computer systems.

One such conflict that the union won involved the "Computerized

Maintenance Administrative System." COMAS stores information on past malfunctions in switching equipment and is used by technicians as a handy reference system to help identify the sources and causes of new problems. By comparing present difficulties with past performance, the job of tracking down breakdowns is simplified. Before COMAS, foremen were never directly involved in the work of tracing and repairing switching equipment malfunctions. But when the system was introduced in 1971, management personnel began using it to do what had always been the work of switching technicians.

The union fought management operation of COMAS through the grievance procedure, claiming it was an unwarranted intrusion into the work of the switching technician. The arbitrator agreed. But the struggle for control of COMAS and other computerized data systems continues. It is a common policy among Bell operating companies to systematically force questions of access to each new workplace computer system into arbitration.

In effect, the job of the switching technician is being split in two. The most interesting and skilled tasks are disappearing into management, leaving behind tasks shorn of skill and reduced to clerical work. This division is occurring throughout the telecommunications industry. It is a new version of the "dual labor market": on one side of the divide, highly skilled work controlled by non-union management personnel; on the other, degraded computerized work relegated to the unionized workers of the CWA bargaining unit.

The skilled crafts are bearing the brunt of the degradation of work made possible by new technology in telecommunications, but no job category is immune. Clerical workers have been deeply affected by the dozen new computer-based administration systems introduced in the past decade. In one sense, computerization has made clerical work easier. Customer records are now at the tip of one's fingers instead of buried in mammoth files. But, as in the case of the switching technician, computerization also brings centralization and a thorough reorganization of work that isolates clerical workers and subjects them to more rigid supervision and control.

LMOS stands for "Line Maintenance Operating System"; it is, essentially, a computerized file. When a repair service attendant answers a 611 customer report of malfunctioning equipment, she types the customer's phone number into the computer and receives on her VDT screen a complete record of past problems, known as the "line card." The clerk adds the information provided by the customer to the line card and gives the customer a "commitment time" provided by the computer indicating when a repair crew will visit the customer's home. Then she transfers the line card, complaint, and commitment time to the repair garage via computer print-out.

Michelle Brooking is a shop steward for repair service attendants in Washington, D.C. She says LMOS may sound efficient but isn't. "It's easier in getting information, but as far as getting your phone fixed, it takes a lot longer now than it ever did before." Before LMOS, repair service attendants worked in local repair bureaus linked to the central offices. They were in frequent telephone contact with repair and installation crews. Often, they would relay messages between the customers and the crews—changes in commitment time, last minute repair jobs, etc. "You knew how many men you had on the street," says Brooking, "how many troubles you could handle." With LMOS, all the attendants are in centralized bureaus. Their only contact with the repair crews is one way, via the computer. They do not know if the crew actually makes it to the customer's house at the appointed time (though, often, they have to listen to the customer's complaints if it does not). "Now, you're doing it blindly," says Brooking. "You're not able to give it the little personal touch you did before."

Michelle Brooking thinks it would have made more sense to have installed LMOS directly into the local repair centers. That way, clerks would have the convenience and efficiency of computerization along with the flexibility of the local work unit. But to use the new computer technology in this fashion would not accomplish the goals the Bell System management has set for it. Thus, computerization is used to facilitate the close supervision of work but at the price of fragmentation, less efficient service, and more customer inconvenience. Telephone workers are aware of the deterioration of service that has, ironically, accompanied the vast reorganization of work inspired by the new technology. Often, it seems illogical to them. But the logic behind these changes is found elsewhere—in management's effort to perfect its control of the labor process.

The computerized control of work has become so pervasive in Bell's clerical sector that management now has the capacity to measure how many times a phone rings before it is answered, how long a customer is put on hold, how long it takes a clerk to complete a call. Charles Sangmeister, district representative of the CWA in Washington, D.C., claims that management has the technical ability to maintain a complete print-out of a clerical worker's output for the day. "They haven't done it yet," he says, "but I can see that this is the next step coming down the road."

In some Bell operating companies, it is here already. FADS (for "Force Administration Data System") is a computer system which, by measuring the pattern of traffic into a particular office each day, can anticipate employment needs for the next. Each morning, workers receive computer print-outs listing their break and lunch times based on the anticipated traffic pattern for the day.

The FADS procedure has been in existence for a long time, but recently it has been computerized. Like other administrative systems, computer control squeezes out what little flexibility workers had before. Before computerization, a worker's morning break normally came about two hours after the beginning of the shift; now, it can come as early as fifteen minutes into the work day. Workers cannot go to the bathroom unless they find someone to take their place. "If you close your terminal," says Jean Miller, a service representative in Washington, "right away the computer starts clacking away and starts ringing a bell." In some offices, where FADS has been computerized, the computer even gives the supervisor a print-out of how many calls each clerk takes per day and how many times she leaves her position. One repair service attendant calls it "being put on the hot seat."

Oversupervision in the telephone company is, of course, nothing new. But just as computerization can build skills into the technology, it can build in supervision as well. "The computer watches you for the supervisor," says Jean Miller. "It is constantly alert to the atmosphere of the office. It's the same idea they've been using in Traffic for years."

The L'Enfant TSPS office is one of three long distance operator offices in Washington, D.C. TSPS stands for "Traffic Service Position System," the computerized successor to the phone company's old "cord-board." The L'Enfant office is about the size of a gradeschool classroom with the atmosphere to match. Magic marker posters with hortatory slogans adorn the walls; the supervisors' desks at the front of the room face about fifty battleship gray TSPS terminals lined up in rows of two.

Operators do not have their own work stations; instead they simply plug their head-sets (kept in lockers in an adjoining room) into any available position at the start of the work day. On the TSPS console, there are rows of square buttons that operators push to connect collect or credit card calls or to calculate time and charges. A small screen on the top of the console gives a digital read-out of this information much like the computerized cash register in a grocery store.

Gloria Newsome, one of seven group supervisors at L'Enfant, demonstrates the TSPS. Calls "drop in" to the console with an electronic beep. Gloria answers them in her professional, slightly sing-song, Bell System voice. Her fingers flash over the console pushing a bewildering array of buttons. In a matter of seconds, the call is completed only to be followed by another beep and another call. The longest time between two calls is about five seconds. Gloria admits that the job can get rough sometimes, "but it also can be very rewarding, if you like to talk with people."

Working a TSPS terminal is easier than the old cord-board. Pushing buttons is faster and less physically demanding than manipulating the plugs of the traditional switchboard. The TSPS also times calls and makes certain computations that previously operators had to do themselves. Not surprisingly, TSPS has sent productivity soaring. Whereas an operator could handle about twenty calls per hour on the old switchboard, now she can average around one hundred.

This does not mean that operators have more control over their work; in fact, they have considerably less. With the cord-board, operators could regulate somewhat the pace at which they responded to calls. And when all the positions on an operator's terminal were occupied, she was free to wait until one opened again. With TSPS, calls drop in one after another, routed by an "automatic call distributor," the moment one's line is free. And ESS means an operator can handle an unending succession of calls. There is no such thing as a full terminal.

Eash TSPS office has a "call-waiting box" which indicates any calls that are waiting to be routed to an individual terminal. If a call is kept waiting for more than ten seconds—what is known as a "high answer"—a light on the box begins flashing. Supervisors can also determine from the call-waiting box whether any of the terminals in the office have been disconnected or put on busy. TSPS groups are evaluated on the number of high answers they have per day.

One of the slogans in neat magic marker attached to the wall of the L'Enfant office reads: "There will be *no* talking until we achieve our 98 Index." A 98 Index is the "outstanding" rating in the Bell System evaluation procedure for TSPS performance. Operators are monitored for courtesy, accuracy, and performance—thus, the evaluation program's acronym, "CAP."

For half an hour, two times each week, every operator is timed by computor to determine her "average working time" or AWT. According to Gloria Newsome, the AWT at L'Enfant is 37 seconds per call, but in some offices it is as low as 30. Operators are also evaluated on their "speed of answer." After the electronic beep, they have three seconds to respond to a call. While most operators admit that working the TSPS is easier and more efficient than the cord-board, many still complain that the pace of their work has increased. "You'd think you were at the Indianapolis Speedway," says a Washington operator. "If you don't keep your AWT down, they start riding you." Willy Leggett, shop steward at CWA Local 2300, the operators' local in Washington, considers AWT a quota to force down the average time spent on calls. When an operator consistently beats the office average, other members of the group are pushed to match her time.

In addition to the twice weekly AWT productivity studies, each operator is monitored by a supervisor for thirty calls per month to evaluate courtesy and accuracy, the other two elements of the CAP program. These tests are called "official evaluatives" or "remote performance observations." Like the AWT studies, operators do not know when they are being observed. "Faults" and "irregularities" are recorded and figured into the individual's Index.

There are still other observations that an operator does know about—"diagnostics," "parallels" (so called because the supervisor sits directly next to the operator), or "performance improvements." There is no limit to the number an operator may receive. Officially, these tests are for training purposes only, but according to Willy Leggett, they are used in employee evaluations. Gloria Newsome confirms this: "If I hear one of my girls doing something she shouldn't do, she *will* be disciplined." In the recent contract signed between the CWA and AT&T last August, language was included expressly forbidding the use of diagnostics for disciplinary purposes. But Willy Leggett claims that it is virtually impossible to prove whether a fault occurred during a diagnostic or an official evaluation.

AWT studies, official evaluatives, and diagnostics are not the only observation that TSPS operators are subject to. There are entirely separate offices which listen in on TSPS groups, evaluate their work, and inform their supervisors. When Elinor Langer wrote about her experiences working at the New York Telephone Company ten years ago, she described similar layers of supervision. "One result of the constant observation (the technology being unbounded)," Langer wrote, "is that one can never be certain where the observation stops. It is company policy to stress its finite character, but no one ever knows for sure." The computerization of work has intensified this sense of powerlessness and spread it throughout the telecommunications workforce.

The "rapid and massive redesign of work" that the AFL-CIO's Dennis Chamot describes also threatens the institutional power of American unions. It is not only that the computerization of work undermines union wages, thins the ranks of union membership, attacks the integrity of the bargaining unit, and provides the techniques for the creation of a strike-proof workplace. By feeding workers' sense of their inability to shape their work lives, new technology is a potent weapon of division and demobilization.

Unfortunately, there is little in the history of the American labor movement to help unions meet this challenge. While union resistance to new technology has always been well-publicized, it has been the rare, and universally unsuccessful, practice of only a few craft unions. For a few years in the mid-sixties, Local 6 of the International Typographers Union won veto power over new automation at New York City newspapers. But the agreement only hurried the slide into bankruptcy of a number of already financially ailing papers and the elimination of typographers' jobs. ". . . A strong and effective union stance in some plants," writes sociologist Andrew Zimbalist, "resulted in failure for New York City typographers as a whole."

In contrast, most American unions have unhesitatingly ceded to management the right to develop and use technology as it sees fit—usually in exchange for generous productivity wage gains, job security provisions, and buy-out clauses. The classic example is the "mechanization and modernization" agreements of the International Longshoremen's Union in the 1960s. In effect, longshoremen gave up their considerable control over working conditions for a $25 million fund providing for early retirement, no lay-offs, and a guaranteed minimum work week. But the plan only protected the most senior workers at the expense of the rest. And like most American union technology agreements, it came "after the fact" when any attempt to influence the actual direction of change was too late. "When it comes to technological change," Harley Shaiken says of America's labor movement, "there has been no more cooperative sector in society."

The new wave of computer-based innovation has inspired some criticism of the defensive practices of the past. Some unionists have begun to call for a more offensive labor strategy, one that would confront the implications of new technology *before* it becomes a fait accompli on the shop floor. This demand for some kind of union influence over technological change is often expressed in the language of "workers' control"; it could be the starting point for the redefinition of labor's approach to new technology. For the present, however, the call for worker control of technology does not so much provide a solution as define an immense problem. The gap between ideas and practical blueprints for union action is extremely wide.

The Communications Workers' union provides a striking example of the dilemmas surrounding the technology issue. Like most unions, the CWA has historically supported management-directed technological change as long as the fruits of increased productivity have been shared with its members. When new technology displaced workers—as it has over 100,000 operators since World War II—the union has negotiated a variety of programs to cushion the blow: attrition plans, transfers, termination payments. This primary emphasis on job security has continued into the present. In the new Bell System contract signed last August, the union won an expanded early retirement plan and a guarantee that workers with fifteen years seniority or more would be exempt from technological downgrading.

However, the widespread effects of computerization have elicited a call from within the union for "reshaping and rethinking" its response to technological change. Union leaders, often under the pressure of the rank-and-file, have begun to perceive the connections between technology issues and deteriorating working conditions like over-supervision and job pressures. There is much talk about "worker participation," "quality of work life," a shortened work week, and more extensive training programs. "Those kinds of things can't be solved in the traditional union sense," says Patsy Fryman, an assistant to CWA president Glenn Watts. "We ought to be willing to experiment, to try some innovative things." The most recent Bell System contract is full of first steps in this direction—six months advance notice of all technological changes, a joint management-labor technology committee to examine the implications of new technology, an industry-wide quality of work life committee, and, most important, a special joint committee to establish the criteria for a comprehensive redefinition of industry job categories.

One central feature of the new direction in union thinking is the idea of "comparable worth." It is an attempt to redefine the standards by which wages are set. "Technology potentially simplifies jobs," reads the pre-negotiation resolution of the CWA Bell System Bargaining Council. "Yet at the very same time, it complicates and adds greater responsibilities to some employees' positions." According to comparable worth, workers should be paid not so much for the skills they possess as for the responsibility their jobs entail. The clerical employee who operates a computer that does the job a skilled craftsman used to do may not have the skills or training of the craftsman; nevertheless, he or she has inherited the responsibility. Likewise the worker who uses new technology to do the same amount of work as three workers did before. In both cases, the job is worth more to the company and the wages of the workers should reflect that increase in worth.

Comparable worth can prevent the use of technology to downgrade jobs.

Under a wage system based on comparable worth, skills may become obsolete, jobs may even disappear, but the wages of the workers who remain behind will reflect their corresponding increase in responsibility for operating technology-intensive equipment. In a modified version of this principle, the union won a wage raise in the last contract for the clerical workers who operate the new MLT computer system. Designed to "get the people off the payroll," MLT has caused major reductions in the skilled category of test-desk technician. But the new contract gives the MLT operator a wage rate that is 80 percent of the test-desk technician's pay.

New ideas like comparable worth or the other technology provisions in the Bell System contract are a first step. But they do not move very far beyond the constraints of traditional defensive protectionism. While they may protect workers' wages, they are incapable of addressing the more intractable problems of the deterioration of skill and the expansion of management control over work. For this reason, Harley Shaiken describes the CWA technology policy as "essentially, to manage an orderly retreat."

Some CWA members have also criticized what one worker calls the union's "band-aid approach" to technology issues. "The bottom line for the union is pay and a job," says Dave Newman. "That's changing. That has to change." Newman and a small group of New York Telephone employees publish an independent newsletter for telephone workers called *The Bell Wringer.* They have urged a more aggressive stance that would protect the job responsibilities of the skilled crafts. "It should be a basic trade union principle," writes Newman in *The Bell Wringer,* "that no matter how the technology of a job changes, the job title and pay rate remain the same (if not get better!). . . . The introduction of new technology cannot be stopped, but it does not have to lead to boring, routinized, less-skilled and low-paying jobs."

The difficulty of protecting skills, however, can be seen in the case of the MLT computer system, the most important struggle to date over new technology in telecommunications and one that the union lost. First introduced in 1976, the MLT system automatically runs tests on certain telephone equipment—tests previously performed by the highly skilled test-desk technician. Management assigned the operation of MLT to a clerical category in order to reduce the labor costs of equipment maintenance. In New York City, CWA Local 1101 disputed this assignment through the industry's grievance procedure.

The union argued that the computer is a tool, just like any other. MLT might be faster and more efficient but it is essentially no different from the meters used by the test-desk technician in his work. Therefore, to assign to a clerk a job traditionally done by a skilled worker was to violate the integrity of industrial occupational categories.

The arbitrator disagreed. Siding with the company, he found that computerization transfers the skill from the worker to the machine. While the output of the two jobs is generally similar," there is "no basic skill parity." Moreover, even if what the union claimed were true, said the arbitrator, the union contract gave management an "absolute right" to assign work to whomever it chose.

The lesson of the MLT case is that to confront the problem of skill, a union must go far beyond protecting traditional occupational categories. Even if the test-desk technician won the right to operate the MLT as Local 1101 had argued (or even if the new MLT operator won 100 percent parity with the test-desk technician's pay, which amounts to the same thing), the problem of deskilling remains. For the idea of protecting workers' skills to have any meaning, it must include the demand for workers' access to new skills. But before unions will be able to do this, they will have to take a giant step: to directly challenge traditional management prerogatives to exclusive control of both technology and the organization of work itself.

At this point, control of workplace technology becomes a political issue. New ideas, experiments in workers participation, joint management-labor committees will be meaningless unless they are linked to the mobilization of a broad cross-section of workers, skilled and unskilled alike, around technology issues. Harley Shaiken has formulated the outlines of an agenda for such a political movement—explicit union rights concerning the alteration of skill and the organization of work (how computers are used to monitor workers, for example); effective union input in design of new technology; recognition of the "social costs" of technological change, and guarantees protecting workers from these costs; and effective procedures to police any labor-management agreement concerning technology issues. But in the absence of rank-and-file mobilization and participation, even the most comprehensive programs are bound to fall short.

There are some countries where such a movement for worker control of technology is being built. In Scandinavia, workers have won extensive rights through both collective bargaining and legislation that guarantees them a say in the design and implementation of new technology. In Norway, unions have negotiated "data agreements" with management on both local and national levels outlining union rights to participate in all technology decisions. A new union position has been created in the workplace called the "data shop steward" who is responsible for analyzing all new systems from the workers' perspective and making sure they contain no features used to restrict workers' freedom and control. The specific provisions of these agreements vary from country to country, but the central feature of the Scandinavian experience has been the rapid expansion in recent years of innovative union education programs that bring pro-union computer technicians and local union members together to learn from each other about the dangers and the opportunities of the new technology. These experiments make clear that any serious worker control of technology requires not only independent union expertise but also an informed, "technologically literate," and politically active workforce.

In the present context of American labor relations and with American unions on the political defensive, it is hard to imagine a major campaign of worker education and rank-and-file mobilization over technology issues taking place in the near future. But there is, perhaps, a historical parallel between the shift from craft production to mass production in the American automobile industry and what is happening in the telephone industry today. During that earlier period of massive technological change, new ways of organizing work were also used to fragment jobs and isolate workers. Traditional forms of worker resistance and traditional structures of trade unionism proved inadequate. At the same time, divisions among workers were worn down by the centralization of mass industry. And the large unskilled workforce of the assembly line—long considered incapable of organizing—began to cohere. Out of the new conditions came a new unity and new forms of political action—industrial unionism and the sit-down strike.

There are similar trends in telecommunications today. The same technology that fragments jobs and destroys skills brings workers from a wide variety of occupational categories together in a way they have not been before. Computerization is also creating whole new categories of workers who are defined as "management" and usually considered unsympathetic to unionism—computer programmers, sales personnel, middle level technicians—who will inevitably face problems and pressures similar to what unionized telephone workers are experiencing today. Most important of all, the computerized workplace, with its concentration of expensive and highly complex technology, is especially vulnerable to disruption by small numbers of centrally located individuals. With increased responsibility come new forms of leverage and potential power.

Of course, what form worker militancy in the Information Society will take—whether sit-down strikes in TSPS offices or worker occupations of Automatic Switching Control Centers—is impossible to predict. But the greatest failure of imagination would be to assume there will be none.

THE HAZARDS OF HIGH TECH WORK

13 *High Tech Hazards: Chipping Away At Workers Health*

ELLEN HOSMER

In the 15 years since General Telephone and Electronics (GTE) opened its plant in Albuquerque, New Mexico, hundreds of young women have worked in the plant producing semiconductors. Dozens of them worked in Department 320, an area where semiconductors were encapsulated and assembled or Department 341 where hybrids were assembled.

Today, an increasing number of GTE employees believe that a legacy of health problems has followed them home from work. Seventy-six employees and former employees, the majority of them women from Departments 320 and 341, are filing suit against GTE alleging that the 21 cases of cancer among them—including uterine, brain, thyroid and skin cancer—dozens of cases of hearing damage, blood abnormalities, menstrual disorders, deterioration of bone and cartilage, infections and disorders that they have suffered were caused by working in the company's high-tech plant. Workers claim that in many cases they were exposed to hazardous chemicals without protection or proper ventilation.

This exposure to hazardous, carcinogenic and suspected carcinogenic substances has had devastating and debilitating results, charges the plaintiffs'

Source: Ellen Hosmer, "High Tech Hazards: Chipping Away at Workers Health," *Multinational Monitor* 7 (January 31, 1986), pp. 1–5. Multinational Monitor is a monthly newsmagazine published by Essential Information, Inc., P.O. Box 19405, Washington, D.C. 20036. $22 individual.

attorney, Josephine Rohr. Forty-nine of the women who worked in Department 320 have had their uteruses removed, she said.

Lawyers in high-tech enclaves across the United States are filing suits similar to Rohr's on behalf of workers who allege they were exposed to chemicals without proper protection. Although Rohr is representing the largest number of employees in suits against any one particular plant, in Silicon Valley, home of the high-tech industry, Attorney Amanda Hawes has filed well over a hundred lawsuits against high-tech companies on behalf of exposed workers.

The cases that flow steadily in represent a disturbing indictment of health and safety provisions in the electronics industry. For years, the high tech industry has capitalized on an image of being a safe workplace and a clean industrial neighbor. In Silicon Valley, where more than 150,000 jobs are provided by the electronics industry, officials in surrounding communities have often overlooked health and safety concerns.

Unfortunately, the electronics industry is neither clean nor safe. It uses thousands of hazardous chemicals. Toxic leaks in the electronic workplace occur frequently, leaving workers debilitated and groundwater contaminated. According to an internal document of Advanced Micro Devices, a large semiconductor company in Silicon Valley, leaking toxics forced the evacuation of workers 18 times within an eight month period. The report was leaked to Attorney Hawes during one of her suits against the company.

The list of toxics used in the industry includes some of the deadliest known. Solvents, acids and gases, ranging from benzene, xylene, arsine and phosphine gas, sulfuric and nitric acids, lead, trichloroethylene, freon, acetone, hydrofluoric and hydrochloric acid are used in many electronics production processes.

In semiconductor manufacturing, workers are expected to dip wafer-thin silicon that has been painted with photoresist into acid baths. The wafers are then heated in gas-filled ovens, where the gas chemically reacts with the photosensitive chemicals. After drying, the chips are then bonded to ceramic frames, wires are attached to contacts and the chip is encapsulated with epoxy. These integrated circuits are then soldered onto boards, and the whole device is cleaned with solvents.

The potential health and safety risks to this process are immense. The silicon in the wafer and the gases used to cook it can lead to lung and respiratory ailments. The acids can cause severe burns as well as tissue and blood vessel damage. Lead can result in poisoning. The solvents and cleaners can lead to liver damage and cancer.

Exposure to solvents such as liquid freon—used to wash circuit boards—causes immediate rashes, drying and cracking of the skin. Freon also acts as a depressant on the central nervous system, and exposure in high concentrations can cause unconsciousness and death. Other solvents, including trichloroethane and methyl ketones, used in making the micro-chips, affect the central nervous system in much the same way that freon does.

According to a recent report by the Office of Technology Assessment, *Reproductive Health Hazards in the Workplace,* "children of solvent-exposed workers are more likely to have congenital malformations and tumors" than those whose parents were not exposed. The report warns that "three studies have implicated solvent exposure in malformations to the nervous system." Epoxies, used to protect circuit boards, are suspected carcinogens.

Lead used in soldering is another potential hazard to the reproductive system. Solder fumes may be absorbed into the bloodstream causing brain damage, paralysis, miscarriage, and sterility. Combined with acid flux which is used to clean metal during soldering and is itself linked to occupational asthma and other respiratory problems, the two make a deadly couple.

According to a 1980 study by the Finnish government, solder fumes can increase the risk of spontaneous abortion among electronics workers. The National Cancer Institute reports that high-tech workers exposed to cyanide and arsenic compounds have a higher than normal rate of oral cancer.

Another dangerous process involves making gallium arsenide microchips. Cheaper and faster than silicon chips, this process uses arsine gas, a combination of hydrogen and arsenic that can be fatal in high doses. Two workers in Massachusetts, one at MIT's Lincoln Laboratory and another at M/A-COM, died after inhaling toxic fumes from the gallium arsenide microchip production. In smaller doses, the gas is the suspected cause of skin and lung cancer.

In 1978, the California Division of Labor Statistics and Research tabulated results for occupational illnesses in the state and found that there were 1.3 illnesses per 100 workers in the electronics industry—while the average number of industrial employees with work-related illnesses was only .3.

In 1980 the number of electronics workers who had become ill while on the job remained disconcertingly high. When widespread doubt about the safety of the electronics industry surfaced, however, the numbers curiously did an about face. Within a year the reported number of worker related illnesses fell dramatically. By 1981, acknowledged worker illnesses in the semiconductor industry had dropped by 70 percent.

What looked at first like a tale of industry responsiveness, however, turned out to be just the opposite. Apparently alarmed by the growing publicity surrounding work-related illnesses, the industry had altered its counting methods so that fewer workers were included in the statistics, according to a special investigative report by KRON-TV in San Jose. "Many companies started using two sets of records on illnesses. The records they sent the state showed far fewer cases of occupational illness than their own records reflected," KRON reporter Vic Lee said in his two-part series aired in May of 1985.

The report documented several examples of companies filing inaccurate information on the number of illnesses that had occurred at the plant. At the Signetics plant in Silicon Valley, a company that has since settled out of court in a number of workers' compensation cases involving chemical exposure, the number of illnesses went from "99 to only two in two years." According to the

KRON-TV report. Intersil, Siliconix and Monolithic Memories no longer reported any work-related illnesses.

Although the Bureau of Labor Statistics did a study into the reason behind the drop, it chose not to challenge the industry's new method of reporting, and instead accepted the companies' revised classification of occupational illness, agreeing that workers who become sick from a one-time chemical exposure should be classified as injured rather than ill.

"Industry throws a very big shadow across a lot of regulatory efforts," explained Attorney Hawes. Changing the definition of illness is a quick, easy way to deal with the ever increasing problem of worker exposure, she said. "It's terrible, but the statistics now look great."

For workers in the high-tech industry, there is little or no health and safety regulation, either by federal or state occupational safety and health agencies or the industry.

"Most companies aren't adequately informing or protecting workers," said David LeGrand, who is in charge of occupational safety and health for the Communications Workers of America. "CALOSHA is still looked upon as the best state operated OSHA—in terms of provisions—but companies [operating in California] did not meet these [provisions]. Companies in other states are probably getting away with murder."

Ensuring compliance with health and safety standards at large electronics plants is difficult, but in the smaller companies where work is contracted out, it is often impossible, LeGrand said. "There's absolutely no protection. Some of the situations we've seen are nightmarish."

The Reagan administration, after cutting back OSHA funding, is asking for a policy of "voluntary compliance" with safety standards from the electronics industry, paving the way for continued worker health and safety risks.

The absence of adequate regulation is compounded by the lack of credible statistics on the health effects of high-tech toxics. "Workers are being exposed in mass throughout the country to chemicals [for which] the health effects are unknown," said Sam Swift, a plaintiffs' attorney in San Jose. "The problem is one of proof."

When the data are collected, Swift said, an epidemic of electronics workers debilitated by high-tech toxics may be already well underway.

The dearth of studies has made the work of plaintiffs' lawyers like Rohr, Hawes and Swift more difficult. With no data to fall back upon, plaintiffs' experts differ with company experts, leaving the judge to decide which side is correct.

In California, New Mexico and many other states, workers compensation is the sole remedy for workers to recover damages in work-related illness or injury. The difficulty of proving the case without epidemiological studies and the relatively small sums that are awarded in most workers' compensation cases has made it difficult for attorneys to find it economically feasible to take the cases of workers alleging chemical exposure.

Lawyers can spend years on a workers' compensation case and at best receive only around $30,000 for their client, said Attorney Hawes. It is not practical for the lawyer or the worker to file the case. As a result, many companies in the electronics industry continue their practices unchallenged.

"Employers have no incentive to be serious about the cases, they can drag them out," said Hawes. "They know exactly what the price is at the end of the line."

For GTE, however, things may not be so easy. Rohr, who represents all 76 of the former and current GTE employees, isn't dissuaded by the economics of the cases. Since August, 1984 when a 37-year-old GTE worker who was riddled with cancer came to her to get legal advice on what to do with $75,000 in medical bills, Rohr began what has turned into a long battle to investigate the toll the electronics industry takes on its workers.

In a conversation about her health problems, the 37-year-old woman had told Rohr that "all the girls I work with" had similar problems. Shocked, Rohr took down the names of several of the woman's friends and telephoned them. "It seemed like every time I called one person she would know of at least three other persons who had had cancer or who had had their uteruses removed," recalled Rohr. "It just kept snowballing. I thought, 'My God will this never stop.' "

Even today, workers continue to come to her law office asking her to represent them in workers' compensation cases against GTE. She has been forced to limit the kinds of cases she will take—all but three of the workers she represents have worked at the GTE plant for more than three years, and several have worked at the plant for more than a decade.

According to Rohr, her battle is fueled by the outrage at what she feels was GTE's handling of health and safety at the plant.

The GTE plant in Albuquerque which opened in 1971 was set up to assemble semiconductors, manufacture optical and data transmission systems, and test digital systems. The plant, originally known as GTE Lenkurt, changed its name to GTE Communication Systems in one of a series of reorganizations in the last 5 years.

Although today the plant employs only 600 people and no longer does semiconductor assembly, in its heyday more than a thousand people in the Albuquerque area were employed at the GTE plant. In its 15 year history, the plant also has used more than 500 kinds of chemicals.

In the printed circuit laboratory, workers regularly dealt with a whole host of acids, ranging from sulfuric to hydrochloric and hydrofluoric. "It was just horrible" Rohr said, "these people were breathing the acids without any masks or any protection."

"They tell me that they used to stick their hands in freon to pick up things and their hands would turn white like chalk," she said.

"It's clearly negligent," Rohr stated in a telephone interview. "All those illnesses could have been avoided with the proper ventilation and information. These women had absolutely no idea what they were working with, it's really sad to hear them talking about it."

Workers were not informed of the health effects of the chemicals they were working with nor were their complaints taken seriously, Rohr charges.

GTE maintains, however, that "there are no indications that materials used at the Albuquerque plant produced the health problems" of the workers who have filed suit. "GTE Communication Systems believes the lawsuits are without merit," said Nancy Colbert, a spokeswoman for the company. "The plaintiffs have not yet presented medical evidence to support their allegations."

The two areas where workers complained of the most health problems are no longer in Albuquerque, however. The company moved Departments 341 to Juarez, Mexico and the equipment for 320 has been shipped to El Paso, claim workers.

The company denies that the moves were in any way related to possible health problems related to working in the departments. GTE moved Department 320 to Juarez, Mexico for "economic" reasons, said Colbert. In the industrial parks along the Mexican side of the U.S.-Mexican border, workers on the average earn less than a dollar an hour.

And in Mexico, the health and safety regulations and enforcement are less stringent than even the most lax of U.S. standards. Mexican workers "have no OSHA, they have no equivalent to the Environmental Protection Agency," said Rohr. "They have absolutely nothing that would monitor or protect those women."

PART III
Structural Transformation and Systems of Inequality

The technological and economically based reorganization of society has created wide disparities in the distribution of economic resources. All people in the United States are affected by the economic changes. The magnitude of structural transformation, however, is different throughout society. The old inequalities of class, race, and gender are thriving. New and subtle forms of discrimination are becoming prevalent throughout society as the economic base shifts and settles.

Four factors are at work here: new technologies, global economic interdependence, capital flight, and the dominance of the information and service sectors over basic manufacturing industries. Together these factors have reinforced the unequal placement of individuals and families in the larger society. They have deepened patterns of social inequality, and they have formed new patterns of domination in which the affluent control the poor, whites control people of color, and men control women.

The disproportionate effects of economic and industrial change are most visible in three trends: (1) structural unemployment, (2) the changing distribution and organization of jobs, and (3) the low income-generating capacity of jobs. The first two trends have their most obvious effects on the changing class structure and the new racial order, while the second and third trends have particular ramifications for women and for gender relations. Still, all three trends have significant consequences for the hierarchies of class, race, and gender.

CLASS

Two major developments stand out when we look at the emerging class structure. The first is the growing gap between the rich and the poor since 1970. The second is the decline of the middle class.

The distribution of income is very unequal and widening. From 1970 to 1986, the income share of the highest quintile rose from 43.3 to 46.1 percent, while the bottom one fifth fell from 4.1 percent to 3.8 percent of all income (Pear 1987). The growing disparity in income and wealth is directly related to the changing job structure as the economy shifts from manufacturing to service.

Many Americans have experienced a sharp slowdown of income growth. This has caused a shrinkage of the middle class that is related in large measure to the erosion of middle-income jobs and the emergence of a bipolar wage structure in high tech and service work. The middle portion of the workforce fell from 52.3 percent of the population in 1978 to 44.3 in 1986 (Rose, 1986:9). To be fair, demographic conditions as well as labor market forces are responsible for middle class decline. Nevertheless, the current period is the first in American history where the rate of downward mobility exceeds the rate of upward mobility. Home ownership illustrates how the middle class is losing ground. The percentage of families owning their homes has declined each year in the 1980s (American Demographics 1986). Furthermore, changes have occurred in the population with a middle-class life style. As Barbara Ehrenreich has observed: "Middle class is a matter of status as well as income and is signaled by subtler cues: how we live, what we spend our money on, what expectations we have for the future. Since the post-war period, middle class status has been defined by home ownership, college education (at least for the children), and the ability to afford such amenities such as a second car and family vacations" (Ehrenreich, 1986:50). Using this colloquial understanding of "middle class," it is clear that fewer Americans will achieve this status because many of the old avenues to social mobility no longer exist in the new society.

RACE

Technology and the changing distribution of jobs are having devastating effects on minority communities across America. The employment status of minorities is falling in all regions. It is worse however, in areas of industrial decline. The labor market status of whites has also been lowered in these areas, but the level of racial inequality has increased to scandalous levels. "In cities such as Detroit, Buffalo, Chicago, and Cleveland, the gap between the labor market position of blacks, especially black males, and whites probably exceeds the highest levels that ever existed in the most racist of the South's cities" (Swinton, 1987:68). Racial inequality is partly a class issue. Prominent sociologist William J. Wilson has argued that the long-term removal of job opportunities for minority skilled and semiskilled workers is the force most responsible for the growth of the Black underclass in America's inner cities (Wilson 1987). Furthermore, the racial underclass is expanding to include Hispanics whose poverty rate now exceeds that of Blacks and who are expected, shortly after the turn of the century, to surpass Blacks as the largest racial-minority group.

Hispanics and Blacks have suffered disproportionately from industrial job loss and declining manufacturing employment. Their concentration in industries that have deteriorated in recent years is well documented in the readings in this section. By every measure including employment rates, occupational standing, and wage rates, the labor market status of racial minorities has deteriorated relative to whites. While the official unemployment rate for March 1987 was 6.5 percent, the rate for Blacks was 13.9 percent and 9 percent for Hispanics. These government rates, of course, are misleading because they count as employed those 5.5 million who work part time because they cannot find full-time jobs, and they do not count as unemployed the 1.17 million discouraged workers who have given up their search for work (Hershey 1987).

GENDER

Women and men are affected differently by the transformation of the economy from its manufacturing base to a base in service and high technology. Industrial jobs, traditionally filled by men, are being replaced with service jobs that are increasingly filled by women. Since 1980, women have taken 80 percent of the new jobs created in the economy. If this pace continues, women will make up most of the work force by the turn of the century (Hacker, 1986:26). Unlike other forms of inequality, sexism in American society has not become more intense as a result of economic transformations. Instead, sexism has taken new forms as women are propelled into the labor market. Women have continued to move from the private sphere (family) to public arenas, but they have done so under conditions of labor market discrimination that have always plagued women. Joan Smith has shown how women have been the principal facilitators of the fundamental shift in the U.S. economy (Chapter 19). Today, the "typical" job is a non-union, service sector, low-paying job occupied by a woman. Despite the growing feminization of the work force, male domination has been more firmly entrenched in the social organization of work. The rise of the contingent work force (including part-time work, temporary agencies, and subcontracted work) offers many advantages to employers but at considerable cost to women workers. Technological innovations in the workplace offer women dubious gains:

> In any number of cases, the outcome is more work opportunities for women, even if that work is less skilled, less autonomous and less rewarding . . . for many other women, new office machines have made them develop greater skills even if the jobs they use them for have become more rather than less fragmented and alienated and are no better than they were in the first place. What was once professional-level work—typically reserved for men—is often eliminated, and the residue or more routine tasks is added to the responsibilities of clerical workers who are, of course, women. The new technologies may develop more skills, but they also introduce clear monitoring and a stepped-up work pace. (Smith, 1987:6)

The full impact of economic restructuring on women must take into account the low wage levels and the limited opportunities for advancement that characterize their work in the new economy.

NEW INTERSECTIONS OF CLASS, RACE, AND GENDER

The hierarchies of class, race, and gender are simultaneous and interlocking systems. For this reason, they frequently operate with and through each other to produce social inequality. Not only are many existing inequalities being exacerbated by the structural transformation of the economy, but the combined effects of class, race, and gender are producing new kinds of subordination and exclusion throughout society and especially in the workplace. For example, the removal of manufacturing jobs has severely increased Black and Hispanic male job loss. Just how much of this is due to class and how much is due to race remains an important question.

In contrast, many women of color have found their work opportunities expanded, albeit in marginal work settings in service jobs and high-tech jobs. Does such growth offer traditionally oppressed race/gender groups new mobility opportunities or does the expansion of new kinds of work reproduce existing forms of inequality? The impact of economic restructuring on race and gender varies considerably. In some cases, it generates no jobs at all and displaces minority women workers. In other cases, minority women benefit by the creation of new jobs. Third world immigrant women provide the bulk of high-tech productive labor force in Silicon Valley. Yet a growing "underclass" in high tech consists of low-paid immigrant women from Mexico, Vietnam, Korea, and the Philippines. These examples reveal new labor systems as well as new forms of racial control based on class, race, and gender. The readings in the last chapter of this section capture the interlocking systems of oppression that are surfacing in the new economy.

REFERENCES

American Demographics. 1986."The Affordable Dream." *American Demographics* (July):12, 14.

EHRENREICH, BARBARA. 1986. "Is the Middle Class Doomed?" *The New York Times Magazine* (Sept. 7):44, 50, 54, 62, 64.

HACKER, ANDREW. 1986. "Women at Work," *New York Review of Books* 33(13) (August 14):26–32.

HERSHEY, ROBERT D. 1987. "Jobless Rate Down but Growth of Jobs Also Falls." *The New York Times* (April 4):7.

MOBERG, DAVID. 1986. "Middle Class May be Losing the Economic War of Attrition." *In These Times.* (Nov. 12, 16).

PEAR, ROBERT. 1987. "Poverty Rate Dips as the Median Family Income Rises." *The New York Times* (July 31):8.

ROSE, STEPHEN J. 1986. *The American Profile Poster.* New York: Pantheon.

SMITH, JOAN. 1987. "Terminal Illnesses," *The Women's Review of Books IV*, 7 (April):6–7.

SWINTON, DAVID. 1987. "Economic Status of Blacks 1986." *The State of Black America 1987.* New York: National Urban League (Jan.):49–73.

WILSON, WILLIAM J. 1987. *The Truly Disadvantaged.* Chicago: University of Chicago Press.

MIDDLE CLASS DECLINE AND WIDENING INEQUALITY

14 What Is Happening to the Middle Class?

Most everyone has an opinion about the decline of the middle class. But until now, little in-depth research has been done to substantiate those opinions. The *New York Times*, the *Boston Globe*, the *Wall Street Journal*, the *Atlantic Monthly*, the CBS Evening News, *Fortune* magazine, and congressional testimony have recently examined the issue of the declining middle class. Many of their conclusions, however, are based on assertions that either lack rigorous documentation or are inconsistent with facts we have uncovered.

We are surprised by the lack of hard thinking and investigation displayed by most commentators on this subject. Is the middle class declining, and if so, by how much? What is causing the decline? Is the decline permanent or transitory? How have the characteristics of middle class families changed over time?

Our analysis shows that the middle class has declined. We have compared data from the March 1969 and March 1983 Current Population Surveys (which ask about income in the preceding year, i.e., 1968 and 1982) to determine whether and why the middle class is shrinking, and to assess the chances that it will ever regain its former strength.

What is the middle class? The best definition would include measures of both social values and income, but the available statistics do not allow for this

Source: McKinley L. Blackburn and David E. Bloom, "What Is Happening to the Middle Class?" *American Demographics* 7, (January, 1985), pp. 19–25. Reprinted with permission.

kind of analysis. Only income statistics are readily available, and that is why most researchers use income as a measure of the middle class.

Previous studies of the middle class have examined total family income. This is not an entirely satisfactory measure because total family income fails to account for family size. A family income of $20,000, for example, goes further if it supports only two family members than if it supports five or more members.

But per capita family income (total family income divided by the number of family members) is not an entirely satisfactory measure either, because it does not distinguish between children and adults in a family, nor does it account for fixed household costs. One person living alone on an income of $10,000, for example, is probably not as well-off as each of two earners who live together with a combined income of $20,000, though their per capita incomes are the same. Since neither total family income nor family income per capita are ideal measures, we examined both in order to assess the sensitivity of our results to the income measure we used.

Other researchers have focused on a variety of criteria for middle class membership. Lester Thurow, in his *New York Times* article (February 5, 1984), examined households with incomes between 75 percent and 125 percent of median household income. *Fortune* magazine (November 1983) defined the middle class as families with incomes between $15,000 and $35,000 in 1982. In *Across the Board* magazine (October 1984), Sar A. Levitan and Peter E. Carlson defined the middle class as families in the middle third of the income distribution.

We believe that the first measure, used by Lester Thurow, is too narrow. It implies that middle-class family incomes ranged from roughly $13,800 to $23,000 in 1982. In our opinion, this classifies too many families in the upper class, when they really are members of the middle class. *Fortune's* measure of middle-class status is not well-suited to the analysis of change over time. For example, would *Fortune* have said that the middle class in 1969 was also in the income range of $15,000 to $35,000 after adjusting for inflation? Or would it have chosen some other range? The third measure suggests that one-third of families are always lower class, one-third are always middle class, and one-third always upper class. This measure can only be used in a round-about way to study the decline of the middle class (for example, by calculating the income cutoffs for the middle class and examining whether they have grown closer together or further apart over time).

The key point to remember when looking at different income classes is that each refers to a certain slice of the income distribution of American families. There is an unavoidable element of arbitrariness, of course, in determining where the cuts are to be made, which is why we have charted the entire distribution of family income in 1968 and 1982.

For our analysis, we have cut the income distribution into five slices, based on fractions of median family income in both 1968 and 1982. The slices are skewed because the distribution of family income is itself skewed.

1. lower class: families with incomes that are less than 60 percent of median family income;
2. lower-middle class: families with incomes between 60 and 100 percent of median family income;
3. middle class: families with incomes between 100 and 160 percent of median family income;
4. upper-middle class: families with incomes between 160 and 225 percent of median family income;
5. upper class: families with incomes greater than 225 percent of median family income.

We used this classification to analyze the distribution of both total family income and per capita family income. We also experimented with alternative definitions of the family so that our results are not dependent on how we define a family. In addition, because the income statistics collected by the Current Population Survey are retrospective (i.e., they refer to the previous calendar year), we determined a family's income class in 1983 by using its income in 1982. Similarly, we determined a family's income class in 1969 on the basis of its income in 1968.

The appropriate unit of analysis in a study of the middle class is the family rather than the household, because many households today consist of several families or unrelated individuals living together who do not share their incomes. The Census Bureau defines a family as two or more related persons who live in the same household. We examined income distributions for Census Bureau families—both primary and secondary.

Using only the Census Bureau's definition of families, however, might yield misleading results, because a large proportion of the adult population lives alone or with unrelated people, and the proportion has increased dramatically since 1969. To circumvent this problem, we also examined income distributions for our own broader concept of family, which includes single-person households, each of which we consider a family, and households in which there are unrelated people living together. We defined each of these unrelated individuals as a separate family. Single-person households and unrelated individuals living in multi-person households represented 22.5 percent of our "families" in 1969, and 32.7 percent in 1983.

THE DECLINE OF THE MIDDLE CLASS

The decline of the middle class appeared in every measure we examined. Whether we used our concept of family or the Census Bureau's definition, middle-class families fell as a share of all families. As the middle-class share has declined, both the lower- and upper-class shares have grown larger, and since 1969 the upper class has expanded more than the lower class.

Using our concept of family, the middle-class share of all families declined from 27.4 percent in 1969 to 23.1 percent in 1983, a 15.7 percent decline in the proportion of families in the middle class. Using the Census Bureau's definition of family, the proportion of families in the middle class fell from 31.4 percent in 1969 to 26.5 percent in 1983, a 15.6 percent drop in the middle-class share.

Using our concept of family and family income per capita as the income measure, the middle class has declined from 23.6 percent of all families in 1969 to 22.0 percent in 1983, a drop of 6.8 percent in the middle-class share of families.

According to our broad definition of family, median family income was $18,426 in 1982, 10 percent less than in 1968, when it was $20,457 (in 1982 dollars). This decline is largely attributable to the drop in family size, meaning that, on average, there were fewer earners per family. The trend to smaller families caused median family income per capita to increase 14 percent between 1968 and 1982.

A middle-class income—defined as an income between 100 percent and 160 percent of median family income—ranged from $18,426 to $29,840 in 1982. In 1968, middle-class family incomes ranged from $20,457 to $32,728 (in 1982 dollars). When we combined the three middle-class categories (lower-middle, middle, and upper-middle), middle-class family incomes ranged from $11,055 to $41,056 in 1982, and from $12,272 to $46,027 in 1968, after adjusting for inflation.

Even using the broader definition of middle class, one that combined the lower-middle, middle, and upper-middle classes, we found that, based on our concept of family, the proportion of families in the three middle-class categories had declined from 62.4 percent of all families in 1969 to 55.9 percent in 1983. Based on the Census Bureau's definition of family, the decline was from 70.6 percent in 1969 to 65.0 percent in 1983. Examining family income per capita using our concept of family, the middle-class share (broadly defined) fell from 61.5 percent in 1969 to 57.3 percent in 1983.

CLASS CHARACTERISTICS

The nation had 89.3 million families in 1983, according to our concept of family, and 20.6 million of them were in the middle-class income category. The lower-middle class had 16.6 million families, and the upper-middle had 12.7 million families. Twenty-eight million families were lower-class, and 11.4 million were upper class in 1983.

The middle class is not concentrated in any one region of the country, nor has any region been immune to the recent trends. The share of families in the middle class has fallen in all regions since 1969, while the share in the upper class has grown in every region. Both the Midwest and the West have experienced a significant increase in the proportion of families in the lower class, while the share of upper-middle-income families has declined in those two regions. In the Northeast and the South the situation is reversed: The share of lower-class families has fallen since 1969, while the upper-middle class has expanded.

A smaller share of both white and black families had middle-class incomes in 1982 than in 1968, although blacks are less likely than whites to belong to the middle class. The share of black families in the middle class fell from 19.8 percent in 1969 to 16.9 percent in 1983, while the share of white middle-class families dropped from 28.3 percent to 23.9 percent.

Over half of black families were in the lower class in 1983, compared with less than a third of white families. Only 3.9 percent of black families were in the upper class in 1983, down from 4.3 percent in 1969. In contrast, the share of white upper-class families grew from 8.6 percent to 13.9 percent during the same period.

The higher a family's economic class, the larger its size, reflecting the greater number of earners in larger families. But family size in all income classes fell between 1969 and 1983. Middle-class families averaged 2.8 members in 1983, down from 3.5 members in 1969. Lower-class families had an average of 2.0 members in 1983, slightly less than their 2.1 members in 1969. The average upper-class family had 3.3 members in 1983, down from 3.9 members in 1969.

The share of multiple-earner families in the middle class declined from 36.6 percent in 1969 to 29.0 percent in 1983, while the share of multiple-earner families in the upper class jumped from 14.1 percent to 25.7 percent. Only 7.7 percent of single-earner families qualified for the upper class in 1983.

One-fourth of families with one earner were middle class in 1983, down from 26.9 percent in 1969, and another one-fourth were lower-middle class. The largest proportion of single-earner families—32.5 percent—were lower class.

For our analysis, we examined the characteristics of the principal males and principal females of each family. We defined the principal male and female as either the family head or the spouse of the family head. Some families have both a principal male and a principal female, while other families have just one or the other. In 1983, for example, 67 percent of our families had both a principal male and principal female, 23 percent had only a principal female, and 10 percent had only a principal male.

For all but one income class, the average age of principal males and females changed little between 1969 and 1983. The average middle-class man was 42.6 years old in 1969 and 43.2 years old in 1983. The average middle-class woman was 40.7 years old in 1969 and 42.9 in 1983. For the lower class, however, the average age of principal males and females fell sharply. For principal men in the lower class, the average age dropped by more than seven years, from 54.9 years of age in 1969 to 47.5 years in 1983. Lower-class women were nearly eight years younger in 1983 than in 1969; their average age fell from 54.8 years to 46.9 years. This decline in the average age of the principal members of lower-class families is the result of household formation by baby boomers, many of whom have low incomes because they have only recently entered the labor force.

In 1969, fully 85.7 percent of middle-class families were headed by a married couple. By 1983, this figure had dropped to 67.6 percent. Similar declines in the share of married-couple families occurred in each of the other classes—except in the upper class, where married-couple families increased from 86.1 percent to 89.5 percent of the total. The smallest proportion of married-couple families was in the lower class, where 27.9 percent of families were headed by a married couple in 1983, down from 37.5 percent in 1969.

Education is directly linked to economic class. The greater the educational attainment of a family's principal male and female members, the higher the family's income. Among families in which the principal male had a college degree, the largest proportion—36 percent—were in the upper class in 1983, up from 28 percent in 1969. About 23 percent of families whose principal male was college-educated were in the middle class, down from 26.6 percent in 1969.

On the other end of the scale, 37.7 percent of families whose principal male lacked a high school diploma fell into the lower class, up from 30.5 percent in 1969; 24 percent were in the middle class, down from 28.5 percent in 1969; 9.2 percent were upper-middle class in 1983, and only 5.1 percent were upper class. Such figures suggest the growing importance of having at least a high school degree in a society in which 75 percent of all principal males are high school graduates, compared with 57 percent in 1969.

Among families in which the principal female had a college degree, fully 31.4 percent were in the upper class in 1983, up from 28.5 percent in 1969. Among principal women lacking a high school education, the majority—51.9 percent—fell into the lower class, up from 42.5 percent in 1969.

Even if the principal members of a family have low-skill jobs, their incomes may still place their families in the middle class. In 1983, 26 percent of families whose principal male held a low-skill job and 22 percent of those whose principal female had a low-skill job were middle-class. But more than one-third of families whose principal members had low-skill jobs were in the lower class.

The largest proportion of medium-skilled workers (e.g., supervisors and low-level managers) were in middle-class families: 29 percent of families whose principal males were medium-skilled and 26 percent of families whose principal females were medium-skilled.

Thirty-six percent of families whose principal males were highly-skilled workers (e.g., professional and upper-level managers) had upper-class incomes, while 29 percent of families whose principal females were highly-skilled earned a place in the upper class. In contrast, only 22 percent of families whose principal male was highly-skilled and 25 percent of families whose principal female was highly-skilled fell into the middle class. Thus, highly-skilled workers earn higher incomes than less-skilled workers, placing their families higher in the income distribution.

The manufacturing industries employ a larger share of principal males in the middle, upper-middle, and upper classes than any other industry. In 1983, 30.0 percent of men in middle-class families, 29.3 percent of those in upper-middle class families, and 28.2 percent in upper-class families worked in manufacturing. The service industry employed the second largest share of men in the middle to upper classes in 1983: 17.1 percent of middle-class men, 17.6 percent of those in the upper-middle class, and 25.1 percent in the upper class. Wholesale and retail trade employed 15.8 percent of middle-class men, 12.5 percent of upper-middle-class men, and 14.0 percent of those in the upper class.

In contrast, 24.2 percent of lower-class men worked in wholesale and retail trade, as did 16.8 percent of lower-middle-class men. The service industry employed 17.6 percent of lower-class men, and 24.7 percent of lower-middle-class men. But 17.2 percent of men in lower-class families worked in agriculture and mining, a larger share than in any other class. Only 5.6 percent of middle-class men and 1.9 percent of those in the upper class had jobs in agriculture and mining.

Compared to principal males, the industry in which the principal female worked had little influence on her family's income class. This is probably because women contribute less to family income than men, and because wages vary less for women in different industries than they do for men.

No matter what her family's class, a larger share of women worked in services than in any other industry. Half of employed upper-class women had service jobs, as did 45.3 percent of those in the upper-middle class, 40.1 percent in the middle-class, 38.2 percent in the lower-middle class, and 39.9 percent in the lower class.

WHY THE MIDDLE CLASS HAS DECLINED

There are five factors that might be responsible for the decline in the middle class: the business cycle, changes in family composition, the growth of two-earner families, the changing industrial structure of the economy, and changes in the age composition of the population.

The civilian unemployment rate was higher in 1982, at 9.7 percent, than in 1968, when it was only 3.6 percent. The fact that a larger share of Americans were unemployed in 1982 than in 1968 is consistent with the fact that the lower class has gained in share between 1969 and 1983. But it is not consistent with the fact that the upper class has grown—except to the extent that higher unemployment results in lower median family income and a lower cutoff for inclusion in the upper class.

Changes in family composition could be partially responsible for the decline in the middle class, because there were more single-person households and households containing unrelated people in 1983 than in 1969. Much of this increase is the result of growth in the number of households containing young workers with relatively low salaries or elderly women with small, fixed incomes. Again, these changes contribute to the growth of the lower class, but do not explain the growth of the upper class.

The growth of two-earner families between 1969 and 1983 may have resulted in the shrinking of the middle class, because as women have gone to work they have raised their family incomes from middle class to upper class. In 1983, for example, 26 percent of families with two or more earners were in the upper class, compared with just 8 percent of families with one earner. In 1969 only 14 percent of families with two or more earners were in the upper income class, suggesting that the number of earners in a family is now a more important determinant of income class than it was in 1969. The growing importance of the second earner is probably due to increased educational attainment and labor force commitment of second earners—most of whom are women.

Changes in the industrial structure of the economy might also have contributed to the decline in the middle class. The popular argument is that job creation has been most rapid in low-paying service-sector jobs and least rapid in high-paying manufacturing jobs. Principal males who work in manufacturing are more likely to be middle-class than principal males who work in the service industries. Thus, a shift from manufacturing to services would contribute to a decline in the middle class. Nonetheless, our calculations show that the size of this effect is small, because many manufacturing jobs pay wages that are high enough to put a worker's family in the upper class, because many service-sector jobs pay quite well, and because there is only a weak relationship industry and income class for women.

The age composition of the population has contributed to the decline of the middle class. The drop in the average age of principal males and females in lower-class families reflects the transition of millions of baby boomers from childhood to adulthood between 1969 and 1983. Large numbers of young workers have formed households, and their low wages classify them in the lower class. Moreover, the changing age composition has had an added impact on the lower class because of the well-documented trend among baby boomers to delay marriage and family formation, creating more single-person households and households containing unrelated people living together. Indeed, 63 percent of single-person households had incomes below 75 percent of the median in 1982. Alternatively, 56 percent of our "families" with incomes below 75 percent of the median were single persons.

While none of these five factors alone explains the decline of the middle class, taken together they shed some light on it. The middle class has declined because of the growth in unemployment, the greater number of single-person households, the growing number of two-earner families, the effect of the baby boom on the age composition of the population, and the changing industrial structure of the economy.

PERMANENT OR TRANSITORY?

There is little evidence that the middle class will continue to shrink. Unemployment rates have been falling and are expected to continue their downward trend. Baby boomers are now in their 20s and 30s and are marrying in record numbers, slowing the growth in single-person households, and as the baby boomers continue to age, their earnings will increase.

On the other hand, the shifts in our economy are not yet complete, nor is it likely that we will soon witness an end to the growth of two-earner families. But the growth of two-earner families is likely to slow in the coming years, because the labor force participation rates of women are getting close to those of men and are not likely to go much higher.

The worst is over, according to the evidence. Any further decline in the middle class is likely to be small, if it occurs at all.

Given that the middle class is not likely to disappear, will it regain its former strength? The answer is no—the middle class will not recoup all of the losses it has suffered during the past 15 years.

Although unemployment rates are falling, it is unlikely that they will ever reach 3.6 percent again, since the work force is so much more heterogeneous than it used to be, and because the structure of the economy has changed so much. Although single-person households will not grow as rapidly as they have in the past 15 years, it is doubtful that the people who live alone today will live in traditional families tomorrow, especially since a large share of single-person households contain elderly women, or young men and women who will probably form two-earner households. Indeed, it is hard to imagine that the growth of two-earner families will reverse itself, because that would mean the return of women to the home.

The upper class has gained from the growth in two-income families; the lower class has been inflated by single-person households and the baby boom generation; the middle class has suffered from high unemployment rates and a changing economy. Yes, the middle class still cuts itself a large slice of the American pie, but the country has moved in the direction of becoming a nation of "haves" and "have nots," with less in between.

MIDDLE CLASS DECLINE AND WIDENING INEQUALITY

15 *U-Turn on Equality: The Puzzle of Middle Class Decline*

CHRIS TILLY

In the decades following World War II, policy-makers proudly noted that the distribution of income in the United States was becoming more equal. The gap between rich and poor seemed to be narrowing and more and more Americans were joining the "great middle class."

Recently, things have taken a sharp turn for the worse. By almost any measure, the distribution of income among families has become much more unequal in the past ten years. Much of this backsliding in equality seems to be due to polarization in wages and salaries; when benefits are included the inequality becomes even more pronounced. The distribution of income among individual workers, which became more equal through the 1960s and the first half of the 1970s, took a U-turn between 1975 and 1978, and has become more unequal ever since. These changes took place before the election of Ronald Reagan, before the passage of the sharply regressive tax act of 1981, and even before the start of the tight money recession of 1979.

The conventional explanations for this explosion of inequality are simple and reassuring. The basic change, say the Washington policy analysts, is a demographic one: Americans have changed the kinds of families they live in. New demographic groups—particularly young people and women—have flooded the labor force. To be sure, four years of recession made the situation

Source: Chris Tilly, "U-Turn on Equality: The Puzzle of Middle Class Decline," *Dollars and Sense* 116 (May, 1986), pp. 11–13.

worse. But all of these problems are temporary and, claim the analysts, do not reflect changes in the overall structure of economic opportunity.

Unfortunately, these conventional explanations are wrong. Recent research by Bennett Harrison, Barry Bluestone, and myself shows that demographic changes, even when combined with the effects of the business cycle, explain very little of the worsening labor market inequality.

INEQUALITY AMONG FAMILIES

How extreme is family income inequality in the 1980s? Consider four facts:

- In 1984, the wealthiest fifth of the population received 43% of national income, while the poorest fifth received less than 5%. The gap between rich and poor is the widest since the government began collecting this information in 1947.
- The most often-used measure of income inequality, the Gini coefficient is also at the highest—most unequal—level recorded since 1947. The Gini coefficient has increased almost continuously since 1968.
- The percentage of U.S. households earning a "middle-class" income (defined as between 75% and 125% of median household income) fell from 28.7% in 1967 to 23.2% in 1983—with an acceleration of that long decline between 1982–83. This is what's often called the "disappearing middle class."
- Americans at the bottom of the income distribution have increasingly slipped into poverty. The percentage of Americans living below the poverty line has been higher during the 1980s than any time since the "War on Poverty" began in the mid-1960s.

In short, the rich are getting richer and the poor are getting poorer. Of course, some groups are more likely than others to end up on the low end. Nearly half of all black families fall into the poorest one-fifth of the population, and black families have fallen further behind white families in income in the past several years. One-third of all persons living in female-headed families are living in poverty, compared with less than one-tenth of people in other families.

SUSPECT #1: CHANGES IN THE FAMILY

What is the reason for the new inequality in the income distribution? The leading conventional wisdom claims that changes in family structure lie at the root of the new trend towards inequality. There are more single-mother families at the low-income end, and more two-earner families at the top end. The low end of the income distribution is also weighted down by increasing numbers of young people and old people living alone; previously these individuals were more likely to live in and share their income with an extended family.

According to this conventional wisdom, then, it is not the economy itself that is generating inequality: Rewards given for individual enterprise haven't changed, just the distribution of individuals among families. The decisions of Americans to live in different kinds of family groups have supposedly caused a re-division of income.

This explanation isn't correct, but even if it were, the changes in family structure would still pose serious economic problems that the free market couldn't resolve. For example, female-headed families sink into poverty because a working woman earns on average 60% of what a man earns, and because welfare benefits leave many families below the poverty line. In a more developed welfare state that provided subsidized housing, a guaranteed health service, and an adequate family allowance per child, an increase in female-headed families would not automatically increase the number of poor families.

There is also mounting evidence that the reshuffling of the family is not the main cause of worsening family income inequality. Two other causes seem to have a role in generating the increased inequality: federal policy shifts and labor market changes.

SUSPECT #2: BUDGET CUTS

What are the effects of changing federal welfare policies? Economists Peter Gottschalk and Sheldon Danziger recently analyzed changes between 1967 and 1982 in the distribution of market income and (government-provided transfer income, and the relation between the two. They concluded that the declining U.S. poverty rate between 1967 and 1979 was equally due to improvements in market income and to increases to transfer income. Between 1979 and 1982, as market incomes suffered from a prolonged recession, increases in transfer income were too small to compensate, and the percentage of Americans living in poverty climbed. Clearly the subsequent cutbacks by Reagan in social service expenditures must bear some of the blame for making the poor get poorer. And although the figures cited up until now are for pretax income, the massive tax cuts initiated in 1981 disproportionately hurt the poor and certainly helped the rich get richer.

Although the policies of the Reagan administration have worsened—the disparities between rich and poor, they are not the fundamental cause of the new inequality. Inequality in individual earnings took its U-turn for the worse in 1978; so did the poverty rate. Inequality in family income has been growing by some measures since 1968. The causes of inequality lie deeper than the policies of any one president.

SUSPECT #3: INEQUALITY IN EARNINGS

People enter the labor market as individuals, then pool their earnings in families and supplement them with government transfers. If the new inequality were *just* the result of altered family structure—the way individuals are distributed among families—and changing transfer benefits, then the distribution of earnings among individuals would not change. But the U-turn that we found in inequality of individual earnings indicates that something more is going on.

We measured inequality by looking at how spread-out annual wages and salaries are around the average level—that is, their variance or dispersion. In research completed early in 1986 at Boston College and MIT, we found that annual wages and salaries of individuals became more equal from 1964 to 1978, although the trend towards equality slowed down during the mid-1970s. Then, starting in 1978, the distribution of earnings abruptly became much more unequal—quickly reaching and surpassing the 1964 level of inequality. This increase in inequality took place in virtually every subgroup of the population: men as well as women, young as well as prime-aged (35–54 years) workers, even full-time, year-round workers.

Three reasons are often used to explain increasing inequality among individual earnings. As in the case of family income inequality, the leading explanation blames demographic changes. It says that increasing earnings inequality results from the entry into the workforce of large numbers of young people (baby boomers) and women—who both have lower-than-average earnings. As these groups enter the workforce, it is thought, not only do they come in with below-average earning power, but they also often compete for the same jobs, "crowding" these job categories and driving wages down further.

A second explanation says that the rocky business cycle of the 1970s, particularly the prolonged recession of 1979–82, was a cause of earnings inequality. It is widely held that wage differentials narrow during periods of macroeconomic expansion and widen during recessions. The reason offered by labor economists is that high wages tend to be relatively "stickier," that is, better protected by contracts (explicit or implicit) than low wages. In recessions, it is therefore the low wages that tend to be eroded relative to the rest, while during recoveries, employers are freer to augment low wages if temporary shortages appear.

Still another widely held view is that the 30% increase in the exchange value of the U.S. dollar between 1980 and 1983, as weighted against the currencies of nine major trading partners, has so decimated U.S. export industries as to dislocate factory workers who are predominantly middle-level wage earners.

But we found that these factors can't adequately explain soaring inequality. First, we looked at the timing of growth in inequality compared with the timing of the workforce entry of young people, the business cycle, and exchange rate fluctuations. After removing the effects of these three variables, we still found a strong U-turn in inequality. In fact, the combination of these three variables only explains about one-third of the U-turn effect.

In a separate analysis, we focused on the demographic explanations. First, we measured the impact of age and sex on the earnings of individuals. Then we compared the age and sex composition of the workforce and the earning power of young people and women (which could have dropped due to labor market crowding) in 1972 versus 1983—an interval in which inequality (variance) rose by 10%. Could the demographic shifts in workforce composition and earning power explain the growth in inequality among earners? Our preliminary results indicate that they account for little or none of the jump in inequality.

Transformations in family structure, new federal policies, more women and young people in the workforce, economic fluctuations—none of these seems to be the major cause of the growing inequality in incomes and earnings. What, then, is the hidden cause? We suspect that the answer has to do with new corporate strategies adopted in response to a profit squeeze.

Throughout the 1960s and early 1970s, corporate profits were squeezed by growing real wages and a growing social wage of government-provided benefits. In the mid- to late 1970s, the problem was compounded by a sharpening of international competition. Corporations reacted by relocating production, accelerating automation, contracting out work to small suppliers, attacking unions and demanding wage concessions, and replacing full-time jobs with part-time ones. These strategies eliminated many blue-collar and even white-collar jobs in the middle pay range and created millions of low-wage jobs, while preserving or even adding to the high-paying professional, managerial, and technical jobs necessary to control the process. This explanation is far indeed from the conventional wisdom—but given the evidence, it just might be right.

RECONSTRUCTING THE RACIAL ORDER

16 Urban Change and Minority Opportunities

JOHN D. KASARDA

Two fundamental yet conflicting transformations pervade the recent history and near-term prospects of older, larger U.S. cities. The first is *functional,* as these cities change from centers of production and distribution of material goods to centers of administration, information exchange, and higher-order service provision. The second transformation is *demographic,* as their resident populations change from a predominantly white European heritage to predominantly black, Hispanic, and other minorities.

Accompanying the functional transformation of these cities have been changes in both the composition and sizes of their overall employment bases. During the past two decades most older, larger cities have experienced substantial job growth in occupations associated with knowledge-intensive service-sector industries. However, selective job growth in these high-skill, predominantly white-collar industries has not nearly compensated for employment declines in manufacturing and other predominantly blue-collar industries that had once constituted the economic backbone of these cities. As a result, the total number of jobs available in most older, larger cities has shrunk considerably over the past two decades.

Likewise, concurrent with the demographic transformation of these cities, there have been substantial changes in the socioeconomic composition and

Source: John D. Kasarda, "Urban Change and Minority Opportunities," in *The New Urban Reality,* Paul E. Peterson (eds.) (Washington, D.C.: The Brookings Institution, 1985) pp. 33–67.

total size of their residential populations. As predominantly white middle-income groups have dispersed from the cities (initially to the suburbs and now increasingly to the exurbs and nonmetropolitan areas), they have been only partially replaced by predominantly lower-income minority groups and a relative trickle of returning urban professionals. The outcome has been dramatic declines in both the total size and aggregate personal income level of the cities' resident populations, while concentrations of their economically disadvantaged continue to expand.

The simultaneous transformation and selective decline of the employment and residential bases of the cities have contributed to a number of serious problems, including a widening gap between urban job opportunity structures and skill levels of disadvantaged residents (with corresponding high rates of structural unemployment), spatial isolation of low-income minorities, and rising levels of urban poverty and welfare dependency. Associated with these problems have been a plethora of social and institutional ills further aggravating the predicament of people and places in distress, such as high crime rates, poor public schools, deteriorating public infrastructures, and the decay of once-vibrant residential and commercial areas of the cities.

The federal government responded to the hardships confronting cities and their inhabitants by introducing a variety of urban programs over the past two decades. Unfortunately, these programs have had little effect in stemming overall urban decline or improving long-term employment prospects for underprivileged city residents. Indeed, mounting evidence suggests that the plight of economically distressed cities and their underprivileged residents is worse today than before the urban programs began.

The inability of federal programs to stem urban decline results primarily from the overarching technological and economic dynamics influencing the locational choices of various industries and the changing roles of major cities in advanced service economies. This chapter will elaborate on these dynamics and changes, especially as they have altered the capacity of America's older cities to offer entry-level jobs and socially upgrade disadvantaged resident groups.

The guiding thesis is that cities always have and always will perform valuable economic and social functions, but changing technological and industrial conditions (both nationally and internationally) alter these functions over time. In this regard, it must be remembered that blue-collar job opportunities expanded most rapidly in the largest and oldest cities during a transportation and industrial era that no longer exists. During this era, cities had comparative advantages over other locations that substantially reduced the costs and raised the efficiency of firms concentrating in these cities. These advantages included superior long-distance transportation and terminal storage facilities; an abundance of immigrant labor willing to work for extremely low wages; many nearby complementary businesses; and availability of private and public municipal services such as police and fire protection, sewage systems, and running water.

Modern advances in transportation and communication technologies, the spread of population and public services, and changing modes of production organization have virtually wiped out the comparative advantages of major metropolitan cities as locations for large-scale manufacturing and warehousing facilities. The precipitous decline of blue-collar jobs in the cities during recent decades reflects, in part, their loss of comparative locational advantages in these industrial sectors . . .

THE EVOLVING STRUCTURE AND FUNCTIONS OF AMERICA'S CITIES

This country's oldest cities originated during the colonial era as mercantile centers of commerce and shipping. Since their chief function was that of funneling raw materials from their immediate hinterlands to the nascent metropolitan economy of eighteenth century Europe, these cities evolved at deep-water sites on long-distance riverine and oceanic routes. By today's urban standards, the cities of colonial America were minuscule—not a single place would have met the current metropolitan criterion of 50,000 residents.

As America entered the nineteenth century, it remained a rural-agrarian nation. Fewer than 4 percent of the U.S. population resided in places of 8,000 or more. Urban expansion during the first two decades of the nineteenth century progressed at a sluggish rate, as externally oriented commerce and shipping continued to dominate the economic bases of fledgling cities. Regional and national markets were constrained by limited transportation systems, while the bulk of urban manufacturing and goods production was of a cottage industry and handicraft form.

The next fifty years witnessed a dramatic development of railroad, highway, and canal systems that substantially broadened urban access to the country's rich raw material resources, spun off new inland towns and cities, and created a nationally integrated exchange network among cities, smaller trade centers, and hinterland villages. At the same time, a revolution was occurring in food production technology and organization that both generated the agricultural surplus that supported a growing nonagricultural population and displaced large numbers of subsistence farmers. Steam-powered machinery and mass production technologies were introduced that, together with entrepreneurial innovations in credit, banking, and finance capitalism, gave rise to large corporate business organizations and their primary units of production—factories. With high domestic fertility rates and increased immigration from abroad providing both an expanded domestic market for manufactured goods and abundant, cheap labor to staff the factories, all the requisites for the rise of America's industrial cities were in place.

Viewed spatially, the industrial cities developed as highly concentrated forms around rail and waterway terminal points within access of northern coal fields. The reasons for such concentration rested in transportation costs and the mechanized application of steam power. Because steam cools quickly and dissipates easily, it must be used where it is produced. Since coal—the fuel used to produce steam—had high transportation and terminal costs, business expenses could be minimized by factories clustering around the terminal and sharing bulk carriage costs. Because the terminal was also the point where other raw materials utilized in the production process were received and finished products distributed, factories concentrating there accrued further cost advantages.

Moreover, during the latter half of the nineteenth century raw materials used in the manufacturing process and finished products tended to be far heavier and bulkier than they are today, and short-distance transportation technologies were, at best, primitive. Under such conditions, agglomeration became cumulative and mutually reinforcing. Complementary units that serviced the factories or utilized their by-products located as close to the factories as possible, as did storage warehouses and wholesalers distributing finished products. Workers employed by the factories, service shops, and warehouses, constrained by pedestrian movement, clustered tightly within walking distance of their place of employment. As late as 1899, the average commuting distance of workers in New York City from their home to their place of employment was roughly two blocks.

Thus the industrial cities that evolved in the late nineteenth century were compact agglomerations of production and distribution facilities, centers of corporate capital accumulation, and both the residence and workplace of millions of unskilled and semiskilled migrants. Industrial expansion and concentrative migration converged in space, generating explosive city growth. Chicago, for example, which was incorporated in 1833 with a population of 4,100, grew to over 2 million residents by 1910, the vast majority of whom resided and worked within a three-mile radius of the city's center.

Spurring the dramatic growth of the industrial cities were a rapidly advancing western resource frontier and strengthening commercial markets interacting with a powerful entrepreneurial spirit which held that individualism, competition, the pursuit of profit, and economic growth were uniformly positive and beneficial. In this political-economic climate, urbanization and industrialization surged together, catapulting the entire country into a period of enormous economic expansion. By the end of the nineteenth century, the output of America's industrial cities surpassed the combined total industrial output of Britain, France, and Germany, the world's leaders in 1860.

A point that merits emphasis is that the employment bases of the country's early industrial cities were characterized by entry-level job *abundance*, compared with their current entry-level job *scarcity*. It was this job abundance, with few requisites for entry, that attracted the waves of migrants and offered them a foothold in the urban economy. The rapidly expanding job base accompanying national economic growth in turn provided ladders of opportunity and social mobility for the migrants, most of whom were fleeing nations or regions of substantial labor surplus. . . .

Technological Advance and Blue-Collar Job Deconcentration

During the first half of the twentieth century, a number of innovations occurred in transportation, communication, and production technologies that markedly reduced the locational advantages the older, compactly structured cities had previously held for manufacturing and distributing activities. Manufacturers and wholesalers soon found that older urban street patterns were not conducive to automobile and truck movement. Traffic congestion, lack of employee parking space, and problems of freight transfer greatly increased direct and indirect costs, particularly for those manufacturers and wholesalers located in the old, densely settled sections of the cities. On the other hand, the development of suburban highway systems, widespread automobile ownership, and increased dependence by manufacturers and wholesalers on trucking for receiving and shipping made uncongested suburban sites more attractive. Manufacturers and wholesalers recognized that by locating on or near the suburban expressway they could reduce their transportation costs, tap an adequate automobile-owning labor supply, and solve problems of employee parking and freight transfer.

The changing mode of manufacturing technology from unit processing to mass production and assembly line methods also hastened the urban exodus of blue-collar industries. Early central-city manufacturing facilities had been constructed as multistory loft-type structures that are not adaptable to much of today's mass production technology. The assembly line, in particular, has large horizontal space requirements that are more difficult and costly to obtain in the central city than in the suburban rings. Likewise, large-scale wholesalers shifting to automated freight-transfer systems found it more practical and less costly to build newly designed facilities on relatively inexpensive suburban ring land (that was often more accessible to regional and national markets) than to convert their obsolete inner-city structures. Owners of newer manufacturing and warehousing facilities that had large space requirements would rarely even consider a central-city site, most locating their facilities in the suburban rings and beyond.

A third, yet no less important, factor stimulating the deconcentration of blue-collar industry was the widespread development of suburban public services and external economies that had previously been restricted to the central cities and their immediate built-up areas. The spread of electric and gas power lines, running water, sanitary waste systems, police and fire protection, and highway services throughout the suburban rings released manufacturing and related industries from their previous dependency on central cities. Moreover, rapid suburban development after World War II brought housing, local suppliers, subcontractors, and other complementary services to nearby areas.

By 1960 further advances in transportation and communication technologies and growing industrial competition from nonmetropolitan areas and abroad made the larger, older cities almost obsolete as locations for manufacturing and warehousing facilities. A massive exodus of blue-collar jobs from the cities commenced, and it has accelerated ever since.

Retail and Service-Sector Shifts

Retail trade and consumer services followed their traditional middle- and upper-income patrons to the suburbs, exacerbating blue-collar job declines in the cities. Between 1954 and 1978 more than 15,000 shopping centers and malls were constructed to serve expanding suburban and exurban populations. By 1975 these shopping centers and malls were the locus of over one-half of the annual retail sales in the United States.

For the current generation of suburbanites, enclosed regional malls have become their Main Street, Fifth Avenue, and community social and entertainment center—all wrapped up into one. The malls not only contain retail establishments of every size, price range and variety, but they also offer professional offices, restaurants, movie theaters, public service outlets, and common space utilized for inexpensive patron attractions. The broad spectrum of goods and services offered by the malls along with special customer amenities (such as controlled climate; pedestrian arcades insulated from automobile traffic, noise, and polluted air; good security and lighting for safer and more relaxed nighttime shopping; free parking) and their accessibility to the automobile-oriented suburban population have combined to give the regional malls a distinct competitive edge in capturing the metropolitan retail dollar. Indeed, large enclosed malls have become so successful that many major cities are now trying to simulate them in their downtowns in an attempt to recoup some of their lost share of metropolitan retail sales and employment.

Urban Growth Industries

There are, nonetheless, significant counter trends occurring in certain retail and service sectors as businesses and institutions offering highly specialized goods and services continue to be attracted to downtown areas. The specialized and complementary nature of these establishments still makes it advantageous for many to choose central locations that maximize their accessibility to certain consumers and businesses. Such establishments as advertising agencies, brokerage houses, consulting firms, financial institutions, small luxury goods shops, and legal, accounting, and professional complexes have been accumulating in the central business districts, replacing traditional department stores and many other standardized retail goods and consumer services establishments that were unable to compete effectively or afford the rents at a central location.

The past two decades have also witnessed a remarkable growth of high-rise administrative office buildings in the largest cities' central business districts. Even with major advances in telecommunications technology, many administrative headquarters rely on a complement of legal, financial, public relations, and other specialized business services that are most readily available at the cities' cores. Moreover, unlike manufacturing, wholesale trade, and retail trade—which typically have large space requirements and whose products cannot be efficiently moved in a vertical direction—most managerial, clerical, professional, and business service functions are highly space intensive and their basic product (information) can be transferred as efficiently vertically as horizontally. Thus people who process information can be stacked, layer after layer, in downtown high-rise office buildings with no loss in productivity. Indeed, such stacking and the resulting proximity often enhance the productivity of those who require extensive, nonroutine personal interaction. The outcome has been an office building boom in central business districts. . . .

EFFECTS ON JOB OPPORTUNITIES

The transformation of older cities from centers of production and distribution of material goods to centers of information exchange and service consumption has profoundly altered the capacity of these cities to offer employment opportunities for disadvantaged residents. The increments in urban blue-collar jobs spun off by newer service industries employing or catering primarily to higher-income persons have been overwhelmed by central-city job losses in more traditional goods-producing and distributing industries. Aggravating these blue-collar job losses has been the exodus of middle-income population and general retail trade and service establishments from much of the city beyond the central business districts. These movements have combined to further weaken secondary labor markets and isolate disadvantaged groups in economically distressed areas where opportunities for employment are minimal.

Major cities in the northern industrial belt have been particularly hard hit by changes in their economic base and the loss of blue-collar jobs. Unfortunately, it is many of these same cities that experienced the largest post–World War II migration flows of people whose educational backgrounds do not equip them for employment in new urban growth industries. As a result, their inner-city unemployment rates are well above the national average and inordinately high among educationally disadvantaged urban minorities caught in the web of change. . . .

CHANGING URBAN EMPLOYMENT BASES

The most dramatic employment declines have been in central-city manufacturing. New York City and Chicago each lost more than 300,000 manufacturing jobs between 1948 and 1977, with the most pronounced employment losses occurring in both cities after 1967.

Wholesale and retail trade employment in all cities likewise deteriorated considerably. However, there are temporal differences. Whereas the majority of central-city retail employment losses occurred before 1967, two-thirds of the wholesale employment losses (like those in manufacturing) occurred after 1967. The accelerating pace of central-city job losses in the manufacturing and wholesale sectors reflects the growing diseconomies of central-city locations for production and warehousing activities.

The selective nature of job declines in most large cities is indicated by the data for their service industries. Only Detroit exhibited a net loss in service-sector jobs between 1948 and 1977, and this is entirely accounted for by losses since 1967. St. Louis and Newark joined Detroit in experiencing net job declines in service industries between 1967 and 1977.

On the other hand, Chicago and New York City have shown substantial vitality in their service industries since World War II. Even in these cities, though, service job growth was overwhelmed by employment declines in manufacturing, wholesale trade, and retail trade. Thus between 1948 and 1977 Chicago lost a total of nearly 400,000 jobs and New York City lost more than 500,000. In each case, the vast majority of overall job losses resulted from blue-collar employment declines. . . .

The most severe job losses by far were experienced by cities that achieved metropolitan status before 1900 and were in metropolitan areas with 1970 populations exceeding 1 million. Conversely, younger, smaller central cities experienced aggregate growth in manufacturing, retail, and wholesale employment.

Regional differences are also striking with regard to more recent central-city changes in manufacturing, retail, and wholesale employment. Central cities in the northeast and north central regions experienced aggregate employment losses in manufacturing, wholesale, and retail employment, while central cities in the South and West registered aggregate employment gains in all three industrial sectors.

By contrast, in the selected services category there were aggregate central-city employment gains in all census regions. More interestingly, selected service employment gains were greatest in the largest and oldest central-city classifications—exactly opposite the patterns found in manufacturing, retail, and wholesale employment. Because of the growing importance of service-sector employment in structurally changing cities, more detailed and more current analysis of this sector is in order.

The Rise of Information-Processing Industries

New York City gained more than 650,000 jobs in its information-processing industries between 1953 and 1980 while losing more than 525,000 jobs in its manufacturing and construction industries. Notice further the selective nature of service employment expansion, especially during the past decade. From 1970 to 1980, employment in information-processing industries continued to mushroom in New York, Philadelphia, and Boston, whereas employment in other, predominantly blue-collar, service industries (such as barbershops, car washes, and domestic services) declined considerably in each city. In just ten years, New York City lost more than 100,000 jobs in its predominantly blue-collar service industries, Philadelphia nearly 30,000 such jobs, and Boston nearly 10,000. All three cities experienced even larger absolute employment declines in their traditional retail and wholesale industries. . . .

CHANGING DEMOGRAPHIC COMPOSITIONS

A rather clear picture emerges from the above analysis. There have been precipitous declines in employment in northern central-city industries that traditionally sustained large numbers of less-educated persons. The job losses have been only partially replaced by growth in white-collar service industries with substantially higher educational requisites. This has implications for the expansion in large northern cities of population groups whose typical educational backgrounds place them at a serious disadvantage as the economies of these cities transform.

Table 16-1 summarizes the racial and ethnic compositional changes between 1970 and 1980 within the four largest northern cities (New York City, Chicago, Philadelphia, and Detroit). New York City, which experienced an overall population decline of over 823,000 during the decade, actually lost 1,393,000 non-Hispanic whites. Thus, in just ten years, New York's non-Hispanic white population (that is, its nonminority population) dropped by an amount larger than the total population of any other U.S. city except Los Angeles, Chicago, Philadelphia, and Houston. Approximately 40 percent of the loss of non-Hispanic whites in New York City was replaced by an infusion of over 570,000 Hispanics, blacks, and other minorities during the 1970s. By 1983 New York City's minority population had climbed to 52 percent of the city's total.

Chicago's demographic experience during the 1970s was similar to New York City's. More than 50 percent of Chicago's minority population increase during the decade consisted of Hispanics. However, it is important to point out that Chicago also experienced the third largest absolute increase of non-Hispanic blacks of any U.S. city. By 1980, 57 percent of Chicago's resident population was composed of minorities, with projections showing the minority proportion climbing to nearly 70 percent during the 1990s.

Philadelphia had the smallest aggregate population decline of the four cities. The number of both non-Hispanic whites and non-Hispanic blacks declined in Philadelphia between 1970 and 1980, while Hispanics and non-Hispanic and other minorities increased. Philadelphia's substantial decline in non-Hispanic whites, together with its net increase of 23,000 minority residents during the 1970s, raised its minority percentage to 43 percent in 1980.

Detroit experienced the highest rate of non-Hispanic white residential

Table 16-1 Population Changes in Four U.S. Cities, by Race and Ethnicity, 1970–80
Figures in thousands

		Non-Hispanic				
Central City	*Total Population*	*Whites*	*Blacks*	*Others*	*Hispanic*	*Percent Minority*
New York						
1970	7,895	5,062	1,518	113	1,202	36
1980	7,072	3,669	1,694	303	1,406	48
Change, 1970–80	−823	−1,393	176	190	204	. . .
Chicago						
1970	3,363	1,999	1,076	40	248	41
1980	3,005	1,300	1,188	96	422	57
Change, 1970–80	−358	−699	111	56	174	. . .
Philadelphia						
1970	1,949	1,247	646	11	45	36
1980	1,688	963	633	28	64	43
Change, 1970–80	−260	−283	−13	17	19	. . .
Detroit						
1970	1,511	820	652	9	30	46
1980	1,203	402	754	18	29	67
Change, 1970–80	−308	−418	102	9	−1	. . .
Total change, 1970–80	1,749	−2,794	377	271	396	. . .

SOURCE: U.S. Bureau of the Census, *Census of Population* summary tapes, 1970, 1980. Figures are rounded.

decline of any major city in the country. Between 1970 and 1980 Detroit lost more than one-half of its non-Hispanic white residents. Concurrently, Detroit had the fourth largest absolute increase of non-Hispanic blacks of any city in the country, falling just behind Chicago in black population increase. Combined with modest increases in both Hispanics and other minorities, Detroit's large increase in black residents and sharp drop in non-Hispanic white residents transformed the city's residential base from 46 percent minority in 1970 to 67 percent minority in 1980.

The bottom row of Table 16-1 provides a summary of demographic transformation and residential decline in the four cities. Between 1970 and 1980 they experienced an aggregate loss of 2,794,000 non-Hispanic whites, while their Hispanic residential bases increased by nearly 400,000. Added to the Hispanic increase during 1970 were substantial cumulative increases of non-Hispanic blacks and other non-Hispanic minorities, resulting in a total increase of more than 1,040,000 minority residents in the four cities during the decade.

Educational Distribution of Blacks and Whites

Racial minorities are at a serious structural disadvantage in central cities that are losing entry-level industry jobs because substantially larger proportions of city minority residents lack the formal schooling to take advantage of information-processing jobs that are expanding in the cities. . . .

The modal education category for central-city black males and black females in the Northeast, the north central region, and the South is "did not complete high school." Conversely, for white male central-city residents in these three regions, the modal education category is "attended college for one year or more." For white females in cities in these three regions, high school completion is the modal category. The statistics of note, however, relate to the much smaller proportion of black central-city residents with schooling levels appropriate to the expanding industries in northern cities and the higher proportions with education levels appropriate primarily to the declining industries.

Black residents in western cities, however, have more favorable educational levels to take advantage of new urban growth industries. In 1982 the number of blacks in western central cities who attended college for at least one year was greater than the number who did not complete high school. The educational distribution for whites shows western cities to have, by far, the highest share of educated nonminority populations of the four regions. Nearly one-half of the white male and female central-city residents aged 16–64 in the West have attended college for at least one year.

In sum, it is apparent that large proportions of black central-city residents still do not possess the education to participate in information-processing industries beginning to dominate urban employment bases. This is particularly the case for blacks residing in larger cities in the northeast and north central regions. With entry-level jobs rapidly disappearing from these cities, their less-educated minority populations are especially disadvantaged.

Transportation and the Isolation of Urban Minorities

An additional structural impediment inner-city minorities face is their increased distance from current sources of blue-collar and other entry-level jobs. As industries providing these jobs have dispersed to the suburbs, exurbs, and nonmetropolitan peripheries, racial discrimination and inadequate incomes of inner-city minorities have prevented many from moving with their traditional sources of employment. Moreover, the dispersed nature of job growth sites makes public transportation from inner-city neighborhoods impractical, requiring virtually all city residents who work in peripheral areas to commute by personally owned automobiles.

The combined costs of maintaining, operating, and insuring an automobile in major cities are substantially higher than elsewhere. This is particularly the case in older, larger, densely settled cities. In fact, automobile ownership in the core areas of these cities is so expensive relative to the actual or potential

Table 16-2 Central-City Minority Households with No Automobile or Truck Available, 1980
Percent

Central City	*Black Households*	*Hispanic Households*
Philadelphia	50.9	50.8
Boston	51.3	54.9
New York	69.3	71.3
Bronx	71.1	72.9
Brooklyn	72.2	73.9
Manhattan	84.7	83.5
Queens	43.4	49.4
Staten Island	54.2	38.6

SOURCE: U.S. Bureau of the Census, *Census of Housing, 1980, Detailed Housing Characteristics.*

incomes of their disadvantaged residents that most cannot afford this increasingly essential means of securing and maintaining blue-collar employment.

The severity of the problem of access to deconcentrating blue-collar job opportunities is documented in Table 16-2. This table presents the proportions of black and Hispanic households in New York City, Philadelphia, and Boston who do not have an automobile or truck available. More than one half of the minority households in Philadelphia and Boston are without means of personal transportation. New York City's proportions are even higher, with only three out of ten black of Hispanic households having a vehicle available. More detailed breakdowns of data show that in New York City's core boroughs (Brooklyn, Bronx, and Manhattan) fewer than three out of ten black or Hispanic households had an automobile or truck available in 1980.

It should be pointed out that the figures in Table 16-2 refer to *all* black and Hispanic households in each city or borough. It stands to reason that black and Hispanic households whose members have lower income and educational attainment have even less access to personal means of transportation. Thus, with automobile or truck ownership becoming increasingly necessary for blue-collar job access by poorly educated inner-city minorities, many find themselves confined to areas of declining opportunity.

CONSEQUENCES OF MINORITY CONFINEMENT

Confinement of minorities to inner-city areas of blue-collar job decline has had three serious outcomes: rising unemployment, increased labor force dropout rates, and growing welfare dependency.

Rising Central-City Unemployment

For all regions, central-city unemployment rates rose steadily for both races, with the exception of better-educated white males. Within racial groups, those who did not complete high school had substantially higher unemployment rates, with the absolute gap in unemployment rates between the lowest and highest educational categories widening between 1969 and 1982. The disproportionate concentration of central-city black males in the lowest category of completed schooling in every region but the West no doubt contributes to the widening gap over time between overall white and black urban unemployment rates. For white male central-city residents at all educational levels unemployment rates rose from 2.6 percent in 1969 to 9.5 percent in 1982, whereas the same rates for black males rose from 5.4 to 23.4 percent during this period. . . .

It is important to note, however, that substantial racial differences in central-city unemployment rates remain even when controlled for schooling completed. Indeed, black male central-city residents in the northeast and southern regions who attended college had higher rates of unemployment in 1982 than did white central-city residents in these same regions who did not complete high school. Putting aside possible racial differences in the quality of schooling, such discrepancies do suggest that problems of racial discrimination may be compounding the structural disadvantage central-city blacks face given their overall educational distributions.

Discouraged Workers and Other Labor Force Dropouts

Rising central-city unemployment rates for minorities illustrate only a portion of the problem of urban economic displacement, because these rates cover only persons who have actively sought employment during the four weeks before being surveyed. Excluded entirely from the labor force and unemployment statistics are disabled persons, those who have simply given up looking for work because they believe no jobs are available (discouraged workers), and those who want to work but cannot hold employment for a variety of physical or personal reasons. . . .

For every region and age group, black males have markedly higher rates of labor force nonparticipation than white males. Moreover, whereas central-city white male nonparticipation rates increased modestly between 1969 and 1982, the rates for black males climbed substantially. Just as the central-city racial gap in unemployment rates is widening, so is the racial gap in urban labor force nonparticipation.

Urban Employment Decline and Subsistence Surrogates

At the same time that blue-collar jobs have disappeared from larger northern cities, with corresponding rises in minority unemployment and labor force nonparticipation rates, the minority populations in many of these cities have continued to expand. This seemingly dysfunctional demographic expansion, which contrasts sharply with that predicted by market equilibrium models, raises some fundamental questions: What continues to attract or hold minority and other disadvantaged persons in central cities undergoing severe blue-collar employment decline? How are economically displaced inner-city residents able to survive? What, in brief, is the economic substitute for traditional blue-collar jobs?

Answers to the above are provided, in part, by the data presented in Table 16-3, which displays the absolute numbers of central-city household heads in each labor force category along with the respective proportion who receive assistance from at least one of three major welfare programs: public or subsidized housing, aid to families with dependent children and other cash assistance, and food stamps. The differentially high rates of welfare recipiency of minority householders who are unemployed or not in the labor force is clearly revealed. This is especially so for black female household heads. . . .

Public assistance rates, along with the much larger proportions of central-city black male householders who are unemployed in northern cities, illustrate the differential scope of welfare economies in northern cities, which serve as partial subsistence surrogates for their declining production economies.

TARGETING, ANCHORING, AND DEMOGRAPHIC DISEQUILIBRIA

From a policy perspective it is important to realize that while the expanding urban production economies of the past offered large numbers of disadvantaged residents entry into the mainstream economy and opportunities for mobility, today's urban welfare economies often have the opposite effects—limiting mobility and reinforcing the concentration and isolation of those without access to the economic mainstream. Many well-intentioned public assistance programs, for example, have been specifically aimed at inner-city areas of greatest economic distress, thereby providing the subsistence infrastructure that keeps disadvantaged people there. By relying increasingly on place-oriented public housing, nutritional, health care, income maintenance, and other assistance programs, a large minority underclass has become anchored in localities where additional welfare aid is directed because of severe employment loss and other indicators of distress.

Serious problems of racial discrimination, inadequate transportation, and insufficient low-cost housing in areas of employment growth further obstruct mobility and job acquisition by this underclass, as do their frequent deficiencies in technical and interpersonal skills so necessary to obtain and hold jobs. The outcome is that growing numbers of potentially productive minorities find themselves confined to segregated areas of urban decline where they subsist, without job opportunities, on a combination of government handouts

Table 16-3 Welfare Recipiency Rates of Central-City Heads of Households Aged 16–64, by Region, Race, Sex, and Labor Force Status, 1982
Figures in thousands

	Number of Heads of Households			*Percent Receiving Welfare*[a]		
Region, Race, and Sex	*Employed*	*Unemployed*	*Not in Labor Force*	*Employed*	*Unemployed*	*Not in Labor Force*
All regions						
White males	7,969	536	901	4	21	16
White females	2,834	190	1,015	8	28	46
Black males	1,434	229	306	9	39	40
Black females	1,121	228	900	28	73	82
Northeast						
White males	1,781	133	250	5	19	21
White females	661	47	373	9	23	58
Black males	321	48	72	12	41	39
Black females	290	53	313	29	75	84
North central						
White males	1,845	174	205	4	27	15
White females	677	55	219	8	42	48
Black males	330	99	98	9	52	35
Black females	255	90	241	27	72	82
South						
White males	2,502	92	240	4	16	16
White females	843	44	215	7	26	35
Black males	605	63	90	8	19	50
Black females	452	58	268	30	76	79
West						
White males	1,842	136	207	3	20	11
White females	653	43	208	8	20	35
Black males	178	20	46	4	32	38
Black females	124	26	79	18	63	86

[a] Welfare recipiency defined as receiving assistance from at least one of three programs: public or subsidized housing, aid to families with dependent children, or food stamps.

SOURCE: U.S. Bureau of the Census, *Current Population Survey* tape, March 1982. Figures are rounded.

and their own informal economies. Such isolation, dependency, and blocked mobility breed hopelessness, despair, and alienation, which in turn foster drug abuse, family dissolution, and other social ills that disproportionately afflict the urban disadvantaged.

This is certainly not meant to imply that targeted government aid to people and places in distress is unnecessary or without solid merit. Most urban welfare programs have had imperative palliative effects, temporarily relieving some very painful symptoms associated with the departure of blue-collar jobs (such as the inability of the unemployed to afford housing payments or adequate nutritional and health care). However, while some success has been achieved in relieving such pains, the disequilibrium in distressed cities between low-skill labor availability and low-skill labor needs grows worse.

SPATIAL INEQUITIES AND EQUALITY OF OPPORTUNITY

This demographic disequilibrium, sustained in part by government subsidies, may work against the long-term economic health of distressed cities and their structurally unemployed residents. Imagine, for instance, what would have happened in the first half of this century if the great numbers of structurally displaced southerners who migrated to economically expanding northern cities in search of jobs and a better life had been sustained in their distressed localities by public assistance. It is possible that many would never have moved and that the significant advances in income levels and living standards that both the South and their out-migrants experienced would not have occurred.

Circumstances today are analogous, but regionally the reverse. Many economically changing northern cities are now characterized by excesses of structurally displaced workers as their blue-collar job bases wither. For example, the March 1984 unemployment rates in Gary, Indiana, and Pontiac, Michigan, were both over 20 percent, whereas in the Raleigh-Durham-Chapel Hill area and in Austin, Texas, they were under 4 percent.

A more equal and, it may be contended, socially equitable territorial distribution of unemployment (say, no place continuously carrying more than 12 percent unemployment) could prevent the downward economic spiral that persists in some distressed cities. It is possible that this downward spiral continues because the social overhead burdens and problems associated with exceptionally high unemployment rates create negative externalities that dissuade new businesses from locating in such cities. If this is so, then under conditions of national economic growth a locality with excessively high unemployment has much poorer prospects of attracting new employers and economically rebounding than a locality with a moderate unemployment rate.

The conclusion to be drawn is that greater spatial equity in sharing the nation's unemployment burden may be essential to greater equality of opportunity for distressed cities and their disadvantaged residents. As long as disproportionately large numbers of unemployed persons remain concentrated in these cities, the economic future of both will be bleak.

NEW URBAN POLICIES FOR NEW URBAN REALITIES

To address the problem of inequities resulting from insufficient private-sector investment in areas of severe blue-collar employment decline, some have suggested a national development bank, a reconstruction finance corporation, enterprise zones, or government-business-labor partnerships, which might "reindustrialize" these areas or otherwise rebuild their historic employment bases. Such suggestions are as unrealistic as they are nostalgic. Government subsidies, tax incentives, and regulatory relief contained in existing and proposed urban policies are not nearly sufficient to overcome technological and market-driven forces redistributing blue-collar jobs and shaping the economies of the major cities. Nor would reversing these redistributional trends and inhibiting urban economic transformation necessarily be in the longer-term interest of either the cities or the national economy.

On the contrary, encouraging the return of older blue-collar industries to urban core areas may additionally saddle urban economies with stagnating or declining industries and further weaken the cities' competitive economic position. Indeed, efforts to assist distressed urban areas through policies that direct older industries to the inner cities where these industires could experience greater costs or lower productivity may well conflict with efforts to strengthen the national economy upon which the health of the cities is inextricably dependent. Economic advancement of cities and maximum job creation can best be accomplished through government programs and private-sector initiatives that promote urban service industries whose functions are consistent with the roles computer-age cities most effectively perform. The most promising new growth industries for cities include headquarter offices, marketing, finance, research, communication, and other information-processing activities. . . .

HELPING THOSE CAUGHT IN THE WEB OF CHANGE

Cities that improve their social and physical environments and adapt to their emerging service-sector roles should experience renewed demographic and economic vitality. However, many urban residents who lack appropriate skills for advanced service-sector industries are likely to remain on the bottom rungs of the socioeconomic ladder. Indeed, their economic plight could further deteriorate. For example, New York City, capitalizing on its strength as an international financial and administrative center, experienced a net increase of 167,000 jobs between 1977 and 1981. Yet, while the city's total employment base was expanding, its overall minority unemployment rates continued to climb. This is because virtually all of New York's employment expansion was concentrated in white-collar service industries, whereas manufacturing employment dropped by 55,000 jobs and wholesale and retail trade employment declined by an additional 9,000 jobs during the four-year period. These figures, together with the demographic and employment data presented for New York City earlier in this chapter, testify that the mismatch between the urban residential composition and job opportunities can worsen even under conditions of overall central-city employment gains.

The stark reality of rising urban minority unemployment under transforming (and possibly even growing) city economies, together with the improbability that government programs can stimulate sufficient numbers of blue-collar jobs in these cities, calls for a shift in policy emphasis. Politically popular (but ineffective) jobs-to-people strategies and essential urban welfare programs must be better balanced with serious efforts to upgrade the education and skills of disadvantaged city residents *and* with people-to-jobs strategies that would facilitate the migration of the structurally unemployed to places where job opportunities appropriate to their skills are still expanding. Contrary to conventional wisdom, there have been massive increases in entry-level jobs nationwide during the past decade. Nearly 1.8 million nonadministrative jobs were added in the food and drink industry alone between 1974 and 1984. This employment growth, the vast majority at the entry level in fast food and drink establishments, is more than twice the total number of production jobs that exist in America's automobile manufacturing and iron and steel industries combined. Unfortunately, entry-level job expansion in such rapidly multiplying service establishments has occurred almost exclusively in the suburbs, exurbs, and nonmetropolitan areas, far removed from concentrations of low-skill urban minorities.

The inability of most inner-city minorities to follow decentralizing entry-level jobs (either because of racial discrimination, inadequate knowledge and resources, or subsidized anchoring) is among the chief reasons for the widening gap in black-white rates of unemployment and labor force nonparticipation. It is also a major contributor to rising demographic disequilibria in the transforming cities and their correspondingly high social overhead burdens. In this regard, smaller central cities, populated largely by those who actually desire to live in them (minorities as well as whites) and by those who have appropriate employment opportunities available, would create far more vibrant cities than those of today, in which millions of disadvantaged persons are confined without jobs and without much hope for jobs.

To increase the mobility options of the urban disadvantaged, revised policies should be considered that would partially underwrite their more distant job searches and relocation expenses. Additional policies must be aimed at further reducing housing and employment discrimination and other institutional impediments to the mobility of minorities who wish to leave distressed urban areas. Finally, existing public assistance programs should be reviewed to ensure that they are not inadvertently attracting or bonding large numbers of disadvantaged persons to inner-city areas that offer limited opportunities for employment.

All of the above, of course, must be complemented by broader economic development policies fostering sustained private-sector job generation. Programs assisting the retraining or mobility of disadvantaged urban minorities will prove fruitless unless new and enduring jobs are available at the end of the training programs or moves.

The vexing dilemma of the urban underclass will certainly not be entirely solved by these and other strategies proposed here. However, without an expanding national economy, improved education and technical training programs for the urban disadvantaged, stricter enforcement of civil rights legislation, and the mobility of the underclass from economically distressed ghettos, the permanence *and* growth of the underclass will be assured.

RECONSTRUCTING THE RACIAL ORDER

17 Deindustrialization and Racial Minorities in the Great Lakes Region, USA

RICHARD CHILD HILL AND CYNTHIA NEGRY

The world economy is going through a profound transition. Manufacturing companies are reorganizing their activities in the face of stagnant demand, excess capacity, and stiff international competition; and that economic reorganization has brought severe dislocation to industrial communities in many nations. Particularly traumatic has been the crisis and reorganization in the world's metal-bending industries: auto, steel, and machine tools. In those industries employment tends to be highly concentrated in one or two regions in each producing country. So when the crisis hit auto and complementary metal-bending industries in the 1970s, job loss was abrupt, it occurred in large production units, and it had sharp regional impact.

There is reason to believe that racial minorities and women have suffered most from the great slump in the world's metal-bending industries. Minorities are more concentrated in the kinds of jobs, in the kinds of factories, in the kinds of neighborhoods that are at the epicenter of today's industrial earthquake. And although men outnumber women in manufacturing, one must not overlook the possibility that women have suffered more from the crisis in the manufacturing sector. The number of households headed by a woman continues to rise, and even in households consisting of two adults, the woman's earnings

Source: Richard Child Hill and Cynthia Negry, "Deindustrialization and Racial Minorities in the Great Lakes Region, USA," Paper presented at the International Conference on Racial Minorities, Economic Restructuring and Urban Decline. University of Warwick, Coventry, England (Sept. 1985).

are usually a necessity. Today, more than ever, women can't afford to lose their jobs. This is especially true of minority women, who are disproportionately represented among the ranks of women who are single heads of household (Sidel, 1986).

Minorities are less protected against economic adversity now since global economic crisis has enabled conservative political coalitions to combat shrinking profit rates by contracting the social regulation and welfare programs fought for and won by civil rights and feminist groups in the 1960s and early 1970s. Even more troubling, the very reasons that indicate racial minorities and women may have suffered most from the current economic crisis also suggest they may benefit least from the reorganization of production now under full steam. The better paying manufacturing jobs in the factories of the future aren't likely to match up well with the skills minorities and women now possess, and the factories themselves aren't likely to be built in the areas where most minorities now live.

DEINDUSTRIALIZATION AND RACIAL MINORITIES

Great Britain and the United States number among the nations which have experienced contracting manufacturing industries, declining industrial productivity, plunging corporate profits, and persistent unemployment since the late 1960s. There are scholars in both countries who argue that these trends indicate a process of national deindustrialization. British political economists have used the deindustrialization concept to designate the declining ability of British manufacturers to earn enough foreign currency through exports to pay for the materials and capital equipment they must import to keep their industries growing (Blackaby, 1979). In the United States, analysts have extended the term to cover other sources of harm to a nation's industries besides declining international competitiveness. Bluestone and Harrison (1982:1), for example, have also placed under the deindustrialization mantle (1) various forms of business disinvestment in new productive capacity ranging from corporations that milk the coffers of new acquisitions to running down plant and equipment to outright plant closure; (2) the reallocation of profits from productive to speculative or merger activity; and (3) unproductive transfer of capital between economic sectors, regions, or nations.

When Bluestone and Harrison investigated Dun and Bradstreet data on plant openings, closings, and relocations, they uncovered extraordinary capital mobility within and between regions in the United States. For example, 39 percent of the jobs listed by Dun and Bradstreet in 1969 had disappeared by 1976 (Bluestone and Harrison, 1982:29). Readily in evidence too was the powerful role conglomerates play in the U.S. economy. For example, when the authors surveyed a list of new manufacturing plants owned by the nation's 500 largest corporations, two out of three turned out to be not really new at all;

instead, they were plants big companies had acquired from other owners (Bluestone and Harrison, 1982:40–41). U.S. business investment abroad has also outpaced the rate of investment at home. Between 1950 and 1980, for example, direct foreign investment grew 16-fold while domestic investment increased 8-fold. Overseas profits are now one-third or more of the total profits of the biggest 100 multinationals, and nearly one-third of all U.S. imports actually come from overseas subsidiaries of American companies (Bluestone and Harrison, 1982:42–44).

Deindustrialization—via runaway shops, plant closings, conglomeration, or investment in other regions or nations—has had a harsh impact on many workers and the communities in which they live. Based upon the analysis of Dun and Bradstreet data, Bluestone and Harrison estimated that various kinds of disinvestment caused between 32 and 38 million jobs to disappear in the United States during the 1970s. In a more recent follow-up investigation, which covered manufacturing plants with 100 or more employees between 1978 and 1982, Candee Harris (1984) discovered that, on average, 900,000 jobs were lost through plant closings each year. And in a special Bureau of Labor Statistics household survey about job loss between January, 1979, and January, 1984, Flaim and Sehgal (1985) found that plant closings or relocations were responsible for about half of the 5.1 million jobs lost by workers who had been at their jobs three years or more. Besides the immediate loss of employment, the human consequences of deindustrialization often include long-term unemployment, loss of family income, decline in occupational status, loss of health insurance and pension rights, deterioration in physical and mental health, family violence, lost tax revenue, and shrinking social services.

There is reason to believe that deindustrialization shrinks employment opportunities most for minorities. People of color, particularly black men, are more likely than whites to work in industries experiencing long-term employment decline: primary metals, auto, apparel, and lumber. For that reason, it has been argued that black men are more likely than white men to lose their jobs due to the restructuring of the economy (Bluestone, Harrison, and Gorham, 1984). Findings from a 1980 study of plant shutdowns by the Illinois Advisory Committee to the U.S. Commission on Civil Rights (1981) support this conclusion. According to Gregory Squires (1981), author of the government report, "Minorities accounted for 20 percent of total employment in Illinois companies that closed between 1975 and 1978, compared with only 14 percent of the statewide labor force."

Overall company employment often drops when businesses move to new plants, but in Illinois the employment drop was more severe for minorities. According to Squires (1981), "Among companies relocating from major Illinois cities to suburban rings, minority employment declined 24 percent and white employment declined nine percent. For companies relocating from Illinois to the Sun Belt, minority employment dropped 26 percent and white employment dropped 17 percent." This suggests that dislocated blacks have a harder time finding new jobs than do whites, and that conclusion is supported by a U.S. Bureau of Labor Statistics (BLS) study of workers dislocated from their jobs between 1979 and 1984. The BLS found that 42 percent of black workers found new jobs after layoffs in comparison to 63 percent of whites (Flaim and Sehgal, 1985). It stands to reason, too, that young blacks, newly entering the labor market, are more likely to find their traditional sources of employment aren't hiring, although there may be gender differences since young black women may be experiencing employment gains in service industries which are less subject to international competition. . . .

Considerations like these led us to undertake a study of deindustrialization in the Great Lakes region, the industrial heartland of the United States. Our research objectives are to assess the magnitude of industrial job loss and the way declining manufacturing employment has been distributed among social groups in the region. But before we begin that assessment we must explain how we have defined the Great Lakes region and why we selected this area for investigation.

THE GREAT LAKES

> . . . a region has no more objective existence than an event; we divide it at will . . . the region will be what we make of it by the questions we choose to put to it. (Veyne, 1984:35)

We chose the Great Lakes, in the first instance, because the crescent of land that bounds the lakes is the traditional heartland of North American industry. Our past work suggests that the states and cities which front this massive waterway are bound together by an interlinked chain of metal-bending industries in crisis and transition (Hill and Negrey, 1985).

Three arguments suggest why we think it is defensible to treat the Great Lakes states as a region and as an object of investigation. The first argument is *ecological.* The Great Lakes states share a natural ecosystem—a fragile web of water and air, minerals and sediments, fish, birds, insects, and countless other species of flora and fauna. And the hazards now posed to this ecosystem by toxic pollutants and the diversion of water to outlying, more parched regions implicates all Great Lakes states in a shared fate (Pierson, 1984; Washington, 1985).

The second argument is *economic.* The bridge from ecology to economy is often short and fairly straight, and so it is with the Great Lakes. One crossing is through the soil, for beneath this industrial crescent lie limestone, coal, and iron ore for the production of steel—the structural fabric of an industrial society. A second pathway is along the St. Lawrence Seaway which links Great Lakes states to one another, to the Atlantic Ocean, and to Europe. Completed in 1959, the seaway was a massive undertaking. Bodies of water were connected, passages dug down as deep as 27 feet, locks built to raise and lower vessels by the height of 60-story buildings—all to allow ships to journey the 2,343 miles from Duluth on the Minnesota side of Lake Superior to other industrial entrepots along the Great Lakes, then into Canada, along the St. Lawrence River, and finally into the Atlantic Ocean (Bailey, 1984).

The Great Lakes and the seaway system allow cargo ships to move iron ore mined in Northern Minnesota to steel mills along Lake Michigan and Lake Erie; to move coal from the eastern fields of Pennsylvania to power plants in industrial metropolises like Buffalo, Cleveland, Detroit, and Milwaukee; and to move grain from Chicago docks to domestic and overseas markets. Just how critical a role the seaway plays in linking the economies of the Great Lakes states to one another and to international markets is suggested by the amount of cargo carried on the seaway each month: it is greater than the tonnage transported on all of Michigan's highways each year (Bailey, 1984).

The third argument is *political.* Facing common ecological and economic problems, public officials in the Great Lakes are cooperating more and constructing a firmer regional political identity. The Regional impact of industrial decline, the consistent effort by the Reagan administration to shift the federal government's domestic responsibilities back to the states, and the threat that water diversion could damage the Great Lakes ecosystem led to the formation of the Council of Great Lakes Governors in June, 1983 (Herbers, 1983). Sometimes this fledgling regional alliance brings together governors from six states, as in the Council's summit conference on cooperative economic growth (Council of Great Lakes Governors, 1983); at other times all eight of the states that border the Great Lakes have joined forces, as in the Great Lakes charter agreement to fight water diversions (Pierson, 1985).

In any case, the Council of Great Lakes Governors is now considering a number of steps toward further cooperation, including (1) putting together a charter, organizational goals, task forces, and priorities; (2) drafting a long-term regional economic blueprint; (3) mounting a coordinated strategy for dealing with Washington; (4) developing a regional natural resource policy; and (5) promoting the image of a unified region through public relations and by designating the region as a common market (Council of Great Lakes Governors, 1983:19–22).

Regional definitions shift with the problem at hand. The Bureau of the Census, for example, does not group the Great Lakes states into one regional category; rather the Bureau spreads them across the Middle Atlantic, East North Central, and West North Central regional census categories. But we think our ecological, economic, and political arguments justify grouping the eight states bordering the Great Lakes into one object of investigation. From west to east the states composing the Great Lakes region are: Minnesota, Wisconsin, Illinois, Indiana, Michigan, Ohio, Pennsylvania, and New York. . . .

DEINDUSTRIALIZATION IN THE GREAT LAKES

Just how severe an industrial crisis has the Great Lakes region experienced? To shed light on this question, we put together a time series from data provided in various U.S. Bureau of Labor Statistics employment and earnings reports. This time series is for (1) total nonagricultural employment, total manufacturing employment, and employment in five basic metal-bending industries; (2) for the years 1960, 1970, 1979, 1982, and 1985; and (3) for the United States, the Great Lakes states, and the Lakefront cities. Deindustrialization in the Great Lakes has been palpable since the Great Slump began in 1979; metal-bending industries have been hit the hardest; and, unlike the nation as a whole, job loss in the Great Lakes has not been counterbalanced by job growth in other sectors. To the contrary, we need just the reverse sort of logic to make sense out of the calamity that seems to have befallen the Lakefront cities. There, as metal-bending jobs disappeared, other jobs left, too. . . .

DEINDUSTRIALIZATION AND RACIAL MINORITIES IN THE GREAT LAKES

Occupation

Production workers (operatives and laborers) were hardest hit by the slump in heavy industry. Between 1979 and 1984, production job losses ranged from −29.7 percent in the USA to −45.3 percent in the Lakefront cities. Pink collar (office and clerical) and service workers also experienced consistent job losses although at a lower rate than production workers. White collar workers suffered fewer job losses than other occupational groups, and some white collar categories, including professional and sales workers, actually added employment while the rest of the manufacturing labor force was sharply contracting. Evidently, during economic slump periods large companies cutback their production facilities more severely than their largely white collar, regional, and central headquarters' operations.

Race

Black workers suffered considerably higher rates of industrial job loss than whites and they did so at every geographical level.

Hispanic workers consistently experienced lower rates of job loss than blacks, but the employment experience of Hispanics in comparison to whites is less uniform. At the national level, Hispanic workers had a lower rate of job loss than whites. In the Great Lakes region, Hispanics and whites lost industrial jobs at about the same rate. In the Lakefront cities, Hispanic workers experienced a bigger percentage job decline than whites. These findings—that Hispanic workers suffered lower rates of industrial job loss than whites in the USA and higher rates than whites in the Lakefront cities—probably reflect the migration of northern industrial capital to low-wage, right-to-work states in the southwest. Most Hispanics of Mexican heritage live in the southwestern United States, and Mexican-Americans are disproportionately employed in low-wage, nonunion jobs.

Gender

Contrary to our expectations, between 1979 and 1984, women workers in durable goods manufacturing suffered considerably lower rates of job loss than did men. This finding is particularly distinct at the national level where the percentage job loss among women was just over half that experienced by men, a similar, though less pronounced disparity is also evident in the Great Lakes states and Lakefront cities.

The distribution of employment losses among men and women in durable goods manufacturing is linked to the kinds of jobs that contracted *and* the kinds of jobs that expanded during the economic slump. Since most women in the manufacturing sector work in clerical and office jobs, lower job loss rates among women as compared to men is partly explained by lower job loss rates among pink collar as compared to blue collar workers. But that is not the whole story.

Nonunion, low-wage, part-time, and temporary employment grew during the economic crisis even while unionized, full-time, high-wage industrial jobs sharply declined. Gender differences in job loss reflect this trend. Of the 7.3 million jobs added to the U.S. workforce between January, 1980, and March, 1986, an amazing 84.3 percent went to women (Serrin, 1986). The percent of the labor force working part-time since 1980 has increased and women are 57 percent of part-time workers (Serrin, 1986). Most part-time jobs filled by women are in retail trade and miscellaneous services, but there is evidence that part-time positions also have grown in durable goods industries. Office automation also encourages part-time work, so some of the decline in office and clerical work may be offset by increasing part-time work in that job sector.

Race and Gender

When race and gender categories are combined, it is black men who consistently experience the highest rate of job loss followed by Hispanic men and black women. Nearly a third of all black men working in durable goods manufacturing in the United States lost their jobs between 1979 and 1984 and that figure rose to 45 percent in the five Lakefront cities. Hispanic women experienced the lowest rate of job loss at the national level, and as noted above, we think this indicates the disproportionate concentration of Hispanic women in expanding low-wage industries in the sunbelt.

Race, Gender, and Occupation

In the final analysis, the group hardest hit by the industrial slump were black male production workers in the Lakefront cities—fully 50 percent lost their jobs in durable goods manufacturing between 1979 and 1984. Black male production workers also suffered the greatest rate of job loss in the region and in the nation as a whole.

Place

The EEOC data, like the findings from the Bureau of Labor Statistics reviewed above, indicate an uneven distribution of industrial job loss across the nation, the Great Lakes region, and the Lakefront cities. With stunning regularity the worst job loss and inequities are most concentrated in the Lakefront cities. Between 1979 and 1984, jobs in durable goods manufacturing shrunk by 18.5 percent in the USA, by 27.8 percent in the Great Lakes states, and by 34 percent in the five Lakefront cities. The decline in production jobs was even more pronounced. Between 1979 and 1984, 29.7 percent of the nation's durable goods production jobs disappeared, that figure rose to −36 percent in the Great Lakes states, and to −45.3 percent in the five Lakefront cities.

Why have blacks suffered the most from the industrial crisis besetting the Great Lakes region? The answer is largely given in one word: segregation. Black workers tend to be concentrated in production jobs and that is where the biggest industrial losses have occurred; black production workers tend to be concentrated in older industrial plants and those are the ones most frequently closed; and black city dwellers tend to be concentrated in older urban neighborhoods and that is where plants and production jobs are disappearing the fastest. . . .

Black employment in the auto factories began in 1916, but until the eve of World War II black auto workers were largely confined to foundry work or the lowest paying maintenance positions. Blacks and women joined the assembly lines during World War II, and by the late 1940s the number of black production workers at Ford and Chrysler was sizeable. But as blacks entered production jobs, the company and the union reacted with dual seniority lists, dual assignment of jobs and other caste-like divisions of the labor force (Denby, 1968; Hill, 1969). And even today, when blacks account for 25 percent of the labor force at Chrysler, 15 percent at Ford, and 14 percent at General Motors, they are still struggling against a racially biased evaluation system administered by whites who want to preserve skilled craft jobs for themselves.

As blacks migrated to Detroit during the postwar period, inner city auto plants were engulfed by black neighborhoods. White workers sought employment in the suburbs while the inner city auto operations became increasingly black. By the mid 1960s, some downtown Chrysler plants had a majority of black hourly employees. On the second shift at Chrysler the ratio was as high as 85 percent (Northrup, 1968). By 1979, Chrysler had 21 production facilities in the Detroit metropolitan area, 15 of those were in the central city. One-third of Chrysler's entire blue collar work force and one-tenth of its white collar employees were black, and 50 percent of Chrysler's black employees resided in central Detroit. The Chrysler payroll to black workers alone amounted to $800 million—an estimated one percent of total black income in the United States (Northeast-Midwest Institute, 1979:11–13).

Between 1960 and 1980, racial segregation in the Detroit region hardly budged an inch. The index of dissimilarity, a good measure of residential segregation, indicates the percent of blacks that would have to move to achieve a distribution of blacks and whites among neighborhoods that corresponds to their percentage in the metropolitan population. In 1960 the index registered 87 percent; in 1980 it was 86 percent. During those two decades, the central city went from under 40 percent to over 60 percent black while the suburbs stayed 94 percent white (Darden, 1985:12–13). The Motor City was choked off from surrounding suburbs by a white noose. And it was in the surrounding suburbs that the auto companies built their new plants.

Car factories are built next to railroad lines, in open space, but not far from an available labor force. Once built, an auto plant attracts metal-bending supply industries, then residential subdivisions. So, as the auto industry expanded, the Motor City sprawled, farther and farther out. Detroit's first mass production facilities were located on the near eastside of the city, the industrial and residential expansion to the northwest followed. Ford later built the Rouge plant to the southwest, and around it grew one of the world's largest manufacturing complexes. After World War II, the Big Three auto companies constructed some 20 new plants in the Detroit region, all in the suburbs (Jacobs, 1981).

Little wonder, then, that when the crisis hit the auto industry in the 1970s, it hit minority workers and inner city minority neighborhoods in Detroit the hardest. . . .

New employment policies are critically needed; job training, to be sure; but the most serious mismatch is not between skills and jobs, but between jobs and places. Between 1973 and 1982, for example, metropolitan Detroit lost 108,000 jobs, but the area's labor force grew by 105,100 (State of Michigan, 1984:108). This is the structural basis for 'deepening economic despair.' Since the supply of jobs is now so desperately inadequate, new job creating policies which complement advanced technology strategies must have priority. More resources need to be put behind work sharing, community-based economic development schemes, and alternative use strategies for abandoned or underutilized plants. The problems of uneven development can only be addressed by targeting places as well as industries and that means regional political cooperation and more state planning. Our findings suggest there are some hard political challenges ahead for those who value racial justice in America.

REFERENCES

BABSON, STEVE with RON ALPERN, DAVE ELSILA, and JOHN REVITTE. *Working Detroit* (New York, Adama Books, 1984).

BAILEY, CHAUNCEY. 'The Seaway: At 25, it looks toward larger link to the world.' *Detroit News* 24 June 1984.

———. 'Black workers and the auto industry—an uncertain future.' *Detroit News* 9 June 1985.

BLACKABY, FRANK (editor). *De-industrialization* (London: Heinemann, 1979).

BLUESTONE, BARRY and BENNETT HARRISON. *The Deindustrialization of America* (New York: Basic Books, 1982).

BLUESTONE, BARRY, BENNETT HARRISON, and LUCY GORHAM. *Storm Clouds on the Horizon: Labor Market Crisis and Industrial Policy* (Brookline, MA: Economic Education Project, 1984).

Council of Great Lakes Governors. *Great Lakes Governors Economic Summit Report* (Ann Arbor, MI: Great Lakes Commission, 1983).

DARDEN, JOE T. 'The housing situation of blacks in metropolitan areas of Michigan,' in Urban Affairs Programs, *The State of Black Michigan, 1985* (East Lansing, MI: Urban Affairs Programs and The Council of Michigan Urban League Executives, 1985).

DENBY, CHARLES. 'Black caucuses in the unions.' *New Politics* (Summer, 1968), pp. 548–577.

Dollars and Sense. 'Hard times for black America.' (April, 1986), pp. 5–7.

FLAIM, P. and E. SEHGAL. 'Displaced workers of 1979–83: How well have they fared?' *Monthly Labor Review* (June, 1985), pp. 3–16.

GREGORY, KARL D. 'Toward a strategy for economic development in the black community,' in Urban Affairs Programs, *The State of Black Michigan, 1985* (East Lansing, MI: Urban Affairs Programs and The Council of Michigan Urban League Executives, 1985).

HARRIS, CANDEE. 'The magnitude of job loss from plant closings and the generation of replacement jobs: Some recent evidence.' *Annals of the American Academy of Political and Social Science* 475 (1984), pp. 15–27.

HERBERS, JOHN. 'Burdens shed by Washington fall unequally on states.' *The New York Times* 3 July 1983.

HILL, HERBERT. 'Black protest and the struggle for union democracy.' *Issues in Industrial Democracy* (1969), pp. 19–29.

HILL, RICHARD CHILD and MICHAEL INDERGAARD. 'Downriver: Deindustrialization in Southwest Detroit,' in Scott Cummings (editor), *Business Elites and Urban Development: Case Studies and Critical Perspectives* (Albany, NY: State University of New York Press, 1987).

HILL, RICHARD CHILD and CYNTHIA NEGREY. 'The politics of industrial policy in Michigan,' in Sharon Zukin (editor). *Industrial Policy: Business and Politics in the United States and France* (New York: Praeger, 1985).

———. 'Deindustrialization in the Great Lakes.' *Urban Affairs Quarterly* (forthcoming).

Illinois Advisory Committee to the U.S. Commission on Civil Rights. *Shutdown: Economic Dislocation and Equal Opportunity.* Washington, DC: Government Printing Office, June, 1981.

JACOBS, JANE. *Cities and the Wealth of Nations* (New York: Random House, 1984).

JACOBS, JIM. 'Suburbs and disinvestment.' Unpublished manuscript, Macomb Community College, Center for Community Development, Detroit, MI, 1981.

MARKUSEN, ANN. *Steel and Southeast Chicago* (Evanston, IL: Center for Urban Affairs and Policy Research, Northwestern University, 1985).

Northeast-Midwest Institute. *The Regional Impact of the Crisis at Chrysler* (Washington, DC, 1979).

NORTHRUP, HERBERT. *The Negro in the Automobile Industry* (Philadelphia, PA: University of Pennsylvania Press, 1968).

Office of Management and Budget. *Standard Industrial Classification Manual* (Washington, DC: U.S. Government Printing Office, 1972).

'Part-time workers: Rising numbers, rising discord.' *Business Week* 1 April 1985, pp. 62–63.

PIERSON, DUDLEY K. 'Michigan fears 'raids' on lakes.' *Detroit News* 2 December 1984.

———. 'Great Lakes charter signed but it is nonbinding.' *Detroit News* 5 February 1985.

SCHWEKE, WILLIAM and LEE WEBB. 'A national industrial policy? States seize the initiative.' *Ways and Means* 7 (1984), pp. 1, 6.

SERRIN, WILLIAM. 'Growth in jobs since '80s is sharp, but pay and quality are debated.' *The New York Times* 8 June 1986.

SHULMAN, STEVEN. 'Competition and racial discrimination: The employment effects of Reagan's labor market policies.' *Review of Radical Political Economics* 16(4) 1984, pp. 111–128.

SIDEL, RUTH. *Women and Children Last: The Plight of Poor Women in Affluent America* (New York: Viking, 1986).

SQUIRES, GREGORY D. 'The flight of capital hits minorities the hardest.' *Chicago Sun Times* 24 May 1981.

State of Michigan. *The Path to Prosperity: Findings and Recommendations of the Task Force for a Long-Term*

Economic Strategy for Michigan (Lansing, MI, 1984).

U.S. Department of Labor. *Time of Change: 1983 Handbook on Women Workers* (Washington, DC: U.S. Government Printing Office, 1983).

U.S. Equal Employment Opportunity Commission. *1979 Report: Minorities and Women in Private Industry* (Washington, DC: U.S. Government Printing Office, 1981).

——— . Printout of 1984 data.

WASHINGTON, THOMAS. 'The Great Lakes: A blueprint for their protection.' *Detroit News* 10 March 1985.

Wisconsin Advisory Committee to the United States Commission on Civil Rights. *Business Incentives and Minority Employment* (Washington, DC: U.S. Government Printing Office, 1982).

VEYNE, PAUL. *Writing History* (New York: Wesleyan University Press, 1984).

WOLIN, SHELDON. 'Political theory as a vocation,' in M. Fleisher (editor), *Machiavelli and the Nature of Political Thought* (New York: Atheneum, 1972).

RECONSTRUCTING THE RACIAL ORDER

18 *Runaway Factories Are Also a Civil Rights Issue*

GREGORY D. SQUIRES

The devastating impact of plant shutdowns, and corporate relocations on abandoned communities is well known. Suppliers and other ancillary businesses are hurt, new business is hesitant to move into the area and things generally slow down. Municipal budgets are drained when the demand for social services is on the rise. The mental and physical health of laid off workers, family and friends deteriorate, rates of divorce, alcoholism, depression and suicide climb. Just the threat of a shutdown can forestall implementation of environmental and occupational safety regulations and other needed public welfare measures.

Despite the visibility of the "runaway plant" phenomenon, curiously little attention has been paid to its consequences for the racial minorities who are most adversely affected. Equal employment opportunity and minority economic development have long been civil rights issues, but the impact of plant closings on minorities has not. Yet there are several dimensions to the problems of plant closings and economic dislocation that adversely affect racial minorities.

Over the past 20 years the nation's minority population has increased its concentration in neighborhoods within metropolitan areas (primarily central cities) and in regions of the country (the northeast quadrant) that are either declining or growing relatively slowly. Blacks constituted 16 percent of all central city residents in 1960 and 22 percent in 1975, while the suburban black

SOURCE: Gregory D. Squires, "Runaway Factories Are Also a Civil Rights Issue," *In These Times* (May 14–20, 1980) p. 10.

population increased only from 5 to 6 percent. During those years the black population in the Northeast and Midwest increased from 7 to 9 percent of the total, while it decreased in the South from 21 to 19 percent.

Because of their concentration in lower level jobs, minorities are less likely to be offered relocation assistance when employers move. Minorities are also concentrated in industries that have been hardest hit by recent patterns of development. While relocation overseas hurts American workers in general, it hits minorities the hardest. Minorities are more dependent on wages and salaries as sources of income than whites (80 percent of black income compared to 70 percent of white income is derived from wages and salaries) and minorities own relatively small amounts of wealth producing assets (blacks own less than 2 percent of all stocks and bonds, business and home equity, and money in banks).

A few examples of the discriminatory effects of shutdowns: when a laundry located in the city of St. Louis began to decentralize in 1964 its workforce was 75 percent black. By 1975, after it opened up 13 suburban facilities and reduced the downtown operations, the workforce was 5 percent black. In 1976 a Detroit manufacturer relocated production to its facility in a rural county just over the state line in Ohio. Salaried employees, most of whom were white, were offered assistance in finding new jobs. Hourly employees, most of whom were black, received no such aid. Minorities constituted 40 percent of the workforce at the Detroit facility when it closed compared to 2 percent at the Ohio plant.

No law or regulation explicitly addresses the discriminatory impact of corporate relocations. However, a number of civil rights attorneys including Rutgers law professor Alfred Blumrosen and former EEOC general counsel Stanley P. Herbert have argued, following the lead of the U.S. Supreme Court in the *Griggs* decision, that a relocation that has a disparate impact on minority employment violates Title VII of the Civil Rights Act of 1964 (which prohibits employment discrimination), unless the employer can demonstrate an overriding business necessity for the move.

If an employer moves, for example, from a predominantly black central city to an all-white suburb, current minority employees may be adversely affected because they will tend to face greater difficulties securing housing in the new location, if they try, or will encounter higher transportation costs in commuting to the new location.

Under these conditions, it is argued, the burden is on the employer to prove an overriding business necessity. And even if the employer can do so, it is further argued, steps must be taken to mitigate the disparate impact such as provision of transportation allowances, aid in locating housing, pressure on local realtors and local officials to assure fair housing practices in the new community, direct provision of housing, consideration of alternative sites that would be less discriminatory and reconsideration of the move altogether. Similarly, it is argued that Executive Order 11246, which requires non-discrimination and affirmative action by federal contractors, also calls for measures to correct the disparate impact of corporate relocations. Such interpretations have led civil rights advocates to file lawsuits against employers who have relocated (*Abney v. Budd* and *Bell v. Automobile Club of Michigan*) and to call for the EEOC and Department of Labor to develop and implement specific regulations on relocation.

But such regulations, no matter how effectively implemented, constitute marginal reforms that address only part of the problem. The fundamental issue underlying the "runaway plant" phenomenon and the discriminatory effects of economic dislocation is one of social control of the nation's resources. As one FTC consultant concluded:

"The private decisions of corporate owners and managers impost costs that affect other businesses, employees, and the community at large. The mental and physical well-being of the community is deleteriously affected; increased stress is placed upon the family; the quality of life in the community can be seriously decreased. Decisions with these wide-ranging effects cannot be viewed solely as private prerogatives; the internal decision-making calculations of the firm do not fully reflect actual costs. Public concern and participation is needed to insure that improperly estimated economic gains do not impose major economic costs."

Regulations that assure non-discrimination and affirmative action when a corporate relocation occurs represent one important part of an overall strategy to combat the deleterious consequences of uneven development. Legislative proposals for advanced notice of shutdown or substantial reduction of operations, severance pay to affected employees, contributions to economic development funds in affected communities, and employee or employee-community takeover of operations are prerequisites for an effective fight against the runaway plant. Among federal proposals currently under consideration are Representative Ford's National Employment Priorities Act of 1979, and Senator Williams' Employee Protection and Community Stabilization Act of 1979. State level proposals include Ohio's Community Readjustment Act, New Jersey's Employment Relocation Assistance Act, the Illinois Employer Relocation Act, and many others.

Racial minorities stand to benefit the most from these proposals. While not generally considered as such, these economic development initiatives are also, in fact, civil rights bills. The civil rights community cannot afford to ignore questions of economic rights. Similarly, those concerned about shutdowns and capital mobility in general can gain moral as well as political advantages by recognizing the civil rights implications of these issues and developing closer alliances with civil rights organizations.

PATRIARCHY'S NEW FORMS

19 *The Paradox of Women's Poverty: Wage-Earning Women and Economic Transformation*

JOAN SMITH

Two recent developments have contributed significantly to the growth of women's poverty in the United States. First, more women than ever before must depend exclusively on their own earnings or on welfare payments as the major source of support for themselves and their families. The number of women heading households with children under eighteen doubled between 1970 and 1980. By 1981 close to one-fifth of all families with minor children were headed by women, who were more than three times as likely to have incomes below the poverty line than were their male counterparts. Second, most available new jobs offer workers little chance to climb out of poverty. The vast majority of these positions are in the rapidly expanding service sector of the economy and are occupied primarily by women.

This article examines the relationship between the increase in the number of women who must enter paid employment and the growth of the service sector. Although these developments are responses to somewhat different circumstances, the two have combined to create a paradoxical situation for U.S. women today. Women's labor has been the major contributor to employment growth in the most rapidly expanding sector of the economy, yet the experience of women wage workers continues to be that of the most marginally

Source: Joan Smith, "The Paradox of Women's Poverty: Wage-Earning Women and Economic Transformation," *Signs: Journal of Women in Culture and Society* Vol. 10 (2) (Winter, 1984) pp. 291–310.

employed. Although central to economic expansion, they still receive the lowest pay and are subject to the least desirable employment practices.

The people who employ these women face another side of this contradiction. The fastest growing sectors in the private economy have been able to expand as rapidly as they have almost exclusively because women have been willing to take work that is less than desirable. Historically, such willingness stems from women's economic dependency and traditional exclusion from paid work; yet it is precisely these conditions that are seriously eroded once women become permanent members of the labor force.

The first section of this article examines the character of the service sector, its phenomenal growth, and its operating premises. While this sector resembles others in the private economy in that it must extract profits at a rate acceptable to investors, it employs unique strategies to accomplish that end. These strategies, inherent in the development and structure of service establishments, affect the entire economy—most particularly the female labor force—since they necessarily entail the creation of low-wage, intermittent, and part-time work.

The second section describes the unprecedented rise in women's paid employment over the past ten years. A substantial portion of that unexpected growth is explained by the decreased number of women who depend permanently and entirely on men for their support: women increasingly do not marry or marry later in life; they are also divorcing much more frequently. In addition, married women are far less likely to engage themselves exclusively in housekeeping and child rearing. Nevertheless, and despite women's vastly increased participation in the labor force, the nature of jobs offered them in the growing service sector is shaped by the presumption that women still have access to sufficient support beyond their own earnings and are at best only partially committed to wage labor.

THE GROWTH OF SERVICES

Without question, service establishments have come to dominate economic activities in the United States over the past thirty years. The IBM plant may be more impressive than the local McDonald's, but the local McDonald's and other such service-sector establishments are absorbing more of total personal expenditures than any other sector in the domestic economy. Between 1950 and 1980 spending on services increased almost twelve times, more than twice the increase for durable-goods spending and half again the rate of growth in total private spending. Although the proportion of private expenditures on durable goods remained near constant during the period, by 1980 nearly half of every dollar of personal expenditures was slated for services, up from just a third in 1950.

Paralleling this increase was the growth in the sector's overall output. In 1950 service-sector establishments produced slightly more than 47 percent of the total gross national product (GNP); by 1980 that share had increased to almost 55 percent. In contrast, goods-producing firms witnessed a decrease from approximately 37 percent to less than 32 percent. Between 1950 and 1980, the gross product of the service sector grew by 200 percent, outpacing the growth in total GNP by 43 percent.

These shifts have been accompanied by a constant differential in the rate of return in the form of gross product—a differential that favors service-sector firms. In 1974, for every dollar invested in wages, salaries, and capital goods, the manufacturing sector realized $1.50 in the value of its gross product; by 1980 that had increased to $1.60. In the service sector, a dollar investment goes a good deal further: in 1974 it returned $1.92 in gross product, in 1981 $2.06.

Though notorious for low productivity, many establishments within the service sector rival the traditional manufacturing industries in achieving high sales in proportion to the number of workers they employ. For example, the food chain of Safeway Stores had approximately the same number of employees as did U.S. Steel in 1980 but 17 percent more sales dollars per employee than the industrial giant. Moreover, Safeway accomplished this sales record with three and a half times fewer assets per employee than U.S. Steel. Kroger food stores reached close to the same employee sales as Ford, but Ford had nearly four times as many assets involved in every dollar of sales per employee. For every dollar of sales per employee among the twenty-five largest industrial firms listed by *Forbes,* including the two major U.S. oil companies, the average asset per employee was eighty-seven cents; in contrast, the average for the twenty-four largest retail establishments was just thirty-six cents.

This very low capital-to-labor ratio is one of the principal characteristics, perhaps the identifying one, of service establishments, and their employment practices are absolutely shaped by this factor and by the kind of business environment within which they operate. The two together—labor intensity and operating environment—are therefore crucial for the majority of women wage workers. Before considering how these factors affect employment practices, however, one must examine the phenomenal growth in the number of workers subject to these practices. Between 1970 and 1980 service-sector establishments absorbed close to 86 percent of all private sector employment growth in the country. Nearly all of that new employment—80 percent—went to just two of its subsectors, retail trade and business and personal services. Had the service sector offered the same type of jobs as those offered in the sectors it was rapidly displacing as the major national employer, the effect on the economy in general and on workers in particular would have been a good deal less consequential. But most service-sector employment stands in sharp contrast to that in other sectors of the private economy.

The wage differential between service-sector work and manufacturing, traditionally large, widened dramatically between 1970 and 1980. In 1970, for every dollar manufacturing employees earned in weekly pay, workers in retail trade earned sixty-two cents; by 1980 retail workers earned fifty-one cents for every dollar paid to manufacturing workers. In general merchandise stores—stores like those occupying the largest spaces in local shopping malls—the ratio declined from fifty-seven to forty-eight cents for every dollar in weekly earnings in manufacturing; in eating and drinking places, from forty-three to thirty-three cents. In personal services, the ratio went from seventy-three to sixty-six cents for every dollar. Even the prestigious financial establishment's relative wage rates have deteriorated from eighty-four to seventy-two cents for every dollar paid in weekly manufacturing wages.

Besides differing significantly from manufacturing firms in their wage rates, service-sector establishments have a considerably greater capacity to sustain relatively high labor turnover. In January 1978, for example, close to 30 percent of the 91 million workers on the job in all sectors held positions that they had not held the year before; service industries were the major contributor to this instability. Between 1970 and 1976 the average quarterly turnover rate for manufacturing establishments was 1.1 percent. By contrast, the general merchandising sector had a rate of 3.2 percent, eating and drinking places 4.4 percent.

The contrast in employment practices is even greater. Not only do employees in the service sector have twice the turnover rate of those in manufacturing, but they also have a substantially higher rate of part-time employment. By 1977, close to 37 percent of those employed in retail and wholesale trade worked less than thirty-five hours per week, whereas a decade earlier the figure was only 29 percent. Comparable increases in part-time work can be found in all service-sector establishments. For example, although the total number of persons working in food stores increased by 21 percent between 1958 and 1975, the total number of hours worked rose less than 1 percent. Despite greatly extended weekly shopping hours, the average number of hours worked per week by nonsupervisory personnel in the retail food industry decreased by 11 percent between 1958 and 1975. Although operating hours escalated substantially, the average work week in eating and drinking establishments declined from 36.6 hours in 1968 to 28.0 hours in 1976.

All the developments associated with growth in the service sector—substantial rise in employment, increases in operating hours accompanied by decreases in full-time work, and higher-than average rate of employee turnover along with considerably lower-than-average wages—are not just the unplanned outcome of typical employment practices but are fundamentally tied to the two conditions of operation mentioned above that sharply distinguish service-sector establishments from their counterparts in the rest of the economy.

The first of these conditions is the low capital-to-labor ratio found among service firms. Not only is the amount of capital equipment per employee in the service sector considerably less than in manufacturing, but the disparity has been growing substantially. Similarly, the amount of total investment per employee (though greater in 1960) is now considerably less in the service sector than in manufacturing firms. Because service-sector establishments are highly labor intensive, expanded production rests less on productivity growth (increases in units of output per units of labor) than on absolute increases in the labor force itself.

The second distinguishing feature of service-sector establishments is the highly competitive nature of their business environment. In contrast to the manufacturing giants, which enjoy virtual oligopolistic control over markets, even the largest service-sector firms face far more severe price competition.

The first condition, a relatively high dependence on labor for increased output, leads to a general rise in overall wage costs. The potential impact of this tendency is heightened by the second condition: a greater imperative to cut wages in reaction to relatively stronger-than-average downward pressures on prices (relative, that is, to manufacturing firms). These two distinguishing conditions within which service-sector establishments must operate—a highly labor-intensive work process and a highly competitive business environment—have fundamentally shaped that sector's development and daily operations. Such operating conditions strongly encourage employment practices that allow overall wage reductions and increases in the size of the labor force simultaneously. Further, these practices necessarily include rapid expansion and contraction of the labor force to match fluctuations in demand. While each of these practices would naturally be advantageous to any firm, they have become indispensable for the vast majority of service-sector firms precisely because of their distinctive operating constraints.

As a direct result of such employment practices, then, labor costs have decreased over the past decade while the labor force absorbed by service-sector establishments has vastly increased. Part-time workers constitute a steadily increasing portion of the labor force and are frequently found in firms open twenty-four hours a day; an increasing number of workers float in and out of jobs depending on seasonal demands. The irony is that the increase in such employment practices signals a vastly increased dependence on a wage labor force that, paradoxically, must be treated in all other respects as though it was entirely dispensable. In short, the contemporary economy has moved to center stage a labor force that must be continually endowed with marginal characteristics.

Creating the Poverty of Working Women

This paradoxical situation was facilitated by the unprecedented number of women willing to enter the ranks of paid labor. Between 1970 and 1980 there were over 13 million of them, the greatest increase in the history of the U.S. labor force. This phenomenal growth in women's paid employment accounts for more than half of the increase in employment in each of the major industrial sectors throughout the decade—even though women constituted no more than half the labor force in each sector. Women contributed over three-quarters of the increase in financial, real estate, and insurance firms and more than 60 percent in both the service-producing and retail food store industries. This trend toward increased wage work clearly challenged stereotypic assumptions about women's labor and encouraged many women to rearrange their daily patterns of living.

In 1980 the number of adult women exclusively engaged in the routine activities of keeping house was close to 10 million less than one would have projected given 1970 figures—a momentous shift in social arrangements. Fewer women married—in 1980 there were 5.2 million fewer married women than 1970 marriage rates would have indicated—and a substantially greater number of married women worked for wages. By 1980 2.8 million more married women were in the active labor force than could have been expected according to 1970 conditions. By March of 1980, according to Labor Department figures, half of all married women living with their husbands were either actively seeking work or actually working for wages. More startling is what happened to the two groups of women between thirty-five and forty-four and between twenty and twenty-four: their employment rate exceeded that of women in similar age groups in 1970 by almost 60 percent.

All of these factors represent a drastic redesign in the fabric of women's lives. In one brief ten-year period, expectations firmly held (although perhaps less firmly realized in practice) were radically altered. The question is, Where did women go for jobs, and how were economic arrangements affected by the sudden appearance of so many new workers on the market?

There is no direct way to measure exactly where each new woman worker went when she found paid employment. (There is no way to identify precisely which women expanded the labor force beyond the size predicted by 1970 rates of employment; such a category of women is an analytical rather than an empirical one.) Indirectly, though, the vastly disproportionate growth of some industries relative to others suggests the kinds of new jobs responsible for absorbing this altogether new source of labor.

Between 1970 and 1980 the service sector grew by close to 14 million jobs—a 31 percent growth in the sector's share of the labor market. Close to 4.5 million more jobs appeared here than would have been the case had the service sector merely maintained its 1970 share of total private employment in the nation. Close to 75 percent of that unexpected new employment was provided by women new to the paid labor force.

Consider, for example, the expansion of retail trade. In 1970 this sector was a major employer of women: 46 percent of its total labor force was female. Yet when it added employment beyond just its 1970 share—in short, when it added approximately 2.2 million unexpected jobs—women took close to three-quarters of them. The same pattern persists in all of the subsectors of service industries. Although women made up 52 percent of the 1970 labor force in financial, real estate, and insurance firms, they contributed 87 percent of the disproportionate growth those firms enjoyed in the decade. All of the disproportionate growth in advertising was contributed by women's unexpected employment. In business services women contributed close to 57 percent of the disproportionate growth, although they had constituted just 34 percent of that sector in 1970. Even in the male-dominated engineering subsector, women's unexpected increase in employment accounted for more than 30 percent of the sector's disproportionate growth. Similarly, in legal services, women provided 82 percent of the unexpected growth in the decade between 1970 and 1980.

In summary, two kinds of growth took place between 1970 and 1980. One is simply the absolute increase in employment of women and the corresponding absolute growth in the firms that employed them. The other is the relative growth that represents significant shifts in the dominance of particular kinds of economic activities and the disproportionate growth in women's employment that accommodated those shifts.

To be sure, the economy grew between 1970 and 1980, and despite rising levels of unemployment, more people were employed in 1980, relative to the population, than in 1970. But these new employees found work in altogether different sorts of industries. There was a significant shift in what people could expect to do because of the significant shift in how profits are generated. A new emphasis, at least domestically, has been placed on generating profits through a relative decrease in domestic goods production and a relative increase in what have been regarded until very recently as merely ancillary services.

Women have been the principal facilitators of this fundamental change in the nature of the U.S. economy. And, as we have seen, the availability of that new pool of labor signals a vast rearrangement in the traditional social institutions that are part of women's lives. Two developments—more women working and fewer women marrying—contributed decisively to the growth of the labor force between 1970 and 1980. Whether there is a causal link between these two events is not at issue here. What is at issue is that, unlike the economic climates in other periods, today's economic climate has been transformed by a considerably different kind of labor force. This labor force, although central to key economic developments, is treated as if its work is only marginal to those developments.

In an earlier time, working outside the home was for most women a stopgap between youth and maturity. It was the exceptional woman who found in paid employment the authority conferred on full-fledged members of the labor force. According to reigning ideologies—and women's experiences led them to accept these ideologies—women could expect to find recognition and respect only through a domestic role. Anything that interrupted that role was viewed as, at best, an aberration, a temporary measure rather than a lifetime commitment.

The kinds of jobs available to women in a predominantly sex-segregated labor force were altogether consistent with these expectations: they were the most temporary, the most dispensable, the worst paying, and the least associated with skill and achievement in the public sphere of production. Economic dependence on men was a proper corollary of being only marginally important to major economic processes. And yet women—and especially married women, that traditionally most marginal group of all wage workers—have now become the principal source of energy necessary for the recent recomposition of economic activities.

As they moved so massively into the ranks of paid labor, women had to rearrange their lives in ways that were different from the traditional stereotype. Part of those rearrangements had to include a rethinking of expectations concerning relations to men. In short, a crack widened in the apparently sturdy wall of economic dependency. Women had to give up the idea that they were properly dependent on men for their sole support. And yet the nature of the jobs they were offered during the decade was totally incompatible with the eradication of economic dependence.

Consider, for example, the issue of part-time employment. A dramatic increase in the proportion of part-time work to all forms of work in an economy is undoubtedly both a sign and an effect of relatively important changes in the labor process itself, quite independent of the characteristics of part-time workers. By 1980 close to a quarter of all jobs in the private sector were part-time—an unprecedented 68 percent increase in the rate of part-time work since 1970. In short, there is little doubt that some profound transformations took place in the organization of work to permit such a drastic shift of available jobs away from full-time and toward part-time work.

Despite what amounts to a structural change in the organization of work, the argument is frequently advanced that the increase in part-time jobs reflects the workers' preference rather than changes in the labor process itself. According to this reasoning, part-time work has increased because women are the new employees and because they prefer part-time work, given their domestic responsibilities. Indeed men took considerably fewer of the disproportionately large numbers of part-time jobs to appear between 1970 and 1980: 1.6 million went to men as compared to slightly more than 3 million that went to women. Countering this argument, though, is the fact that men and women increased their part-time work almost at the same rate. Of the total number of women who worked part-time in 1980, 25.8 percent would not have been expected had the 1970 rates for women merely been maintained. Of the total number of men working part-time in 1980, 24.8 percent would not have been expected, given 1970 rates for men). In short, both men and women demonstrated equally a substantial growth in the degree to which they accepted part-time work. This strongly suggests that the growth in overall part-time labor, rather than rising from putative gender-based preferences, resulted from changes in the form and content of available work.

Nevertheless, men apparently have an upper limit on the degree to which they will engage in part-time work, a limit considerably lower than women's. Women are both socially and culturally assumed to be in a position to accept less than full-time work. Between 1970 and 1980, of the close to 6 million women who went into the labor force unexpectedly (that is, women who represent the disproportionate increase in women's labor force participation rate), 51 percent received only part-time work. Very surprisingly, however, married women alone cannot account for this growth. Of the 2.8 million married women who, in joining the work force, exceeded 1970 work force participation rates, just 28 percent took part-time work; of the women workers who did not live with husbands and entered the labor force at a rate that exceeded 1970 levels, however, 73 percent had taken part-time work by 1980.

In sum, the degree to which work has ceased to be a matter of full-time employment cannot be explained exclusively by changes in work preferences. Men took a considerably greater proportion of the new part-time work than part-time labor rates in 1970 would have led one to expect, and women who did not live with husbands took on considerably more, proportional to their 1970 rates, than did married women living with husbands. Yet, of course, it was still married women who contributed a disproportionate share of part-time labor: though they formed just 28 percent of the total 1980 labor force, they constituted 34 percent of the part-time labor force.

The capacity to shift work from full-time to part-time, to be sure, was facilitated by the increased rates of employment of married women—of that there can be no question. Nevertheless, what facilitates an economic process is not to be confused with what initiates it. The substantial growth in part time work reflects significant changes in the pattern of work itself and the terms on which dominant economic activities take place. The stretch out in operating hours accompanied by a relative decrease in wage costs—the service sector's key strategy for increasing production—has been influenced by a vast increase in part-time personnel.

A second feature of most new jobs to appear in the private economy over the past decade is that they yield the lowest wages. The vast majority of women workers labor in precisely those occupations that pay less than the average wage. Of the 30.6 million working women in 1980, 51 percent held jobs paying less than 66 percent of an average crafts worker's wage. Close to 90 percent of the women who unexpectedly entered paid labor between 1970 and 1980 found jobs in establishments that paid on average just sixty-three cents for every dollar paid elsewhere in the private economy—a decrease of almost 23 percent in the relative wage rates paid in those sectors that are the major employers of the new women workers.

Much like the growth of part-time work and periodic employment, the unprecedented movement of women into the ranks of labor within these low-wage sectors facilitates fundamental changes in the character of work. One sign of those changes is the disappearance of the so-called family wage. Two or more jobs are now required to meet family financial needs. Although it is often argued that this is the result of a rapid rise of prices relative to wages, the growth of two-earner families is also an effect of substantial changes in the structure of employment: the new jobs are in low-wage sectors, and there has been a resulting decline in the average wage. As we have seen, virtually all the new jobs between 1970 and 1980 were in sectors that paid less than average wages. Rising prices combined with the scarcity of well-paid work requires a family to do more paid work.

Compare, for example, the wages paid in the most rapidly expanding sectors to what the Bureau of Labor Statistics (BLS) defines as the "lower living standard"—an amount established as a benchmark against which eligibility for public assistance is determined. (Should a family of four fall below this income level, it qualifies for a wide variety of government support programs, including admission to public housing projects, food stamp programs, programs for the distribution of surplus government food, and Medicaid.) In 1979 close to half of all working women found themselves in industries paying an average wage less than the bare minimum set by the BLS for a family of four. The two and a half million women working in eating and drinking places in 1979 found themselves in an industry that paid the average workers $151.37 per week less than the BLS minimum for a family of four. Another million and a half women worked in an industry—general merchandising—that paid its average worker $111 less per week than the amount a family of four needed to stay alive without government assistance, according to the BLS. Even in the financial, insurance, and real estate industries, which are relatively well paid in comparison to other service-sector industries, the average worker earned in 1979 nearly $46 a week less than the BLS minimum. Because the jobs most available to women pay less than an acceptable family wage, many fully employed women who head households remain poor despite their work efforts. In 1980 fully employed women heading households had a poverty rate almost three times that of husband-and-wife households and twice that of men who headed households without a wife present.

To be sure, most of the women working in these sectors were not exclusively responsible for the support of a family of four. That is not the point. The point is that those industries most responsible for employment growth pay a wage rate that absolutely requires the recipient of those wages to find additional resources in order to support a family above the poverty level. And it is this condition that is largely responsible for the "feminization of poverty." This is not an entirely novel set of circumstances; these service-sector industries have historically paid considerably less than the going rate and certainly less than a family wage. Two related developments are what is novel.

First, these low-wage industries are growing at a rate that far outstrips the growth of other higher-waged sectors, and, second, a considerable portion of economic activities have been transferred to them. Thus, these industries have taken on an importance far beyond that which they may have had in the past. Their characteristically low-wage levels and their less than full-time, less than all-year, work now play a role in the private economy far different from that played in an earlier period during which goods-producing firms were expanding.

What might have been taken as aberrations in more normal employment practices in an earlier period have now become the dominant patterns in the

1980s. These patterns in turn depend on the availability of a labor force prepared to continue to accept less than optimal employment. And here is the basic contradiction. The expansion of the economy over the recent decade has had as its major premise the availability of workers who will act as though they perform only marginal, inconsequential labor. Because of the social relationships and expectations that stereotyped them as nonwage workers in the past, women are expected to accept less than standard employment—employment that is frequently less than full time, considerably less stable, and substantially lower paid. Yet once this labor force is utilized as a labor force, the nonmarket conditions that shaped these expectations—conditions most frequently associated with being outside the main currents of economic development—are by definition fundamentally eroded.

Conclusion

Although the availability of women to the labor force has required a massive shift in many women's expectations—they must think of themselves as workers in addition to their other social identities—women wage earners must consider their domestic roles the most salient ones in shaping their experiences at work if they are to play their assigned parts. While this has always been the case for wage-earning women at least during the past century, something entirely new has occurred that has become the central feature for the decade's most decisive economic developments. Now, rather than actually being a marginal sector of the overall labor force, women constitute upwards of 40 percent of it. Much more important than their mere numbers, however, is the fact that women constitute by far the bulk of that labor that was and is necessary to accommodate the massive shifts in employment practices that were necessary for the continued profitable operation of the fastest-growing sectors of the private economy.

Had women not increased the rate at which they worked for wages, these changes in typical practices now characteristic of the private economy could not have taken place—at least in the manner they did. But this is hardly evidence that the presence of women caused such changes unless one wants to accept the untenable position that a consequence of an economic event can be equated with its cause. The growth in the number of jobs in the service sector relative to growth in other sectors and women's greater willingness (or greater need, as the case may be) to work for wages are in all likelihood two separate aspects of deeper social and economic developments over the recent and not-so-recent past.

Nevertheless, the increased rate at which women became available for wage work during the recent decade and the distinctive constraints on employment practices in the fastest-growing sectors of the economy that absorbed this new labor meant that the development of these industries, their daily operating premises, and their characteristic labor processes are all predicated on a supply of workers who are presumably involved primarily in nonmarket institutions. Therefore, to the extent that the growth and relative importance of these industries are central to the transformations in typical economic practices in the nation, the set of social arrangements characterizing women's lives as nonwage workers are incorporated into the very grounds of the economy. These social arrangements are precisely those that support the assumption that women are properly considered a cheaper and more dispensable labor force and are less dependent on their wages than are male workers.

In short, women's poverty and continued economic dependency are now the central operating premises of the most rapidly expanding sectors of the U.S. economy today and form the basis for the profound changes that have characterized that economy over the past decade.

But this situation is highly unstable. Women's earlier economic dependency arose under entirely different circumstances and was strongly related to conditions very different from those facing women today: women then either excluded themselves entirely from the more visible aspects of paid labor or faced severe social degradation. Individual poverty and its companion, economic dependency on men, were the concomitants of the social conditions women were required by social convention to deal with in their daily lives. It is no surprise that necessity was transformed into a virtue. The so-called cult of true womanhood elevated such poverty and dependency to the level of a societal good.

Under contemporary conditions, all that has changed; but the economic dependency remains the same. Poverty and dependency are now the results not of women's relative exclusion from the wage labor force but of the very conditions under which women are actively incorporated into its ranks. The ideology of domesticity previously used to support the nonwaged roles of women must now be invoked to support the conditions of their waged work. But the ideology wears thin under such circumstances. Rather than being obscured behind the veil of a biological destiny that seems to be somewhat easily associated with housekeeping and child rearing, the pauperization of women and their continued dependency on either men or the state are now clearly linked to profound economic developments.

PATRIARCHY'S NEW FORMS

20 Word Processing: Forward for Business, Backward for Women

ANNE MACHUNG

"The era of focusing on automating the secretary and clerical labor force is drawing to a close," David Liddle, office vice-president at Xerox recently announced, "and the emphasis now is on automating white-collar workers" ("Invasion," 1981). That is not how word processing companies see it, and it certainly is not how secretaries, bookkeepers, and bank tellers are experiencing it. In fact, the automation of the secretary, the typist, and the bookkeeper, which began in the 1970s, is growing apace. Almost every large corporation in the country now has installed its own word processing center or is looking into it. Word processors are popping up in small offices and private homes everywhere. Word processing, in fact, is big business. Worth only about $780 million in 1978, predictions are the word processing market will generate $5 billion in revenue by 1984, a growth rate of over 500 percent in five years (Morner, 1979: 76; Schatz, 1982: 69).

But word processing constitutes only a fraction, though a sizeable fraction, of the whole office automation market, a market that spawns such prosaic products as electronic copiers and electronic filing cabinets, as well as more exotic items, such as local area networks (LANs) and Ethernets which link it all together. Moreover, while word processing is big, office automation is so

Source: Anne Machung, "Word processing: Forward for Business, Backward for Women," In *My Troubles are Going to Have Trouble with Me,* Karen Brodkin Sacks and Dorothy Remy, (eds.) (New Brunswick, New Jersey: Rutgers, The State University, © 1984) pp. 124–139. Reprinted by permission of Rutgers University Press.

much bigger. With hardware costs dropping and sales increasing 40 to 45 percent annually, that market could exceed three hundred billion dollars through 1987, a rate *Time* claims "dwarfs almost every other sector of U.S. business" ("Fighting," 1981: 66).

Word processing, however, is not quite the same as real office automation. Zisman (1978), in fact, distinguishes between word processing (or "mere mechanization") and "true automation." With mechanization, hand labor is replaced by machine labor, but with automation, mental processes are displaced and machines themselves exercise discretion. In the truly automated office of the future, the "paperless office" as it is often called, the computer not only will control office activities but will exercise judgment by performing tasks according to a preprogrammed logic. So, while it is often claimed that word processing is paving the way toward the truly automated office, with word processing we are still only halfway there.

All this is very exciting for business, equipment vendors, and systems analysts alike, but its consequences for clerical and managerial workers alike are somewhat less exciting. Computerization is visibly reshaping office work in America today, but how is not really known.

Historically, office work has been labor intensive. Indeed, the technology involved in the office changed little between the invention of the typewriter in 1883 and the development of microchip technology in the 1970s. A few simple machines were added, some for adding, some for xeroxing, some for affixing postage; but the typewriter, in the main, remained the backbone of office technology. The office, as a consequence, it is often said, is a remnant of the nineteenth century.

But the development of microchip technology in the 1970s and the emergence of computerized typewriters (or word processors, as they are usually called) promises to change all this, to bring the office into the twentieth century. In their simplest form, word processors are electronic typewriters that can store texts, display whole paragraphs on television screens, play back pages at high speeds, and print out multiple versions almost instantaneously. Moreover, by permitting small editorial changes—commas added, sentences deleted, paragraphs shifted around—to be made without retyping the entire document anew, word processing vastly speeds up the typing function. All this technology is fascinating to those weaned on the old-fashioned typewriter, where each new revision required the slow and tedious job of retyping the entire letter or document.

THE TECHNICAL-MANAGERIAL MENTALITY

By eliminating tedious retyping, word processing is also eliminating the tedium from many jobs, many believe. Indeed, showcase companies, such as Ball Corporation, a diversified manufacturer of food and beverages, claim to have eliminated many tedious jobs altogether. With word processing they needed only 82 secretaries/clerk-typists in 1980; but without word processing they would have needed 126—a "savings," that corporation estimates, of 44 employees. At the same time, productivity rose 643 percent; while a secretary averaged 143 pages a month in 1974, a word processing operator in 1980 averaged 1,062 pages a month. This, the company furthermore claims, is saving them $475,000 in labor costs each year (Jones, 1981).

Other trade journals are likewise filled with such success stories. Each month, for example, *Modern Office Procedures* carries a story showing how word processing pays off, increasing productivity while reducing labor costs. In 1975, before word processing, for example, the 108 secretaries at Michigan Gas and Electric served 310 principals and produced 8.3 million lines of type. Three years later 215 secretaries there served 821 principals and typed 25.9 million lines. Word processing made the difference. "To date," *Modern Office Procedures* concludes, "the word processing operation has obviated hiring some 70 secretaries at a total cost of $27,000 each, which incorporates the cost of salaries, fringes, furniture, and equipment, saving Michigan Consolidated a cost-avoidance figure of about $1.6 million" ("1979 Word Processing," 1979: 76). Exactly what it cost that company to install and maintain expensive computer equipment, redesign offices, and hire computer programmers and systems analysts to oversee the whole automation process is not reported.

Promoted by a vast media blitz, the business rationale for automating the office is clear: to increase white-collar productivity and lower white-collar labor costs. Numbering nearly fifty million workers in 1980 and proliferating faster than any other segment of the labor force, business sees the growth of the information sector contributing significantly to an overhead cost squeeze in the nation's offices ("America's Offices," 1981; "Information Processing," 1979). But by investing huge sums of money in automated equipment, business hopes to curtail this growth. Underlying this are huge advertising and sales campaigns by computer manufacturers to sell their products; the aim of these campaigns is equipping offices everywhere with desk-top terminals, first supplanting clericals, next "supplementing" managers.

Profit, of course, is the motive. By increasing the productivity of each office worker (clerk and manager alike), business hopes to restrict the size of the white-collar labor force and control aggregate wages. To managers, go personal computers; to secretaries, word processors. Word processors therefore constitute but one of the many kinds of computers designed to meet emerging business needs, but also one of the earliest. Word processing is thus helping transform the office from a labor-intensive to a capital-intensive venture.

What business journals do not make clear is that automated technology at the clerical level is also, in part, a response to management fears that clericals are about to unionize. While only about 19 percent of the clerical labor force is currently unionized (Gifford, 1982: 50), those who are earn about 30 percent more than their nonunion counterparts (Working Women, 1982:4). Unionization would threaten the very low wage labor base that has long characterized clerical work and would set up countervailing organizations within business that oppose managerial interests. By automating the office before unionization occurs, management hopes both to restrict the number of clericals who might, in the future, have access to substantially higher wages, and to curtail any future limits unions might place on their right to automate as fully as they wish.

WORD PROCESSING AND THE DE-SKILLING OF CLERICAL WORK

Word processing as such also needs to be distinguished from continuous-process technologies. In some cases, automation in the office is proceeding by reassembling the components of different jobs into one. One worker is put in control of a computer which has recombined into a single task a number of different and highly routinized procedures that used to be performed by different people. This is apparently how Citibank in New York City, another showcase company, automated its letter-of-credit department, transforming clericals into "front line professionals" (Matteis, 1979).

But most of the automation now going on in the office does not, however, involve the reintegration of broken-down, de-skilled jobs into a single, more coherent entity. Just the opposite. Word processing technologies, through which most office automation has been introduced, differ from the automation of wholesale procedures in that they enable a relative uncomplicated task, that of typing a letter or manuscript, to be broken down into yet smaller components. In word processing centers, for example, the different components of typing a letter are divided up and assigned to different people: a supervisor to schedule and allocate the work; a word processing technician to key the material into the central memory; a printing operator to monitor the output process; a proofreader to catch mistakes; and a clerk typist to pick up and deliver the work. In this sense, word processing closely resembles classical de-skilling: a whole craft is broken down into ever smaller components, mental work is separated from manual work, and bit pieces of the remaining jobs are parceled out to "detail workers" (Braverman, 1974). Because of this, word processing centers in large organizations clearly resemble assembly lines in the factory.

RESTRUCTURING SOCIABILITY

From the vendors' point of view, the major problem involved in introducing word processing and fully automating the office is not raising the level of technology but gaining acceptance of that technology from managers and clericals alike. Achieving the high productivity gains promised by equipment vendors requires, it is often said, restructuring the office, transforming the traditional secretarial-managerial relationship, and eliminating the "inefficiencies" of human relationships, interpersonal contact, and irrelevant conversation. Indeed, the equipment is often sold as a way to "end the social office," as if sociability among office workers impairs productivity.

The classic (and now discounted) grand IBM plan of the early 1970s for introducing word processing, for example, called for dividing the job of a secretary into two separate portions: typing and administrative support. All typewriters were to be banned from the secretary's desk, and all typing was to be consolidated into a single word processing center. Secretaries then became administrative assistants, typists became word processing technicians, and bosses became principals. It was anticipated that this would dramatically reduce the number of secretaries needed; some would leave by normal attrition, others "voluntarily" would enter the word processing center, and those left would double up on the number of bosses they worked for.

While some organizations still follow this plan, many have abandoned it, mainly because of the adverse effects it had on personal relationships between secretaries and bosses. The plan has not worked well, but reasons for its failure are obvious only in retrospect. Having worked their way up the corporate hierarchy, managers have resented losing the convenience and status associated with a personal secretary. Accustomed to having her there, many have complained their typing is not done as well in an impersonal center—that turnaround time is slow, that important versus unimportant matters are not distinguished from each other, that personal idiosyncrasies are not understood or appreciated.

Secretaries, on the other hand, have disliked losing some of the protectiveness that secretarial–managerial relationships offered as well as the chance to "advance" up the corporate ladder with their bosses. They have missed their typewriters, so it is said, along with the flexibility that a one-on-one secretarial–managerial relationship provides. So, at least, goes the business explanation for their resistance. The real story is somewhat different; and this, untold, lies with word processing operators. The following observations, collected by interviewing word processing operators working in centralized word processing centers in the banking and insurance industries of San Francisco between 1978 and 1980 (Machung, 1983), indicate a quite different situation.

Working face to face with a TV screen all day long, operators, especially those in large centers, have few social contacts except with each other. Unlike secretaries, who are highly visible at front desks, word processing operators are virtually invisible. Isolated from the rest of the corporation by their dress (frequently more casual), their age (frequently younger), and their race (frequently minority), they work anonymously under fluorescent lighting in crowded back offices and windowless rooms. A supervisor usually handles all contact with users of the center; operators often never see the person whose paper or letter they have typed. Bosses are known only by their names, not by their faces, and sometimes operators are even assigned arbitrary initials to prevent favoritism from developing between bosses and operators. Unlike secretaries, who are known by their first names and friendly manners, word processing operators are virtually unknown. Many have taken these jobs to escape the stigmatization and subordination of secretarial work, only to find themselves even more faceless and nameless. Gaining autonomy has cost them sociability; some miss it, some do not.

Though first introduced in this centralized fashion, more recently word processing terminals have been installed at the individual secretary's desk rather than in one central location. Decentralized word processing is beyond the scope of this essay. Vendors claim, however, that decentralization eliminates many of the problems associated with centralized word processing centers. How true this is remains to be seen. While decentralization may reduce depersonalization, the woman–machine relationship still seems to remain paramount, as secretary–boss relationships increasingly come to be mediated through contact with computer terminals.

Word processing introduced in such a way indeed destroys much of the social fabric that formerly wove the office together, undermining bases of loyalty and interaction between clerical and managerial staff, dividing their social worlds from each other, and further segmenting these relationships, all in the name of increasing productivity.

RESTRUCTURING PRODUCTIVITY

In word processing jobs, the semblance of craft in the classic sense of that term is returned to clerical work. All documents produced in the center come out looking beautiful, and operators, initially at least, feel quite pleased with their work. But there the resemblance to craft ends. In fact, it takes fewer skills than ever before to produce a beautiful-looking document; skills formerly belonging to secretaries are now built into word processing machines. Editing facilities, for example, make typos so easy to correct that both typing speed and accuracy atrophy over time; dictionary-checking facilities eliminate the need to spell well.

Moreover, to speed up production (which is the main justification for using word processing equipment) all work, especially that produced in large centers, is standardized. Formats are determined in advance and precoded onto the machine; margins are automatically set at a certain width; and signature lines are placed just so. There is little variety in the task at hand, and few decisions to make, except which keys to press, which codes to call up. Discretion is eliminated from the job, and so are surprises. Interruptions in the routine are experienced as aggravating and disruptive, rather than as challenges to master new skills and techniques. People become problems that operators escape by facing the machine, and bosses become "users" or even "dictators."

The job also takes little time to learn. Most word processors fully master their machine within three months; most tire of it within six months—then go looking for another job and another machine to learn. Operators describe themselves as having learned a new skill yet respond toward their jobs like only the most unskilled and unattached workers. Interest in the job quickly peaks, then declines; and as interest declines, errors go up and productivity goes down. Supervisors face the dilemma of either encouraging top management to invest in new, more sophisticated equipment in order to retain the interest of their staff, or encouraging worn out operators to leave. Most often, however, it is not even necessary to encourage operators to leave; they have already quit, voluntarily, long ago.

But this is commentary not on women and their motivation for working but on the nature of their jobs. Word processing equipment is deliberately designed for simplicity of use and for speed of production, not to enhance the dexterity, skill, or self-esteem of the craftsman. Operators are seen as easily trained, replaceable, and interchangeable, not as highly trained, especially competent, or invaluable workers.

That this is so is made clear not only in interviews with word processing operators but also in reports of leading equipment manufacturers. Recently Thomas Kavanagh, president of NBI, quipped that his corporation manufactures word processors so easy to learn that "users can take someone off the street with typing skills and put them into word processing" ("How not to Exploit," 1980: 92). WANG laboratories also design their terminals with such "human engineering" in mind. That it now takes someone with minimal typing skills only a day or so to learn how to use a word processor attests to its increasing simplicity. Manufacturers are here speaking with forked tongue: simplifying their equipment so thoroughly that it takes almost no time to learn and selling it to large corporations on that basis, but encouraging women to leave secretarial jobs and enter word processing on the expectation of upgrading their skills and career opportunities. Large corporations are buying, not only the equipment, but the line as well.

PRODUCTIVITY COUNTS, QUOTAS, AND CONTROLS

Word processing operators also work under constant surveillance, but the surveillance comes not so much from their supervisor as from their terminal. "An additional advantage of word processing systems," explains Donald Vadnais of 3M Corporation, is that "business can finally monitor and measure the productivity of clerical function" (Thoryn, 1979: 74). This basically means that word processing operators work under heightened production pressures; each line, page, and document they type is counted and tallied, along with each error, day by day, month by month. Almost all centers maintain such counts, ostensibly to justify the cost of the equipment. Sometimes these measurements are expressed in number of pages typed per day per operator, sometimes in number of clericals needed to support the managerial staff (the "secretary–principal" ratio). But the purpose of such measurement is control: mechanized, depersonalized, industrialized.

Operators, though, are told such counts are kept solely to justify the cost of the equipment: is it producing as much as was promised? Driscoll (1980), however, reviewed office automation investments and found that not one of forty large office automation users had ever gone back to evaluate the economic benefits of using automated office equipment once it had been installed. Cost justification, that is, occurs only before installation, never after; productivity counts and quotas, it seems, are kept only to speed up the work.

Needless to say, word processing operators know they are working under intense speedups disguised as production quotas; keeping their jobs means keeping up with these quotas; getting a raise means exceeding them. Thus the job is extremely pressured, perhaps the most highly stressed job in the country today. The National Institute of Occupational Health and Safety recently found stress among some clerical VDT operators to be greater than that ever reported before for a single occupational group, including air traffic controllers. Of the VDT operators at Blue Shield in San Francisco, 80–90 percent experienced eyestrain or muscle strain; anxiety, depression, and fatigue were also common (National Institutes for Occupational Safety and Health, 1981; see also Smith, 1981; Working Women, 1981). Social skills are also eliminated ("It doesn't take much to talk to a machine"), and machine skills (the willingness to withstand boring, repetitious, and routine work for long periods of time) are highlighted. Such constitutes the freedom of the new clerical worker.

Turnover is accordingly quite high (running as much as 100 percent a year in many centers) as operators change jobs looking for variety, slightly higher pay, and a little less pressure. Nobody expects to receive a pension—the thought is ludicrous—and many operators contemplate working for temporary agencies where job security is even lower but pay slightly higher. Word processing has indeed become the vanguard of the six-month worker; rare is the operator who has stayed at the same job for more than a year or two. Such high turnover, of course, is saving business even more money that might otherwise be expended in raises, job benefits, profit-sharing plans, or pensions.

It is difficult to understand vendors' claims, widely repeated as verities throughout the business and trade journal literature, that automation is creating "stimulating careers" and "eliminating the drudgery" normally found in offices (Eddinger, 1980; "Office 'Miracles'," 1978; Wohl, 1981; for example). At issue is whose career is being stimulated and from whose job drudgery is eliminated. Under office automation, some jobs are indeed gaining in autonomy and creativity, but many are losing. For managers, those who design, manufacture, sell, and administer the new technology, automation does mean more control over labor costs, faster and more complete information, and higher levels of expertise, responsibility, and salary. But for clericals, automation means something completely different: more routinized jobs, more standardized work, greater pressure, and less sociability. And most clericals are women.

SEXUAL STRATIFICATION AND TECHNOLOGICAL TRANSFORMATION

The effect of automation and computerized technology upon the labor force, that is, is sharply stratified by gender. True when data processing was first introduced into corporate and government offices in the 1950s, this is even more true today. Comparing, over time, keypunch operators (who do the least skilled and most routine computer jobs for the lowest pay) with systems analysts (who have the most challenging computer jobs and the highest pay) illustrates this point (Table 20-1). In 1960, for example, 73 percent of all keypunch operators were women, while only 22 percent of all systems analysts were; but in 1980, 95 percent of all keypunch operators were women, and still only 22 percent of all systems analysts were. Despite extensive affirmative action efforts, things hardly changed in data processing—except to grow more stratified.

But what about word processing? Word processing in some sense

Table 20-1 Growth of Computer-Related Jobs, By Sex, 1960–1980

	The Professional Line			
	Computer programmers		*Systems analysts*	
Year	*Number*	*Percentage female*	*Number*	*Percentage female*
1960	8,140	26.7	3,280	22.0
1970	161,382	22.5	79,961	14.7
1980	341,000	28.7	201,000	22.4
	The Clerical Line			
	Keypunch operators		*Computer-equipment operators*	
Year	*Number*	*Percentage female*	*Number*	*Percentage female*
1960	161,067	72.9	1,928	65.2
1970	272,921	89.8	117,313	29.2
1980	266,000	95.3	522,000	61.6

SOURCES: Figures for 1960 and 1970 are taken from the U.S. Bureau of the Census (1973: Table 221). Since 1980 census data are not publicly available at this writing, the closest detailed approximations to census data are from the U.S. Bureau of Labor Statistics (1981: Table 23).

represents the intersection of repetitious typing with repetitious computer work. At present, it is impossible to say exactly how many word processing operators there are in the country or what proportion are women, since neither the Bureau of Labor Statistics nor the Bureau of the Census classifies word processing operators separately; they are supposedly counted as secretaries, typists, or computer operators. But in 1980, 61 percent of all computer operators were women, 99 percent of all secretaries were women, and 98 percent of all typists were women. If the tendency for women to get slotted into the most routine, repetitious, and least skilled clerical computing jobs continues, more and more word processors will be women.

Right now the seeming neutrality of the new word processing technology conceals this gender divide. Just as the earliest typists were called "typewriters" after the name of their machine, so many word processing operators today are called "word processors" just like the machine they operate. But just as the novelty of typewriting wore off and women rapidly got slotted into these jobs (Davies, 1974), so, I suspect, as the novelty of word processing wears off (and the work is seen for what it actually is), women, especially minority women, will be slotted into these new jobs.

I am not here advocating that more men become word processors and keypunch operators to equalize the invidious and widespread effect of job segregation on the labor force. Rather, the nature of work itself must be changed. Bad jobs are bad jobs, unskilled labor is unskilled labor, whether done by women or by men, by minorities or by whites. But if my observations of San Francisco banks and insurance companies and my analysis of long-standing labor-force trends are accurate, word processing—and its consequences for the destruction of jobs, the health of the worker, and the loss of job satisfaction—are pre-eminently a woman's issue.

WORKER RESISTANCE AND ORGANIZATIONAL RESPONSE

There is no doubt that breaking up traditional working relationships and destroying jobs is provoking deep resistance from executives and secretaries alike, and the word processor industry has long realized it is people who form the biggest barrier to swiftly automating the office. To many, the willingness of office workers to accept such changes in jobs and office organization and to "discipline themselves" to the new system is the major factor determining how quickly word processing catches on ("Office of the Future," 1975: 50). Accordingly, the industry is devoting itself to a two-pronged attack: "educating" top management on the advantages of the new technology and devising ad hoc procedures for undermining resistance from below.

Computerworld, for example, explains resistance to technological change as a normal reaction to change of any kind; adjustment to that change therefore follows a "five phase predictable process": "First is self pity (why me?), then denial (not me!), then anger (I'll get even.), then bargaining (I'll cooperate if you will . . .), finally comes acceptance (what change?)"(Gaffney, 1981). Ironically, each of these stages almost parallels Kübler-Ross's (1969) five stages of death and dying—denial and isolation, anger, bargaining, depression, and finally acceptance—but no mention of Kübler-Ross (or of death and dying) has been made. "Each individual," Gaffney (1981) instead continues, "moves through this process at his own speed and some may get stuck in one phase if not helped to move on." Rather than "helping" clericals to move on, most of the trade journals, though, devote themselves to selling the equipment to top management.

Still there is some concern, however limited, with the jobs of lower level clericals, for resistance from these people (expressed as high turnover rates, high drug abuse, and high absenteeism) can significantly lower productivity. Even a carbonated drink innocuously spilled on a terminal effectively kills it for a day or two. So each month *Modern Office Procedures* describes how one showcase company solved the "people problems" in its word processing center by retraining the staff, redesigning the center, establishing career ladders, and hiring an experienced supervisor. None of these solutions in themselves is fallacious, but none of them really improves the job either.

Take career ladders for example. Establishing career ladders in most centers means little more than creating minute gradations among word processing jobs. Operators enter as "word processing trainees" and move to "word processing technician I" and then "word processing technician II"; but for most, career advancement peaks when they become senior word processing technicians or specialists. While a few operators do aspire to becoming word processing supervisors, this job leads nowhere, pays little more than the operator's job, and is far more cross-pressured; most operators do not want it. Other operators hope to become marketing service representatives for various equipment vendors or computer programmers, but to do so often requires skills, expertise, and college degrees not acquired as a word processing operator. There really is no career advancement possible within the field, and most operators know this.

Hiring an experienced supervisor also means little more than bringing in someone with smooth personal and organizational skills to replace the social contact operators qua secretaries formerly had with bosses. An experienced supervisor is somebody who restores sociability to the center, but in limited amounts; someone who oversees the entire process without directly interfering; someone who praises and compliments her operators for a job well done; and someone who "runs interference" between her operators and the professional and managerial staff. Most often the first-line supervisor is also a woman. But she too, like her operators, is under corporate pressure to produce, and this in itself points to the amount of organizational control that has been gained by installing word processing equipment. Not only is the unstructured secretarial–managerial relationship controlled by reducing the number of secretaries hired and doubling up on the number of bosses served, but the amount of personalism possible in the supervisor–operator relationship is restricted and bounded by the productivity pressures of the organization to which the supervisor herself is subject. She has to meet her quotas too.

WHAT NEXT? AUTOMATING MANAGEMENT

What has happened, however, is that word processing systems have not lived up to their promises. Office costs have continued to rise despite the introduction of word processing, and few offices have achieved the productivity gains promised by vendors. For every showcase company there are untold numbers of failing centers. As a consequence, management efficiency experts and vendors alike are beginning in the 1980s to look at the real source of soaring office costs: not clerical salaries but managerial salaries, not clerical productivity but managerial productivity.

Almost all efforts at introducing office automation technology, Abraham (1981) has noted, have been directed at improving the productivity of the clerical staff; yet productivity gains in the typing area account for only about 1.2 percent of all U.S. white-collar business costs. Moreover, in 1979, U.S. businesses spent three times more on the salaries of "principals" (managers, professionals, and executives) than they did on the salaries of clericals: $435 billion was paid to principals, $135 billion to clericals.

Noticing this, vendors are shifting from selling office automation equipment exclusively as a device to increase clerical productivity and are now selling it as a device to increase managerial productivity. Celebrating this technological revolution, *Time* in 1982 named the personal computer "machine of the year." "TIME's Man of the Year," the editors announced, "is not a man but the computer itself" ("Machine of the Year," 1983: 3). Meanwhile vendors fear the word processing market is drying up; essentially they consider the nation's 5.5 million secretaries nearly fully automated and the word processing market "pennies" compared to the one hundred billion dollar potential of the business professional market (Schatz, 1982: 66). Moreover, by 1986 low-cost word processors will have "penetrated 30% of the U.S. heavy duty installed typewriter base," its estimated saturation point ("White Paper," 1982: 52). Whither next is into management. But nothing really is different on the executive terminals, Driscoll (1981: 20) has quipped, except for their mahogany paint.

But something is fundamentally different in the way computerized equipment is used at the managerial level and the way it is used at the clerical level. When the First National Bank of Boston installed minicomputers for both secretaries and professionals, some secretaries lost their jobs and others were reclustered into a "secretarial services center"; but professionals kept their offices, their desks, and their jobs (Synott, 1981).

Rationales for using computerized equipment also differ between the two levels. At the clerical level, word processing is cost justified—does it reduce the number of secretaries needed and increase the number of pages typed each day? But for professionals, no such justification is wanted. Listen to William Synott (1981), senior vice-president of the First National Bank of Boston, tell it:

> When savings are expressed in "soft" terms affecting professionals, such as getting better information faster and more reliably, the benefits usually do not translate into savings in jobs; thus, the results become more difficult to measure with anything approaching reliability. Justification in these cases can only be in the form of testimonials given by satisfied customers.

How typical the Boston bank case is cannot really be known yet since automation of the managerial sector has barely begun. I have already argued that reports of how word processing affects clerical productivity often overstate the truth, that word processing does not increase clerical productivity nearly as much as vendors claim. Synott's comments that cost savings at the managerial level need only be expressed in "soft" terms (such as getting him better information faster) rather than in "hard" terms (such as job elimination) may likewise overstate the truth. Privately, vendors are predicting that automation will eliminate as much as 30 percent of all midmanagerial jobs by 1990. Whether this is so also remains to be seen. But since the impact of office automation on managerial jobs is indeterminate at this point in time, any remarks at this stage on how automation divergently affects managerial and clerical jobs can only be tentative and exploratory, meant to raise issues rather than resolve them.

But these differences between clericals and professionals suggest, however, that professionals and managers do not have to work under the same kinds of speedups and face quite the same threats of job loss or skill reduction that secretaries and clericals experience as their jobs are automated. They further suggest that the purposes for automating the office are different at the professional and managerial level from those at the clerical level. At the managerial level, the purpose of automated equipment is to please and satisfy the "business professional"—to get him better information faster and more reliably. But at the clerical level, the purpose is something else.

Astute observers speculate that the real reason for automating the office is not so much to control clerical labor costs as to control clerical labor. Noble (1979) argues that technology is a means of social control in the conflict between classes in society; numerical control was developed over record-playback techniques in the machine tool industry, for example, not because it was more cost efficient but because it enhanced managerial control more. Likewise, Barker and Downing (1980) see word processing in the automated office as a mechanism to replace looser patriarchal forms of control (characteristic of secretarial–managerial relationships) with more direct capitalists forms: the move, they label it, toward the real subordination of office workers.

WHITHER NEXT?

So while vendors claim that automation is needed in the office to control rising labor costs, it is mostly jobs in the clerical labor force that are being controlled; managerial needs for information are being well fed and attended to. In the capitalist economy as developed in late twentieth-century America, corporate needs for information (regardless of the ability of managers to digest it, and control take precedence over human needs for interesting, challenging, and dignified work. "How not to Exploit the Hardware" (1980), the title of a recent article in *Business Week,* sums it up; the viewpoint that it is possible to exploit human beings with the new technology disappears from industry.

The defining characteristic of the managerial–technical perspective, then, is that it takes for granted the "needs" of the corporation—for higher profits, more efficiency, greater productivity, and tighter control—and does not examine how these conflict with and undermine more basic human needs—for meaningful, dignified, challenging, and rewarding work. It is therefore up to clerical women themselves to define these as basic human rights and claim them as their own.

REFERENCES

ABRAHAM, STEVEN. 1981. "Is Office Automation the Best Darned Thing You've Ever Seen? Maybe." *Computerworld,* September 28: 21, 23.

"America's Offices Enter a New Age." 1981. *Nation's Business,* July: 48–54.

BARKER, JANE, and HAZEL DOWNING. 1980. "Word Processing and the Transformation of the Patriarchal Relations of Control in the Office." *Capital and Class* 19:64–97.

BRAVERMAN, HARRY. 1974. *Labor and Monopoly Capital.*New York: Monthly Review Press.

DAVIES, MARGERY. 1974. Woman's Place Is at the Typewriter: The Feminization of the Clerical Labour Force. *Radical America* 8:1–18.

DRISCOLL, JAMES. 1980. "Office Automation: The Dynamics of a Technological Boondoggle." Working Paper for the Alfred P. Sloan School of Management, Massachusetts Institute of Technology.

———. 1981. "Is Office Automation the Best Darned Thing You've Ever Seen? No!" *Computerworld,* September 28: 20, 22.

EDDINGER, JOHN. 1980. "21st Century Office: A Paperless Wonder." *Nation's Business,* July: 65–66.

"Fighting the Paper Chase." 1981. *Time,* November 23: 66–67.

GAFFNEY, CAROL. 1981. "Crisis in the Workplace: Selling the Staff on Office Automation." *Computerworld,* September 28: 11

GIFFORD, COURTNEY C., ed. 1982. *Directory of U.S. Labor Organizations.* 1982–1983 ed. Washington, D.C.: Bureau of National Affairs, Inc.

"How not to Exploit the Hardware." 1980. *Business Week,* March 24: 90, 92.

"Information Processing and Tomorrow's Office." 1979. *Fortune* (special advertising suppl.), October 8: 31–62.

"Invasion of the Computers." 1981. *Newsweek,* December 28: 57.

JONES, REBECCA. 1981. "Food Firm Finds Automation Fortifying." *Computerworld,* September 28: 69–70.

KÜBLER-ROSS, ELISABETH. 1969. *On Death and Dying.* New York: Macmillan.

"Machine of the Year." 1983. *Time,* January 3: 3, 12–32.

MACHUNG, ANNE. 1983. "From Psyche to Technic: The Politics of Office Work." Ph.D. diss., University of Wisconsin.

MATTEIS, RICHARD. 1979. "The New Back Office Focuses on Customer Service." *Harvard Business Review* (March-April): 146–159.

MORNER, MAIMEE. 1979. "Why Word Processing Companies Are the Talk of Wall Street." *Fortune,* December 31: 76–77.

National Institute for Occupational Safety and Health. 1981. *An Investigation of Health Complaints and Job Stress in Video Viewing.* Cincinnati.

"1979 Word Processing Executive: Leadership, Cost-Effectiveness, Career-Enhancement." 1979. *Modern Office Procedures,* June: 74–79.

NOBLE, DAVID. 1979. "Social Choice in Machine Design." In Andrew Zimbalist, ed., *Case Studies on the Labor Process.* New York: Monthly Review Press.

"Office 'Miracles' that Electronics Is Bringing." 1978. *U.S. News and World Report,* September 18: 76–78.

"The Office of the Future: An In-depth Analysis of How Word Processing Will Reshape the Corporate Office." 1975. *Business Week,* June 30: 48–82.

SCHATZ, WILLIE. 1982. "A Terminal on Every Desk?" *Nation's Business,* February: 64–69.

SMITH, MICHAEL. 1981. *An Investigation of Health Complaints and Job Stress in Video Display Operations.* Cincinnati: National Institute for Occupational Safety and Health.

SYNOTT, WILLIAM. 1981. "Boston Bank Takes Office Automation via Well-Travelled Word Processing Route." *Computerworld,* September 28: 12–14.

THORYN, MICHAEL. 1979. "Office Equipment: Electronics Empties the Typing Pool." *Nation's Business,* February: 70–78.

______. 1980. "Tomorrow's Office: Wired for Words." *Nation's Business,* July: 56–64.

U.S. Bureau of Labor Statistics. 1981. *Employment and Earnings,* January, 180.

U.S. Bureau of the Census. 1973. *Census of the Population: 1970,* vol. 1, *Characteristics of the Population,* pt. 1, "United States Summary," sec. 1. Washington, D.C.: GPO.

"White Paper: Word Processing in Transition." 1982. *Modern Office Procedures,* June: 51–60.

WOHL, AMY. 1981. "'Office of Tomorrow': What to Expect." *Computerworld,* September 28: 5.

Working Women. 1980. *Race against Time.* Cleveland.

______. 1981. *Health Hazards of Office Work.* Cleveland.

______. 1982. *Office Work in America.* Cleveland.

ZISMAN, MICHAEL. 1978. Office Automation: Revolution or Evolution? *Sloan Management Review* 19: 1–17.

PATRIARCHY'S NEW FORMS

21 *Integrated Circuits/Segregated Labor: Women in Computer-Related Occupations and High-Tech Industries*

MYRA H. STROBER AND CAROLYN H. ARNOLD

We are just beginning to see the repercussions in all of our lives of the technological feat of fitting the electronic wiring and switches of what was a room-sized computer in 1946 onto a less than fingernail-sized piece of silicon and metal by the end of the 1970s. This silicon chip is the core of a technological revolution, the result of many attempts over a century to produce a "computing machine" that is small, fast, and cheap. Now, as the chips and, thus, the computers they make possible get smaller, faster, and cheaper, their applications in both old and new products are spawning a new high-technology industry. We see changes in a multitude of workplaces and homes, the expansion of opportunities in existing industries and occupations, and the creation of new industries and occupations that were not even imagined just a few years ago.

The occupations most involved in the computer revolution are engineers, computer scientists/systems analysts, programmers, electronic technicians, computer operators, and data-entry workers; these occupations are expanding both within the computer industry and in other industries as well. In addition, the computer industry provides new opportunities for managers, clerical workers, and production workers. Traditionally women have been sharply segregated into different occupations from men and have been paid less

Source: Myra H. Strober and Carolyn H. Arnold, "Integrated Circuits/Segregated Labor: Women in Computer-Related Occupations and High-Tech Industries," in *Computer Chips and Paper Clips,* Heidi I. Hartmann, (ed.) (Washington, D.C.: National Academy Press, 1987) pp. 136–182.

than men (Gross, 1968; Lloyd and Neimi, 1979; Blau and Hendricks, 1979; O'Neill, 1983; Bielby and Baron, 1984; Strober, 1984; Treiman and Hartmann, 1981). This study poses several questions. Are there better opportunities for gender integration and earning equity in these new occupations that are growing rapidly, are exhibiting labor shortages, and are supposedly not locked into past traditions and stereotypes? Are there better opportunities for gender integration and pay equity in high-technology (high-tech) industries? What does the growth of these occupations imply for women's employment?

This paper is divided into four sections. The first discusses the details of the occupations we analyze. The second section uses published data as well as the 1/1000 Public Use Samples (P.U.S.) from the 1970 and 1980 U.S. Censuses to look at how women are faring in the six major computer-related occupations and in high-tech industries.

In brief, we found that although high-tech in general and computer occupations in particular are often seen as the great equalizers, especially for those with higher education, in fact, there is considerable gender segregation in both high-tech industries and computer-related occupations in all industries; there is also considerable male-female earnings differentiation. We suggest that one possible cause of the earnings differentiation is that men and women in computer occupations are not employed equally across industries; women tend to be employed more frequently in the lower-paying end-user industries.[1]

[1]We define "end-user" industries as those that use the products of the computer industry and make only minor changes to the products to accommodate their needs, rather than making basic new developments in these products. The companies developing computers and/or their software are part of the computer industry. The companies in all other industries, which will use these computers and/or software, are part of the end-user industries. It is true that even within the computer industry the administrative divisions of the companies are end-users of computers. However, census data do not permit us to make such fine distinctions within industries, and we are interested here mainly in any broad differences between industries.

COMPUTER-RELATED OCCUPATIONS

Descriptions of Occupations

The development of semiconductors, computers, and computer languages spawned several new occupations and expanded several others. There are six major computer-related occupation groups—engineers, computer specialists, engineering and science technicians, production workers, computer operators, and data-entry operators. For three of these groups (engineers, engineering and science technicians, and production workers), we have restricted the analysis to those employed in the computer industry. For the other three groups (computer specialists, computer operators, and data-entry operators), we have examined employment in all industries. The following descriptions of the occupations in these groups are based on definitions in the *Standard*

Occupational Classification Manual, 1980 (U.S. Department of Commerce, 1980), California Employment Development Department publications (ABAG, 1981), and interviews with workers and employment counselors.

Engineers. *Engineers* design hardware for computers, including the electronic circuits. The largest group of engineers is electrical engineers, but mechanical and industrial engineers also work in the computer industry. Sometimes they incorporate software designs into the circuits. Engineering is the highest-status and highest-paid computer-related occupation, with engineers generally having at least a B.S. degree in engineering; many have advanced degrees.

Computer Specialists. As we look beyond designing hardware to designing software, the sets of instructions that tell the computer which operations to perform, we encounter the *computer specialist* occupations. While computer engineers tend to be employed largely by computer companies, computer specialists are employed in virtually every major industry group. These jobs involve a hierarchy of tasks that used to be done by one person with the title of computer programmer.

When the first computer was unveiled in 1946 (it was room-sized because the circuits were made with glass vacuum tubes), the engineers who designed it thought that the main task of arranging the circuits had been done, and that giving instructions to the computer to perform calculations would be a simple clerical task. So they hired people who usually do clerical tasks—women. In this case the women were recent college graduates with math backgrounds. However, these women found that in order to get the computer to do calculations, they (the programmers) had to know all about the design of the circuits and the way those circuits worked in the computer; they had to tell the computer not only *what* to do, but *how* to do it. The simple operation of performing calculations (in this case for Navy shell trajectories) became a high-level task that involved a knowledge of logic, mathematics, and electronic circuits. These women programmed the necessary calculations and went on to do others. However, those who watched the programming process began to realize that programming was a high-level, challenging, and creative occupation. As the occupation grew, it became largely male (Kraft, 1979).

Ironically, some programming today is akin to the type of clerical job that computer designers (mistakenly) thought it would be in the late 1940s. Over time, with the development of higher-level languages (closer to human languages) and more routine applications, programming tasks that were previously highly skilled, highly paid, and concentrated among highly educated workers have been broken down into more routine tasks and distributed among less-skilled workers. Kraft (1977) has suggested that, as this "deskilling" has occurred, it is women who have moved into the less-skilled jobs. Greenbaum (1979) found that the lowest-level programming jobs were disproportionately occupied by racial and ethnic minorities.

This history of the developing hierarchy in computer programming is reflected in the designations given by the Bureau of the Census to the computer specialist occupations. In 1960 and 1970, computer specialists, including programmers, were included in the professional category. By 1980, the Census put the three-digit occupational category of computer scientists/systems analysts in the professional category and the three-digit category of computer programmers in the technical category.

The following descriptions attempt to capture the current hierarchy and educational requirements among computer specialists. Some workers in these occupations do not have formal credentials, having been trained or self-taught on the job. These job titles and descriptions continue to change and overlap.

Computer scientists and some systems analysts work with engineers to design the overall hardware and software systems and sometimes know just as much about the hardware, although their training is more concentrated in the logic and mathematical models of computer systems, rather than on electronic principles. They also develop new languages to be used by other programmers. Generally, they have an M.S. or Ph.D. in computer science (CS) or electrical engineering (EE) or both (CS/EE).

Computer systems analysts conceptualize and plan how a business or industrial task, such as automating a payroll or an assembly line, will be solved by computerization. Systems analysts do not write the programs but make flow charts to show the subtasks that need to be done by people and computers and their sequence and timing.

Computer programmers are often promoted into systems analysts positions because these positions represent higher-level skills, responsibility, and pay than do programming positions. If systems analysts were not previously programmers, their education is either in business or data processing. There is a debate in this field about whether systems analysts need programming skills or not.

Systems programmers maintain and modify operating systems—systems of programs that coordinate all the hardware in a particular computer so it will run according to certain high-level languages. They are also responsible for updating the high-level software on the system in their particular company. They generally require a B.S., M.S., or Ph.D. in CS, EE, or math.

Programmer/analysts update operating systems and write programs that tailor the computer's uses to each individual work-place. Although ready-made software is available for many purposes, most firms need programmers to modify or write programs that reflect their own computing needs. These programmers need to know both operating systems and high-level languages. Education requirements are a B.A. in related subjects with some programming experience, a B.S. in CS, or an M.B.A.

Software engineers, as some programmers are increasingly called, design and write programs in high-level languages specifically for certain computers. These programs, often called packages, are sold with the computer to make it easier for nonprogrammers to use. Packages can include such items as games, accounting programs, and instructional programs. Producing these programs requires the creativity to conceptualize and design new ways to use the computer. People writing these software packages require a good knowledge of the language used to program the software and good ideas about marketable packages. Acquiring knowledge of programming and having creative ideas are more important for job success than are educational degrees. Consequently, a range of people, from high school students to Ph.D.s, are designing and writing software.

Programmers, sometimes called coders or applications programmers, are mainly translators. They translate instructions for a certain application in one language into the programming language that their particular computer will use to produce the same results. The job category itself encompasses a range of skill and creativity from routine coding of sections of an application program to a task more like a programmer/analyst, depending on their industry or firm. These jobs can be done with less than a B.S. in CS. However, if those with a B.S. degree are available, employers often prefer to hire them.

A third major group of computer-related workers is *engineering and science technicians.* This group is found in the computer industry and is mainly made up of engineering technicians trained in electronics, although there are also some science technicians working within the industry. The electronic technicians have enough specialized knowledge of electronics to be able to construct, test, and repair the circuitry and components of computers that the engineers design and to understand engineering specifications and problems. Although they do not do original design work, they operationalize designs, test them, and then advise engineers on possible modifications. They work both in research design and in production to test and troubleshoot both new and existing products. Also included in the engineering technician category are drafters, who, using both manual and computer-assisted drafting tools, make drawings of the circuit boards and components that the engineers design. The standard degrees are the 2-year associate of arts (A.A.) or associate of science (A.S.) for an engineering or science technician and an A.A. in drafting for a drafter.

Computer Operators. The fourth computer-related occupational group is *computer operators.* They are employed in all industries. Computer operators run the external operation of the computer; ensure that the computer receives the programs and data; and coordinate disks, tapes, and printing connections to the computer, either manually or by supervising automated systems. This occupation ranges from active high-level interactions with the programs to routinized supervision of automated systems. It is sometimes an entry-level job leading to low-level programming. Education needed is simple knowledge of the equipment from a short training course and/or from experience. The occupation is rapidly being deskilled as more of its functions become automated.

Data-Entry Operators. *Data-entry operators,* the fifth computer-related occupation, put information into a form that can be read by a computer. This information used to be keypunched onto cards but is now almost always put onto tapes or disks from terminals. The operators, who basically type numbers and letters into a terminal, require training in typing. They are also employed in all industries.

Production Workers. The sixth major computer-related occupation is *production worker.* While many of the production jobs are similar to those in other industries, the following jobs are unique to semiconductor and computer production. There are generally no specific educational requirements for these jobs, although people with some knowledge of electronics are generally preferred. Often hazardous chemicals are used in the production process.

Semiconductor processors put materials through chemical and mechanical processes to create semiconductor integrated circuits on chips. They work either manually or, as these tasks become mechanized, with processing machines.

Semiconductor assemblers assemble chips into wired devices which become the complete integrated circuit. This includes bonding wires to circuits, a task which is done under a microscope, and cleaning the circuits with chemicals.

Electronic assemblers assemble the integrated circuits and other electronic components into a frame that becomes the finished product (e.g., a computer). Electronic assemblers can be promoted to *electronic testers,* who test chips, boards, and components as they go through assembly, or *electronic inspectors,* who examine the components for errors and specification requirements.

A skilled occupation that is sometimes part of production and sometimes part of customer service is *data-processing repair,* which involves installing and repairing data-processing machinery in offices and on production lines. This job category includes a range of workers from electronic mechanics to assembly and wiring technicians.

There are also, of course, managerial, professional/technical, sales, clerical, and service occupations within the computer industry. These jobs tend to be similar to such jobs in other industries.

GENDER SEGREGATION

Computer-Related Occupations

In this section we begin the analysis of gender segregation by examining trends in total employment and women's employment in computer-related occupations. Between 1970 and 1980, employment in computer-related occupations grew about 80 percent—from about 1.5 million to 2.4 million. However, although the growth of these occupations is widely heralded, it is important to note that they represented only 2.0 percent of all employment in 1970, and 2.5

percent in 1980. (Among all women workers, those in computer occupations represented 2.3 percent of employment in 1970 and 2.9 percent in 1980. The corresponding percentages among men workers were 1.8 percent in 1970 and 2.2 percent in 1980).

This growth took place in the context of increasing participation in the labor force by women. In 1970, women were 38 percent of the U.S. labor force; by 1980, they were 43 percent. However, in both those years, women's representation in computer-related occupations was either considerably below or considerably above their representation in the labor force as a whole, depending on the specific occupation.

Despite the fact that the computer-related occupations are of relatively recent origin, they are already remarkably segregated by gender. In 1970, women were 2 percent of all engineers in the computer industry; in 1980, that figure had risen to only 5 percent. Thus, in the highest-paid, highest-prestige computer-related occupation, women are virtually absent.

Among computer specialists in all industries, the situation is somewhat better, although women are still below their proportion in the overall work force. In 1970, women were 15 percent of all computer scientists/systems analysts. This occupation more than doubled from 1970 to 1980 (from 93,000 to 201,000), but by 1980, women were still only 22 percent. Among programmers, the proportion of women were also increased, but they were also still underrepresented. The number of programmers almost doubled (from 161,000 to 313,000), and the proportion of women grew from 23 percent in 1970 to 31 percent in 1980.

Women were better represented among engineering and science technicians in the computer industry than among engineers, but were less well represented than among computer specialists. In 1970 in the computer industry, women were 11 percent of engineering and science technicians; in 1980 they represented 17 percent. They were similarly represented among electronic/electrical technicians—11 percent in 1970 and 15 percent in 1980. As drafters, women did less well in 1970 than other technicians—only 7 percent were women—but did better in 1980, when 16 percent were women.

Initially, there was no clear indication as to which gender would be assigned to the occupation of computer operator. In 1960, when there were only 2,000 computer operators, women held 65 percent of the jobs (Dicesare, 1975). Between 1960 and 1970, the jobs in this occupation increased more than 50-fold, to 117,000. More of these new jobs were filled by men than by women so that in 1970, women were only 29 percent of all computer operators. In the period from 1970 to 1980, however, while the occupation increased fourfold, to 408,000, more of the new additions to the occupation were women, so that in 1980, women were 59 percent of all computer operators.

Like most clerical occupations, data entry is preponderantly female. In 1970 women were 90 percent of data-entry operators. Between 1970 and 1980 the occupation became even more segregated so that by 1980 women represented 92 percent of such operators.

Of all production workers in the computer industry, women were only slightly over their representation in the labor force as a whole: 46 percent in 1970 and 49 percent in 1980. However, when we look more closely at the less-skilled production occupations, women's representation is much higher. Of all the operators, fabricators, laborers, and transportation workers, a group which includes the semiconductor processors and assemblers and all other lower-level production workers in the computer industry, women represented 58 percent in both 1970 and 1980. Among assemblers, a subset of operatives, women were about 73 percent in 1970 and 1980. Of electronic assemblers, a group identified only in 1980, women were an even higher proportion: 77 percent. However, in the occupation "data-processing machine repairers," we find again the extraordinary gender segregation we often see in technical occupations: women held 3 percent of these jobs in 1970 and 8 percent in 1980. . . .

GENDER, RACE, AND ETHNIC DISTRIBUTION IN FOUR COMPUTER-RELATED OCCUPATIONS

If we look more closely at the four computer-related occupations that are present in all industries—computer scientists/systems analysts, computer programmers, and data-entry operators—we can see how women and men in the three largest racial and ethnic groups are represented. In the employed labor force as a whole, white men are 50 percent of the workers, white women are 36 percent, black men are 4.8 percent, black women are 4.8 percent, and other racial groups make up 8.4 percent of the workers. Men and women of Spanish origin, who can be of any race, represent 3 and 2 percent of the workers, respectively.

In the occupations of computer scientist, systems analyst, and computer programmer, women of all groups and minority men are underrepresented compared to their representation in the labor force as a whole, while white men are overrepresented. Within each racial and ethnic group, men are better-represented than women.

Among computer scientists/systems analysts, the highest paid of these four occupations, white men hold 71.3 percent of the jobs, which is much higher than their representation of 50 percent in the labor force. White women occupy 19.5 percent of the jobs, only half as high as their representation in the labor force. Black men are less represented in this occupation than in the labor force (3 percent versus 4.8 percent). Black women's representation is even poorer, 1.7 percent in computer scientists/systems analysts versus 4.8 percent in the labor force. People of other races are only about half as well represented in this occupation as in the labor force. Like black men, the representation of men of Spanish origin is just over half that of their presence in the labor force (1.8 versus 3 percent). Women of Spanish origin are virtually unrepresented; they are only 0.1 percent of computer scientists/systems analysts, though they are 2 percent of the labor force.

Among computer programmers, the next lower paying occupation, white men are still overrepresented, and the other groups are still underrepresented, but somewhat less so. White men are overrepresented by 12 percentage points (62.2 percent versus 50 percent) and white women are underrepresented by 10 percentage points (26.5 percent versus 36 percent). Black men's representation is still about 3 percent, but black women's representation has increased to 2.5 percent. However, both percentages are below their share of the labor force. Other racial groups have also increased their representation, but it is still below their labor force representation. Men and women of Spanish origin have virtually the same representation among computer programmers as among computer scientists/systems analysts—1.9 and 0.1 percent, respectively.

This situation changes and in some cases reverses itself in the lower-paying occupation of computer operator. White men are now only one-third of the workers, 16 percentage points *less* than in the labor force, and white women are now half of the workers, 14 percentage points *above* their representation. Black men's representation is about the same as in the labor force, and black women's representation is higher among computer operators (6.9 percent) than in the labor force (4.8 percent). People of other racial groups are still underrepresented, however. Men of Spanish origin have a representation among computer operators below their labor force percentage, while the percentage of women of Spanish origin is above their percentage in the labor force. Thus, in this lower-paid computer occupation, white and black women and women of Spanish origin are overrepresented, and white men are underrepresented, while the percentage of black men and men of Spanish origin reflects their percentage in the labor force as a whole.

When we look at the lowest-paid computer occupation of data-entry operator, a clerical occupation, we see that all women's representation is dramatically higher and all men's representation is dramatically lower than for the other occupations listed in Table 4. White women are 71 percent of data-entry operators, twice their labor force representation. Black women and women of Spanish origin are represented in this occupation with a frequency three times greater than their representation in the labor force—15 and 6 percent, respectively.

Like gender, race and ethnicity of workers in computer occupations is associated with pay and status of the occupation. The higher status and pay of an occupation, the more white men are overrepresented and the more minority men and all women are underrepresented. In occupations with lower pay and status, the presence of white men drops to much below their percentage of the labor force, the percentage of women of all races becomes higher than their labor force percentage, and the percentage of minority men approaches their labor force representation. In an occupation that is clerical, women's representation doubles and triples above their labor force percentage, and men virtually disappear.

HIGH-TECH INDUSTRIES

This section looks at how women are faring not just in computer-related occupations but in the group of industries known as "high-tech," of which the computer industry is one part. We are interested in whether the computer industry and other high-tech industries, because they are growing rapidly and are relatively new, are, therefore, perhaps less gender stereotyped and more hospitable to women than are non-high-tech industries.

The Bureau of Labor Statistics recently reviewed all current definitions of high-tech industries and developed a range of definitions based on three factors (Riche et al., 1983:51): "(1) the utilization of scientific and technical workers, (2) expenditures for research and development, and (3) the nature of the product of the industry."

Although employment in high-tech industries grew over the 1970-1980 decade from 4.6 million to 6.1 million, these industries employed only about 6.5 percent of all employed workers, about 5 percent of all women workers and about 7.4 percent of men workers.

We find also that these industries are no less segregated by gender than are other industries and, in fact, may be even more so. In both 1970 and 1980, most men in high-tech industries were in management, professional/technical work, or production, while most women were in clerical work or production. Compared to women in non-high-tech industries, women in high-tech industries are less likely to be in managerial and professional/technical jobs and are much more likely to be in production work. High-tech industries provide more low-status occupations to women, as a group, than do non-high-tech industries. By contrast, men in high-tech industries are more likely to be in managerial or professional/technical positions than are men in non-high-tech industries. . . .

CONCLUSIONS

Findings

1. Although computer-related occupations are of recent origin, they are not gender-neutral. The computer field was sired by the fields of mathematics and engineering, and the newly born prestigious and technical jobs quickly took on the gender designation of the parent fields. Computer engineering and electronic technical work employ very few women. On the other hand, data entry, which quickly took on the characteristics of clerical work, became a virtually exclusive female preserve. Production work, too, is preponderantly female.

   Computer programmers were female when the occupation first emerged, but very shortly after the computer was introduced, men began to fill the emerging jobs. Although women have increased their representation in the jobs of both programmer and analyst, women remain less than a third of the incumbents of these occupations. While the occupation of computer operator did not seem immediately to be gender typed, in that it was preponderantly female in 1960 and male in 1970, it was becoming preponderantly female again by 1980.

2. When the difference between the percentage of women in the total labor force and the percentage of women in each computer-related occupation is used as a measure of occupational segregation, the level of segregation was approximately the same in 1970 and 1980, except for computer operators.
3. Among four computer-related occupations found in all industries—computer scientists/systems analysts, computer programmers, computer operators, and data-entry operators—the higher the status and pay of the occupation, the more white men were overrepresented, compared with their representation in the labor force as a whole, and the more minority men and women of all racial and ethnic groups were underrepresented. In occupations with much lower pay and status, the presence of white men dropped to much below their percentage in the labor force, the percentage of women of all racial and ethnic groups became much higher than their percentage of the labor force, and the percentage of minority men approached their labor force representation.
4. Within high-tech industries, most men were in production, professional/technical, or managerial occupations, while most women were in production or clerical occupations. Women and men were equally likely to be in production occupations. However, men were more likely to be in managerial and professional/technical occupations in high-tech industries than in non-high-tech industries; women fared worse in these occupations in high-tech industries than in other industries.

These findings dispel the myth that high tech is automatically a great equalizer. High tech may produce integrated circuits, but it does not necessarily produce an integrated work force.

REFERENCES

ABAG (Association of Bay Area Governments) 1981. Silicon Valley and beyond: high technology growth for the San Francisco Bay Area. *Working Papers on the Region's Economy,* 2. Berkeley, Calif.

AMSDEN, ALICE H. 1980. Introduction. pp. 11–38 in Alice H. Amsden, ed., *The Economics of Women and Work.* New York: St. Martin's Press.

BIELBY, WILLIAM T., and JAMES N. BARON. 1984. A woman's place is with other women: sex segregation within organizations. pp. 27–55 in *Sex Segregation in the Workplace: Trends, Explanations and Remedies.* Committee on Women's Employment and Related Social Issues. Commission on Behavioral and Social Sciences and Education, National Research Council. Washington, D.C.: National Academy Press.

BLAU, FRANCINE D. 1977. *Equal Pay in the Office.* Lexington, Mass.: D.C. Heath.

BLAU, FRANCINE D., and WALLACE E. HENDRICKS. 1979. Occupational segregation by sex: trends and prospects. *The Journal of Human Resources* 14:197–210.

BLAU, FRANCINE D., and CAROL JUSENIUS. 1976. Economists' approaches to sex segregation in the labor market: an appraisal. pp. 181–199 in Martha Blaxall and Barbara B. Reagan, eds., *Women and the Workplace: The Implications of Occupational Segregation.* Chicago: University of Chicago Press.

BRAVERMAN, HARRY. 1974. *Labor and Monopoly Capital.* New York: Monthly Review Press.

Bureau of the Census. 1971. *Census of Population:*

1970. Alphabetical Index of Industries and Occupations. U.S. Government Printing Office. Washington, D.C.: U.S. Department of Commerce.

Bureau of the Census. 1972. *Census of Population: 1970. Occupation by Industry.* Final Report PC(2)-76. U.S. Government Printing Office. Washington, D.C.: U.S. Department of Commerce.

Bureau of the Census. 1982. *Census of Population: 1980. Classifed Index of Industries and Occupations,* Final edition. U.S. Government Printing Office. Washington, D.C.: U.S. Department of Commerce.

Bureau of the Census. 1983. *Data User News* 18 (June): 8–9.

Bureau of the Census. 1984. *Census of Population and Housing: 1980. Subject Report, Occupation by Industry.* PC80-2-7c. U.S. Government Printing Office. Washington, D.C.: U.S. Department of Commerce.

CAIN, GLEN G. 1976. The challenge of segmented labor market theories to orthodox theory: a survey. *Journal of Economic Literature* 14 (December): 1215–1257.

CHATTERJEE, SAMPRIT, and BERTRAM PRICE. 1977. *Regression by Example.* New York: Wiley and Sons,

DEBOER, GEORGE E. 1984. *Factors Affecting the Science Participation and Performance of Women in High School and College.* Hamilton, N.Y.: Colgate University, Department of Education.

DICESARE, CONSTANCE BOGH. 1975. Changes in the occupational structure of U.S. jobs. *Monthly Labor Review* 98(March):24–34.

GREENBAUM, JOAN M. 1979. *In the Name of Efficiency: Management Theory and Shopfloor Practice in Data Processing Work.* Philadelphia: Temple University Press.

GROSS, EDWARD. 1968. Plus¸ca change . . .? The sexual structure of occupations over time. *Social Problems* 16:198–208.

GROSSMAN, RACHAEL. 1980. Women's place in the integrated circuit. *Radical America* 14(January/February):29–49.

HACKER, SALLY L. 1981. The culture of engineering: woman, workplace, and machine. *Women's Studies International Quarterly* 4(3):341–353.

KRAFT, PHILIP. 1977. *Programmers and Managers.* New York: Springer-Verlag.

KRAFT, PHILIP. 1979. The industrialization of computer programming: from programming to software production. pp. 1–17 in Andrew Zimbalist, ed., *Case Studies in the Labor Process.* New York: Monthly Review Press.

KRAFT, PHILIP, and STEVEN DUBNOFF. 1983. Software workers survey. *Computer World* XVII (November 14):3–13.

LLOYD, CYNTHIA B., and BETH T. NEIMI. 1979. *The Economics of Sex Differentials.* New York: Columbia University Press.

O'NEILL, JUNE. 1983. The Determinants and Wage Effects of Occupational Segregation. Project Reports. Washington, D.C.: The Urban Institute.

RESKIN, BARBARA F., ed. 1984. *Sex Segregation in the Workplace: Trends, Explanations, Remedies.* Committee on Women's Employment and Related Social Issues. Commission on Behavioral and Social Sciences and Education, National Research Council. Washington, D.C.: National Academy Press.

RESKIN, BARBARA F., and HEIDI I. HARTMANN, eds. 1986. *Women's Work, Men's Work: Sex Segregation on the Job.* Committee on Women's Employment and Related Social Issues. Commission on Behavioral and Social Sciences and Education, National Research Council. Washington, D.C.: National Academy Press.

RICHE, RICHARD W., DANIEL E. HECKERS, and JOHN U. BURGAN. 1983. High technology today and tomorrow: a small slice of the employment pie. *Monthly Labor Review* 106(November):50–59.

ROGERS, EVERETT M., and JUDITH K. LARSEN. 1984. *Silicon Valley Fever.* New York: Basic Books.

RYTINA, NANCY F. 1982. Earnings of men and women: a look at specific occupations. *Monthly Labor Review* 105(April):25–31.

SOKOLOFF, NATALIE J. 1980. *Between Money and Love: The Dialectics of Women's Home and Market Work.* New York: Praeger.

STROBER, MYRA H. 1984. Toward a general theory of occupation sex segregation: the case of public school teaching. pp. 144–156 in Barbara F. Reskin, ed., *Sex Segregation in the Workplace: Trends, Explorations, Remedies.* Washington, D.C.: National Academy Press.

STROBER, MYRA H., and BARBARA B. REAGAN. 1982. Sense and nonsense in measuring employment discrimination in higher edcuation. Unpublished paper.

TREIMAN, DONALD J., and HEIDI I. HARTMANN, eds. 1981. *Women, Work, and Wages: Equal Pay for Jobs of Equal Value.* Washington, D.C.: National Academy Press.

U.S. Department of Commerce, Office of Federal Statistical Policy and Standards. 1980. *Standard Occupational Classification Manual.*

NEW INTERSECTIONS OF CLASS, RACE, AND GENDER

22 *On the Margins of Development*

SOUTHEAST WOMEN'S EMPLOYMENT COALITION

In 1982 the board of the Southeast Women's Employment Coalition (SWEC) decided to undertake a research project on the economic status of rural women in the South. The decision did not come easily. As an activist organization dedicated to race and sex equity, the coalition at first perceived research as peripheral to its purpose. Academics do research, not activists.

Yet the board recognized the vacuum of basic information about women in the South, especially rural women, and the existence of damaging stereotypes that flourish with such ignorance. Southern rural women have been visualized as petulant "Scarletts," loyal "Mammies," and dim-witted "Daisy Maes," but they have only recently begun to emerge in either history or literature as complete human beings—diverse, complex, historically active.

The SWEC report, excerpted here, contributes to more realistic depictions by examining the economic activities, circumstances, and problems of rural women in three Southern states—North Carolina, South Carolina, and West Virginia. Approximately 700 women—as interviewers, respondents to interviews, staff coordinators, or in some other capacity—contributed their experiences and efforts to Women of the Rural South.

Source: Southeast Women's Employment Coalition, "On The Margins of Development," *Dollars and Sense.* No. 117 (June, 1986) pp. 9–18; excerpted from Southeast Women's Employment Coalition, *Women of the Rural South: Economic Status and Prospects* (Lexington, KY: Southeast Women's Employment Coalition, 1986).

SOUTHERN HISTORY

The abolition of slavery ended property rights in human beings, and renewed a central economic concern of planters: how to secure an adequate reliable labor force, a dilemma soon shared by Southern industrialists. The make-up of various work forces soon reflected different but complementary patterns of discrimination whereby specific kinds of jobs were allocated or denied based on gender and race. In many industries, these patterns remain visible today.

In the rugged mountains of southern West Virginia, for example, the work force for the booming railroads and coal mines of the late 1800s was relatively diverse racially and ethnically, but completely uniform in gender. Despite racism, language, and cultural differences, and the violently anti-union response of employers, male Southern coal miners from Alabama to West Virginia were able to band together, unionize, and eventually secure a decent wage.

In West Virginia today, where the bituminous coal industry still presides over the economy, employment opportunities for women—even low-wage "women's" jobs—have always been extraordinarily scarce. Women in West Virginia have by far the lowest labor force participation rate of any state in the United States.—39% in 1984.

In the ex-slave states of North and South Carolina, large-scale and labor-intensive agriculture left a relatively dense rural population in many areas. The problem here was not so much importing a work force as creating a reliable and disciplined one out of ex-slaves and poor white farmers. After brief trial and error with other systems (including wage labor), sharecropping evolved as the principal means of organizing an agricultural labor force for the large planters.

Thus many blacks in the rural South although emancipated from slavery, were caught in a wageless, coercive labor system well into the twentieth century. For black women, occupational segregation was especially severe. For nearly 100 years after the Civil War, the primary source of employment for black women was the lowest wage job in the United States: domestic service. As recently as 1960, nearly half (45%) of employed black women in the South were domestic workers.

The industries that formed the mainspring of New South growth in the late nineteenth century offer a third set of patterns. These industries, principally textiles, moved to the Southern piedmont areas of North and South Carolina, and Georgia partly for water power, but even more for the cheap, abundant labor available from the poor, white rural population. The resulting work force in Southern textiles was overwhelmingly white, young, and female. Well over half, and in some mills as high as 80%, of the workers were white women and their children. Only during the 1960s, when faced with a labor shortage and the political unrest of the civil rights movement, did textile employers finally lower their color bar.

What stands out about the South is the centrality of race in virtually all areas of economic, social, and political life. Without minimizing the extent of racism in other regions or against other groups, the thoroughness of racial proscriptions against black Americans in the South is unique in the United States.

Contemporary economic scenarios do not bode well for women in the rural South. Textiles and many other light manufacturing industries, the traditional mainstay of their employment, are either automating or moving overseas. Machinery production and other emerging industries, which tend to employ predominantly men, are by no means absorbing women displaced from the mills.

The public sector, a growing source of jobs for women since World War II, is now stagnating. Even the service industries, far and away the leaders in employment growth both regionally and nationally, favor metropolitan centers and offer little opportunity to rural women. Tourism, which generates a particularly low-wage and destructive form of growth, seems to be the major focus of investment and source of new jobs for women in many rural locations.

Thus, far from basking in the benefits of a vibrant Sunbelt, many women in the rural South face seriously declining economic opportunities. Indeed, over the last ten years, many rural women have not found new prosperity, but have fallen into poverty.

DEVELOPMENT FOR WHOM?

Many view North Carolina as a leader in successful economic development, a model for other states to emulate. Over the last 15 years, North Carolina's growth in employment and new industrial investments has outstripped most other states in the country, including many in the Sunbelt. During 1983, the state ranked first in the United States in the number of large, new manufacturing plants and first in the number of large plants financed by foreign capital.

North Carolina offers two persuasive attractions to capital. The first element—a low-wage, non-union labor force—has attracted manufacturers for over 100 years, originally and most importantly in the textile industry. Such policies have paid off, at least for business: North Carolina boasts the second lowest (next to Mississippi) average manufacturing wage in the U.S. ($6.97/hour in 1984) and the second lowest (next to South Carolina) unionization rate of its workers. During the six year period 1978–1983, labor unions in the state lost 65% of their recognition elections and were decertified in 28 other cases.

The second attraction—the aggressive use of North Carolina's higher education institutions and highly educated work force—is a more recent phenomenon. Thirty years ago, under the "businessman's" governor Luther Hodges, the state established the Research Triangle Park. The success of the

"Triangle" in high technology research and development spurred the state recently to invest in a new Microelectronics Center, "a $40 million plus facility for state of the art research and development in semiconductor technology," and a Biotechnology Center for research and development in genetic engineering.

This "favorable business climate" attracts new investments in manufacturing, particularly high technology manufacturing. There are currently 275 plants classified as high technology manufacturers in North Carolina, employing 67,000 workers. During 1984 alone, various corporations announced 25 additional high technology plants to be located in the state. Because of such investments, manufacturing employment in North Carolina is holding steady, running contrary to the dramatic national decline.

Despite impressive new investments and climbing employment levels, however, North Carolina's growth has not been dispersed evenly. A significant portion of recent investment generates no jobs within North Carolina at all; in fact, it is eliminating jobs and displacing workers. This is especially the case with the enormous capital outlays in the textile industry. During 1983, over one-fourth of investment in expansion of existing industries was in textiles, and was largely for automation of the work process. Larger textile companies like Burlington Industries, which can afford expensive automated equipment, replace workers with machinery in order to raise productivity and keep up with worldwide competition. Less productive plants close down; less well-capitalized companies go bankrupt.

As a result, recent economic growth in North Carolina is characterized by a seemingly paradoxical combination of high investment levels and a large number of plant closings and layoffs. North Carolina ranked first in the nation in the number of plant closings in 1982, with much of the job loss due to the "shakeout" in the textile industry and the relocation of apparel plants overseas. Between 1980 and 1982, more than 23,000 jobs were eliminated statewide in these industries.

For more isolated rural communities, long dependent on textiles and apparel as their economic foundation, these trends are disastrous. Rural women have been hardest hit. In 1980 textiles and apparel together accounted for nearly 20% of all female employment in North Carolina, the majority of it in rural areas. Women make up two-thirds of the production operatives in these industries, which are the jobs most affected by automation and relocation.

For black women, who are 22% of textile operatives and who have fewer occupational alternatives than any other group, the implications are particularly grim. In 1983, supposedly a banner year in North Carolina's economy, the unemployment rate for black women was 17.5%.

Official projections suggest employment growth in several new or expanding manufacturing industries, but these opportunities are often not in the geographic areas where old jobs are disappearing. As one analyst concluded, "Recent manufacturing losses are clearly concentrated in small towns and nonmetropolitan counties of the mountain and coastal regions, while manufacturing job gains occur predominantly in metropolitan counties of the piedmont."

The one industry that is growing in certain more remote rural counties is tourism, along with the related development of retirement or second homes. For those with land to sell or with capital to begin a new business, these trends bring an economic bonanza. For those applying for the service jobs generated by tourism, these trends often bring disappointment. Women who are laid off from the mills rarely find a service job that matches even the low wages they received in textiles. Service sector employment is frequently seasonal, part-time, and, above all, low-wage. In 1983 in North Carolina the average weekly wage in services was $128.57.

North Carolina highlights many of the contradictions and problems with current economic trends and development policies. Clearly, growth in and of itself has not automatically brought prosperity or an improved quality of life to all citizens. For some, economic growth has brought unemployment, even impoverishment. For others, it has brought destruction of community and former ways of life.

NEW CHALLENGES FOR WOMEN

The strength of North Carolina's economy offers opportunities for women, but it will require strong and collective advocacy by women themselves to fulfill this potential. North Carolina is undoubtedly a state where nontraditional job strategies still have great relevance, given the continued importance of manufacturing and the projected growth in certain predominantly male industries.

Women make up less than 30% of the work force in growth sectors like machinery, transportation equipment, and fabricated metals. Skilled craft occupations carry some of the highest wages for nonprofessional work in the state, and are also among the most difficult jobs for employers to fill. At present, women claim only 10% of the 360,000 precision production, craft, and repair jobs in North Carolina.

The inevitable growth of service sector employment, the fastest growing employment sector in the state, presents a much more difficult strategic dilemma for women. The varied and dispersed industries that make up the service sector are exceedingly problematic to organizers, especially in anti-union North Carolina. One of few potential points of intervention may be the public sector, where women have legal and legislative levers that are not available in private industry.

For rural women in counties far from the piedmont, tucked away from cities, not even near an Interstate corridor, the problem of creating economic opportunity is greatest. Some North Carolina communities in these circumstances have turned to collaborative development projects such as worker-owned cooperatives. The New Bern Bakery, Worker Owned Sewing Company, and United Zipper Company are all examples of this form of community-based economic development. However, most cooperatives remain marginal in at least two respects: they do not significantly raise the standard of living of their low-income workers; and they do not resolve a fundamental economic problem of rural areas—lack of investment capital.

These structural problems can only be attacked effectively on a larger (e.g. statewide) level, but at present there is little organized pressure to do so. Current state policies are blatantly premised on the poverty (i.e. low wages) and powerlessness (lack of unions and strikes) of North Carolina's workers. Unfortunately, such bald neglect of the needs and interests of working-class residents is also an indication of their lack of political voice. If investment is to be encouraged in areas that lack it most, and if development is to serve the needs of ordinary people in North Carolina—including rural women—then this political powerlessness must be turned around.

* political power by workers

NEW INTERSECTIONS OF CLASS, RACE, AND GENDER

23 *The Impact of "Sun Belt Industrialization" on Chicanas*

PATRICIA ZAVELLA

Historically, Mexican American women have had lower labor force participation rates than Anglo or black women. In 1960, only 24.4 percent of Chicanas were employed as compared to 34.5 percent of Anglo women. Explanations of these disparities in labor force participation rates have relied on a model of cultural determinism that identified traditional cultural values and norms as the causal factors. Many authors assume that traditional family values alone determine whether Chicanas will enter the labor market, and that Chicanas violate those values when they work outside the home. I call this the "familism" or "Hispanic traditionalist" explanation. According to this model, Chicano "machos" prefer that their wives not work, and Chicanas, who value their homemaker roles over taking jobs, thus submit to their husbands' wishes. A corollary notion is that acculturation will occur when Chicanas are gainfully employed and that, consequently, they will lose traditional family values.

Over the last two decades, however, Chicanas have rapidly entered the labor force. Between 1960 and 1970, for example, married Chicanas entered the labor force nationally at a rate 15 percent higher than their white and black counterparts. In 1970, the gap in labor force participation rates had narrowed; 43 percent of all women worked as compared to 36 percent of Chicanas. By 1980, the gap between Chicanas and other women workers had nearly disap-

Source: Patricia Zavella, "The Impact of 'Sun Belt Industrialization' on Chicanas," *Frontiers* Vol. 8 (No. 1) (1984) pp. 21–27.

peared: 52 percent of all women worked, while 49 percent of Spanish-origin women were employed.

At the same time, Chicanas consistently have had higher unemployment rates that Anglo women. In 1970, women of Mexican origin had an unemployment rate of 14 percent compared to 9 percent for all women. By 1980, their percentage of unemployed declined to 10, while that of Anglo women was 8 percent. Technically, the unemployed are persons who are actively seeking jobs and are available for work. The official unemployment rate, of course, does not include the many "discouraged workers"—those who have completely given up the search for work. Groups with high unemployment rates usually also have higher numbers of discouraged workers. The historically high Chicana unemployment rates, whether official or not, indicate that more Chicanas would work if they could find jobs. In addition, since Chicanas tend to have larger families than Anglo women, the availability of child care facilities may have a greater influence on their ability to remain in the work force.

The traditionalist approach neither predicted nor explained these trends. Instead of challenging the traditionalist perspective, adherents of the model merely modified it and claimed that *declining* family values explain Chicanas' recent entrance into the labor force. Most cultural determinants do not actually examine Chicanas' values and beliefs. A recent empirical study, however, investigated the effect of traditional values on Chicana and Anglo women's labor force participation. Vilma Ortiz and her colleagues found that the cultural argument was not supported; "traditional sex-role attitudes did not have a stronger impact on the labor force behavior of Hispanic females." In addition, these researchers found that the significant factor influencing Chicanas' commitment to the labor force was women's prior work experience. Women with unskilled, low-paying jobs tend to prefer homemaking, while those who have more stable jobs want to continue working. Ortiz argues that the availability of jobs, not traditional culture, determines whether or not Chicanas are committed to the labor force.

Furthermore, the traditionalist approach is ahistorical and does not consider varying or changing regional economies. Yet regional economies are the products of distinct patterns of accumulation and industrial restructuring. The "spatial division of labor"—the regional variation in industrial location—is the product of managerial strategies for minimizing production costs. Managers engage in "global scanning," whereby they seek industrial sites whose lower labor costs and "locational capability" allow for new ways of organizing production and workers. Uneven regional development provides varied work opportunities for women workers. For example, two studies show that even though Chicanas are generally segregated into the lowest paying jobs, there is regional variation in the types of jobs they hold, the labor force participation rates, and discrimination. The traditionalist argument also assumes a mechanistic relationship between values and behavior by viewing the lack of employment as voluntary. In this regard, analyses of unemployment rates of Chicanas also help to further illuminate the model's deficiencies.

Rather than assuming that culture determines whether Chicanas enter the labor force, a more fruitful approach is to examine changes in local labor markets, structural defects in the regional economies, and the varying impact of women's work on their families. Such an approach provides a corrective to the traditionalist perspective that has perpetuated the stereotype of the reclusive, passive Mexican woman who devotes herself to motherhood, housework, and her man. Instead, I believe that Chicanas have always placed a high value on work outside the home.

This paper analyzes data from a research project on women's work and family strategies in the context of "sun belt industrialization" in Albuquerque, New Mexico, for which in-depth interviews with twenty-two married Chicana workers were conducted. The chief criterion for selecting informants was employment in either electronics or apparel factories; the focus was on women with at least one child under the age of six so as to understand working women's child care arrangements. These women were relatively young, with a mean age of twenty-nine, and, on the average, they had two children. Most of the women were entry-level electronics assemblers or sewing operators who had worked at their jobs for about three years.

I will discuss two related processes regarding the relationship between work and family life. Recent industrialization in Albuquerque has created jobs primarily for women. The contours of the local labor market, including the availability of jobs women consider to be good ones and the instability of men's jobs, are the primary reason Chicanas choose to enter and to remain in the labor force. Concurrently, women's work has dramatically altered Chicano families. The available evidence does not indicate a decline in familism, but suggests instead a more complicated adaptation. The impact of women's work on Chicano families varies depending on family circumstances, such as the ages of children in the household, the availability of child care, and the relative need for the woman's wages at home. Furthermore, traditional values and norms should be viewed as family ideology—a set of values and beliefs that prescribe proper behavior for women and men and that are symbolically opposed to women working for wages outside the home. Women's views about being working mothers respond to the prevalent family ideology. All these related individual and familial processes must be viewed in the larger context of regional industrialization.

The state of New Mexico has been slow to industrialize. As late as 1970, manufacturing jobs accounted for only about 6 percent of total employment. The service sector (including trade, finance, and public administration) provided over 60 percent of New Mexico's jobs. Between 1960 and 1970, however, important changes occurred that are turning New Mexico into an industrial center in the Southwest. The number of industrial jobs in the state has grown significantly because of the expansion of two industries: apparel and electrical machinery. The dramatic development in both industries is reflected in the increase in employment in electrical machinery manufacturing, which grew by 353 percent, and in the apparel industry which quadrupled its employment.

Most of the recent industrial growth is located in Albuquerque, the state's largest city, and surrounding Bernalillo County. Prior to industrialization, Albuquerque's economy was based on "guns and butter"—the military and related enterprises—and the public sector. Besides the State of New Mexico, other large employers included Sandia Laboratories, Kirtland Air Force Base, the Veterans Hospital, and the Air Force Special Weapons Center. As the commercial center of the state, Albuquerque also provided a large number of sales and service jobs.

Albuquerque attracted manufacturers for several reasons. Among them was a low wage scale, especially for women's jobs. The median wages for female manufacturing jobs during the 1970's were 44 percent of the wages of male jobs. In electrical machinery, for example, women earned about 44 percent of the wages of men, while in the apparel industry women earned almost 70 percent of men's wages. Taking inflation into consideration, average earnings in Albuquerque were almost 10 percent lower than national levels. That few manufacturing plants are unionized partially accounted for lower wage rates. An additional attraction to manufacturers was the available labor force in Albuquerque, especially women. In Bernalillo County, the unemployment rate of Anglo women was only 5 percent in 1960, but this figure grew to 6 percent in 1970 and to over 7 percent in 1979. During the same years, Spanish-surnamed women had unemployment rates of 5 percent, 8 percent, and 9 percent respectively.

The recent recessions notwithstanding, Albuquerque has become a center of "sun belt industrialization." Beginning in the late 1960's, several companies—GTE Lenkurt, Singer Fridan, Levi Strauss, and Ampex—opened plants in or near the city. Subsequently, other firms such as Motorola, Digital, Honeywell, General Electric, Signetics, Sperry Flight Systems, and Ethicon built plants in Bernalillo County. Between 1969 and 1979, employment in manufacturing in Albuquerque grew over 100 percent, while total employment increased only 60 percent. By 1980, manufacturing was the third largest sector of Bernalillo County's economy, employing over 218,000 workers. Industrial growth continues. Intel Inc., for example, planned to employ about 1,000 production workers by the end of 1984.

Changes in industrial employment affected men and women differently. The bulk of new jobs were "women's jobs." In electrical machinery, the female labor force grew by 680 percent. By 1970, women constituted 46 percent of the total electronics labor force. In apparel—a traditionally female industry—the number of women employed increased fourfold, while male employment increased only 300 percent. By 1970, women formed 82 percent of the apparel labor force. These increases in women's employment occurred during a period when total manufacturing in the state dropped by 4 percent. Male jobs, however, accounted for a good portion of the decrease, declining by 16 percent, while female employment increased by 58 percent. Women grew from 16 percent of the total manufacturing labor force in 1960 to almost 27 percent in 1970. In sum, although manufacturing declined in the state as a whole, the electronics and apparel industries—which employed increasingly greater proportions of women—grew tremendously.

Although the growth of industrial jobs in Albuquerque benefited all women, Chicanas were affected the most. Between 1960 and 1970, the labor force participation rate of Anglo women increased 18 percent while that of Chicanas rose by 66 percent. The labor force participation rate of Anglo women was 43 percent in 1970 and 40 percent for Spanish-surnamed women. By 1980, 52 percent of all white women and 50 percent of Hispanic women were in the Albuquerque labor force. Clearly, a significant proportion of Chicanas entered Albuquerque's labor market once industrial expansion began.

The twenty-two women interviewed for this study first entered the labor force during this time of rapid growth in the labor market; the interviews, however, took place in 1982–83 during a period of economic recession. How has industrial employment affected these women? Have they become less family oriented or traditional in their outlook?

Our informants entered the labor force, as women everywhere increasingly are doing, when their children were under the age of six. Finding satisfactory child care placed great pressure on them. In addition, these women struggled with both the demands of caring for young children after work and the ideology that mothers—especially those with young children—should not work. According to the traditionalist viewpoint, we would expect these women to leave the labor force as soon as they could, but we found instead a more complicated situation.

When the women originally decided to seek wage work, most of them were in difficult financial situations. A few were already working when they married, but most of them started work after marriage. Many entered the labor force when their husbands were either unemployed or making low wages. Thus, the women's primary motivation for seeking work was economic. There was a common refrain: "The way it is now, everything's too high and we can't make it on one check." In the majority of couples, both husband and wife agreed that the woman should work, even though eight of these husbands preferred that their wives stay at home. In these cases, women's wage work is an economic strategy for families that have little choice.

Five of the women wanted a job primarily to get out of the house; generally their spouses supported their decision to get a job. The women were critical of the boredom of housewifery. "I decided it [homemaking] wasn't for me," one woman commented. Sometimes they wanted personal satisfaction. One woman who was very independent said simply, "I'm a self-supporter."

An examination of our informants' work histories and family circumstances sheds additional light on the question of declining familism. Most of them had moved through a succession of low-paying, dead-end jobs—as waitresses, fast food workers, or cashiers—before they acquired their current jobs. The apparel workers usually knew someone, often a relative, already working in the plant who notified them that the company was hiring; some of these women had worked only at the one factory where they were employed. The electronics workers, on the other hand, had longer work histories and therefore a wider base of comparison. They tended to have had prior work experience in electronics production (as well as in apparel factories) and often had some technical training. Electronics workers usually applied for their present jobs without prior knowledge that the company was hiring and without the help of someone who was already employed there. These women heard, either through their training or thorough social networks, that electronics factories pay better and provide good benefits.

Compared to their previous jobs, the new positions provided far better pay, benefits, and job security. The benefits in particular, especially medical insurance and paid maternity leaves, were great improvements over their previous fringe benefit packages. Women's job benefits often were better than their husbands', which meant that couples relied on the women's medical insurance, for example.

Many women believed that, compared to other jobs in the Albuquerque area, these factories were relatively good places to work. Several women compared their present positions to previous ones or to the overall Albuquerque labor market. They were aware that, in general, "Albuquerque's [jobs are] always underpaid," as one informant said. They characterized the wages in electronics as "the best you can find in Albuquerque," and recounted news stories of job seekers numbering in the hundreds waiting hours for the privilege of submitting job applications to new firms. In light of these stories or the experiences of their relatives who were chronically unemployed, one woman summed up the general view: "There's no place else. It's hard to get a job and I feel lucky to have this one."

It is important to emphasize the financial contribution that these women made to their families; their wages were not supplemental. Electronics workers earned, on the average, $5.25 an hour, and generally they were the best paid blue-collar women in Albuquerque. There was considerable variation in wage rates for apparel workers, since they were paid on the piece rate system. Depending on how fast they could sew, apparel workers earned between $4 and $7 an hour. On the average, electronics workers earned about $12,000 a year; apparel workers earned about $9,500 annually. One-third of our informants earned more than their husbands. These women were the economic mainstays of their households, since they provided over 60 percent of the total family income. The other women, who earned between 30 and 59 percent of the total family income, made a significant contribution to their families' support.

But more than the amount that they contributed, these women's wages were pivotal in terms of their families' economic stability. When both husband and wife were working, the family was relatively secure. Unfortunately, many of the husbands—particularly those who worked in construction—were periodically moving between the stability of two incomes and the hardship of one. Therefore, the woman's income saved the family from financial disaster when her husband was laid off, and assured the family of relative stability when he was employed.

The family incomes of the informants were slightly lower than those of other dual-worker families in Albuquerque. Combined with their husbands' wages and other income, the informants' gross annual family incomes were $24,425. (Gross family incomes ranged from $13,500 to $40,000 a year.) The mean income of all families with two or more workers in Albuquerque was $26,846 in 1979. Compared to households in the rest of the state, however, our informants were relatively well off. The median family income for all residents of the state of New Mexico was $16,930 in 1979. New Mexico Hispanics had a median family income of only $13,800 during this time, a significant disparity from other families.

It is important to recall that these were relatively young families with great economic needs. The men had not yet reached their peak job earnings, and many of them received low wages that were eroded by inflation. Several of the women had recently taken maternity leaves which considerably lowered the families' incomes. Furthermore, since most of these couples were buying or constructing their houses, they had major payments to make. When the husbands went back to work after periods of unemployment, the couples had to catch up on previously unmet needs. Since they had to cope with the fluctuations and loss of real wages by husbands, the women's wages were essential for family survival. Rather than diminished familism, changes in the economy pushed these women into the labor force. These couples were part of a national trend, in which the proportion of families with two or more wage earners is increasing. Most of our informants resigned themselves to the fact that, as one woman said, "Nowadays, both people have to work."

That these women worked primarily for financial reasons could be seen as a reflection of Chicana familism. To a certain extent, support of their families is one way to express traditional values. But these women's economic contribution to the families was but one aspect of their commitment to the labor force. Another important factor was the satisfaction they received from their jobs.

In order to gauge job satisfaction our informants were asked a number of questions: what they liked and disliked about their jobs, what the advantages and disadvantages were; they were also asked to rank ten qualities of the job. Generally, the women believed that job security was the most important aspect, and most thought that they had secure jobs. The second most important feature was pay, and all of the women believed that they had good pay. In addition, the majority of women considered their benefits good, as illustrated by one woman who explained, "That's the main reason I went back to electronics." The women were generally satisfied with the economic features of their jobs. Beyond issues of remuneration, though, there were important differences between electronics and apparel workers in their views of their jobs, stemming, of course, from the different production processes. Electronics workers, particularly those who did not work as assemblers, had higher job satisfaction than apparel workers.

In the work process, two-thirds of the women believed that they could work at their own pace. Unfortunately, because of production quotas—which the women generally felt were a disadvantage of their jobs—self-pacing meant that they pressured themselves to work fast. Assemblers lamented the physical discomforts (such as eyestrain, backaches, and headaches) that are a consequence of the fast pace as well as the monotony. Apparel workers who sewed fast and who bettered their production quota had more job satisfaction. These women believed that the piece rate system allowed them to make good money. The women who barely made their quota, or who performed at less than the desired company speed, disliked the pressure of the piece rate system. Those not meeting production quotas were either pressured to quit or, after a series of warnings, were fired. It is not surprising that the women found their work challenging.

Virtually all of the women believed that they had a safe work place. Furthermore, all of them stated that they had good supervisors and ample opportunities to converse with coworkers.

But there were also several disadvantages to the job. Apparel workers had shortened work weeks during the 1982 recession, something most of them disliked. They preferred a forty-hour week so they could make steady wages. In addition, most of the women believed that they were not paid well enough for the work performed, and some of them would have liked more appreciation for a job well done. Only half of the women thought they had opportunities for promotions.

Despite these reservations, however, the majority of the women liked their jobs, but for varied reasons. An assembler said, "I enjoy it. It's easy." A machine operator reported, "I like the work; I think it's interesting." A materials handler liked the autonomy of her job because, in her words, "No one bothers me." An inspector cherished the opportunity to associate with friends and coworkers. "I like going to work," she said. "I get to talk to different people. It makes me forget half of my problems." Another assembler made the analogy between her department and home, saying, "It's like a family." A quality control worker could find nothing wrong with her job, claiming, "I like everything about that job." These women's comments were not unusual: on the whole, most of the women valued their jobs.

Our informants' views on their status as working mothers reflected the contradictions between economic circumstances, on the one hand, and family obligations and ideology on the other. Traditional family ideology asserts that women should stay home and care for their families, especially the young children. Finding the balance between these "competing urgencies" is difficult and frustrating. In an attempt to understand their values and attitudes, we asked the women to indicate whether they agreed or disagreed with several statements about gender roles and women's work. Their responses illustrate both how they construct the meaning of family and how their circumstances temper their beliefs and values. Probably the most striking aspect of their opinions is their variation, ranging from traditional to egalitarian viewpoints, some of which were ambivalent and even inconsistent.

Given the state of Albuquerque's economy, with inflation and relative high male unemployment, it is not surprising that virtually all the women agreed with the following statement: "Women need to work to help their families with the high cost of living." Their responses to other economic issues, however, were less clear. Most of the women disagreed with the notion that it is better for a marriage if a husband earns more than his wife, and of the six who agreed, two earned more than their husbands, while the rest earned less. In addition, the majority disagreed with the more general view that men should get most of the higher paying jobs because they have families to support. According to several women, "Women have to support families, too."

Child care arrangements profoundly affected the women's views on being working mothers. Two-thirds of the women had their husbands or female kin care for their children. Others took their children to either a private baby sitter or a day care center. Many had frequently changed child care providers and had had bad experiences with former baby sitters. At the time we interviewed them, though, most of the women were generally satisfied with their present child care arrangement.

Despite this satisfaction, some of the women felt both sorrow and guilt about leaving their young children in child care. The statement that elicited the most responses was: "Working mothers miss the best years of their children's lives"; most of them agreed. They believed that by going to work while their children were young, they had missed those important "first steps" in the child's development. Several of the women had delayed entering the labor force not only because they did not want to leave infants with a sitter, but also because they wanted to enjoy their children's development. For example, one informant had decided to wait until her daughter was one year old before going back to work, but circumstances forced her into the labor force for three months when her daughter was six months old. Another was defensive on this issue. When her sitter informed her that her son had said the word *cookie,* she replied, "Yeah, I know; I taught him how." She went on, "I felt like I heard it first."

Working while having young children placed certain demands on the women's time. They had to miss work to take their children for medical care, both for periodic checkups and for the inevitable illnesses. Many women indicated that the hardest aspects of a working mother's life are the lack of time to accomplish everything that needs to be done at home and their fatigue from working and parenting the very active youngsters. It is understandable, then, that the majority of women agreed with the statement: "It would be much better all around if a woman can stay home and take care of her family instead of having to work."

We found further variation in the household division of labor. Five of the women had a relatively egalitarian arrangement. Couples in these households either divided up the work (she washed clothes and he took out the garbage, for example), or they both performed the same chore. Several of these couples negotiated a more equitable division of labor after the women complained that they were doing too much. A second group of nine women performed most of the housework. The arrangements of the remaining couples varied between these extremes, with the women usually assuming a greater share of the housework. Three-quarters of the couples, however, shared child care equitably. The differences between the divisions of housework and child care were reflected in the women's agreement with several statements. Every woman agreed that "Taking care of the children should be shared equally between husband and wife." Furthermore, most of the women agreed that "Sometimes I think that I cannot do enough for my family when I work," and "Sometimes I feel it is unfair that I have to work and also spend so much time taking care of my home and my children."

Finally, the women were asked if they ever thought of quitting their present jobs. Most of them had considered it, but for varied reasons. The frustrations or boredom of the work itself, the pressures of the piece rate or of managing home duties and a job, and the desire for higher wages (especially by apparel workers) were the reasons usually given. Yet despite the majority's agreement with the statement that "it would be much better," only a few of the women wanted to stay home and rear their children. The others preferred better opportunities, specifically a job that paid more and allowed some advancement. Most of the electronics workers who wanted better jobs preferred to remain in the electronics industry.

One woman's statement illustrated how most of the women vacillated between their desire to support their families and still care for their young children at home. When asked whether she ever thought of quitting, she stated, "I always wanted to work; it's a a habit I have. I can't be without work, I can't stay home; it's so boring. I have to work or else I won't be happy." Later, however, when asked whether she would quit if she did not have to work, she contradicted herself: "It's kind of hard. I do like to work, but I think my family is more important." She realized she was contradicting herself, so she joked, "Maybe I like to work because I need the money."

Whether or not they held traditional views regarding family, the economic needs of their families often took precedence. Each family had to find its own balance in meeting these competing needs, and the women constructed meanings that took each need into consideration. Thus, despite their misgivings about not being able to spend time with their children, the majority agreed that "Working is an important part of my life that would be hard for me to give up." Furthermore, the majority of women claimed, "Even if I didn't need the money, I would continue to work." Finally, virtually all of the women believed that their work did not conflict with family responsibilities.

The changing local economy—including job opportunities for women—and its multiple effects on women and, in turn, on the family economy must be analyzed before traditional norms can be interpreted as labor force determinants. Contrary to the traditionalist viewpoint, the Chicanas interviewed for this study indicated a strong commitment to their jobs. Their employment had a significant effect on the family economy, but its impact on family ideology was more complex. Once these Chicanas entered the labor market, some of their views regarding family changed, while others remained the same. The extent to which they accepted the prevalent ideology or constructed a new meaning depended on their particular work histories and job satisfaction, their contribution to the family's economic status, and the demands at home. In other words, rather than continuing an ideological separation between work and family, based on the assumption that family values determine women's labor force participation, we must investigate the reciprocal effects of work and the family on women employees. Clearly, a combination of factors determines whether or not Chicanas prefer to continue wage work, and these factors are similar to those that face other American working-class women.

PART IV
Institutional Changes in Family, Education, and Politics

The diffuse sense of "crisis" that Americans express about our institutions is related to the fact that many established social arrangements are inadequate solutions to the unique problems facing a society in the throes of change.

Macro level changes in the American economy are causing changes in virtually every arena of society. The structural rearrangement of global, national, and local economies is causing a vast rearranging of social institutions. Although the institutions of society—family, education, religion, polity, and economy—are interrelated, the economy is the most basic institution. The way that society is organized to produce and distribute goods and sources is the crucial determinant of the way other institutions are organized (Eitzen, 1983: 403).

While there are reciprocal effects among the institutions, the increasing interconnectedness of work and other important areas of society is requiring new solutions to the ongoing business of social life. The readings in this section on the family, education, and politics show how strongly each institution is affected by economic changes in the contemporary United States.

A central theme of this section is that economic restructuring is causing a great institutional turbulence. And while American institutions show many signs of malfunctioning, there are also indications of positive change. People often forge new and liberating social strategies as they are forced to adjust and cope with institutional pressures. Furthermore, growing forms of resistance to capitalist development and domination can offer alternatives to patterned inequalities within institutions.

FAMILY

The deindustrialization of America has profoundly affected families. As capital is shifted outside the United States and as old industries die or are relocated and production methods become automated, large numbers of wage earners are displaced. Besides the immediate loss of employment, the human consequences of deindustrialization include the loss of family income, decline in occupational status, loss of health insurance and pension rights, deterioration in physical and mental health, lost tax revenue, and shrinking social services (Hill and Negry, 1985: 5).

New patterns of downward mobility are caused by the elimination of relatively high-wage jobs and the predominance of low-wage jobs. Throughout most of American history, children expected to do better than their parents did, and many succeeded. Upward mobility in every generation was a higher probability than downward mobility, but at some point in the 1970s the odds shifted. The article by Levy and Michel in Chapter 30 examines the multiple causes of downward mobility. Demographic factors including the baby boom generation and the influx of women in the labor force have crowded the labor market and depressed wages and family income. Changes in job-holding patterns of Americans create important changes in family structure and family composition. The increase in women's labor force participation is associated with dual-earner families, high divorce rates, and female-headed households. Other social and demographic changes, while not causal, are contributing to the redistribution of families within the class hierarchy. Changes in family composition are associated with income inequality. The rise in families with two workers increases the proportion of families with high earnings, while the rise in single-parent families increases the proportion of families with low earnings. We have limited knowledge of how job loss is related to divorce and single parenthood, but it is clear that they are related and that divorce and marriage are factors in emerging patterns of family inequality.

The shifting social placement of families combined with changes in family composition and family structure have imposed new pressures and demands on home life. Women, men, and children are devising a variety of ways to cope. Delayed marriage and childbearing and sending more family members into the work force to work longer are coping strategies. More adult children come home to live. Many of these family strategies have their own costs. Still, Hartmann (Chapter 30) argues that changes can be positive for women when they contribute to women's increased autonomy from men and to women's increased economic independence.

Family life is affected by technology as well as by changes in the workplace. Electronic homework is rapidly increasing. Estimates of the numbers of Americans who will be working at home within the next decade range from five to fifty million (Walsh 1987: 122). Many women opt for homework as a way to balance a job and family. Unfortunately, the home workplace can be fraught with new forms of enslavement for women.

A more privatized home environment will undoubtedly be the result of more personal computers in American homes. How this affects the quality of family life remains to be seen.

EDUCATION

A vital function of education is to prepare individuals for the specialized roles in the occupational system. Rapid changes in the labor market have altered educational requirements. Most education has not kept pace with the shifting job market. Instead, education reacts to economic, social, and technological forces that are restructuring the workplace. New attempts are needed to upgrade schools in ways that can better match jobs and potential employees. Mass education generally fails to provide most Americans with marketable job skills, or prepare them for life in the information age.

Computer-aided learning is one solution to the crisis in American education, but it does not address basic shortcomings in the schools. Furthermore, without uniform access to the new learning technologies, "computer literacy" will only perpetuate the social inequalities of class, race, and gender.

Microcomputers have begun to proliferate on some campuses across the country and are changing teaching and learning for a few fortunate Americans. These powerful tools in the hands of students and faculty can make a difference in higher education. The computerized university campus is but one example of how electronic learning is transforming education. New technologies make it possible to get an education without setting foot inside a classroom.

POLITICS

The relationship between structural transformation and the American political system is neither well documented nor well understood. It is clear, however, that basic changes in American society are shifting the balance of power within national and local political arenas. The forces behind the current social reorganization are political as well as economic. As R. Jeffrey Lustig has astutely observed,

> Contemporary shutdowns, relocations, and dismantelings reveal not simply movements of capital and misfortunes of communities, but more importantly shifts in the relations of power between corporations on one hand and

> communities and unions on the other. . . . The basic story with plant closures, to put it differently, is a political one, and it raises serious questions for citizens even beyond the precincts of the affected community. (1985: 126)

The power of labor unions and other organizations that have traditionally influenced the exercise of political power have declined. Expanding control by the owners of American capital reveal a new political configuration. Will the new power arrangements call forth the new forms of political action? Deindustrialization and its economic consequences have become politicized as more communities and workers organize to resist shutdowns. These political uprisings are not large in number, but they serve as important lessons in learning how to struggle in these changing times. Local struggles to save plants, mills, and jobs take place alongside a national political discussion about capital flight, regional decline, corporate responsibilities, and workers' rights.

Crucially important for the political future will be the new distribution of income. Widening income inequality discussed in the last section could lead to political and social unrest. If class and race polarization inevitably lead to social tensions, the actual forms that collective action might take are difficult to anticipate.

The political future of the "new collars," those middle-level workers in the service economy with economic pressure unknown to their parents, might form an important voting block. Now labeled conservative and politically apathetic, they remain an untapped force that is ripe for political mobilization.

America is at a critical historical moment. Signs of protest, reestablished labor militancy, and growing resistance to plant closings, unemployment, and welfare cuts are gathering new momentum. Will these opportunities for mobilization transform the basic institutions of society into agencies of democracy? That depends on fundamental structural changes in the present control of capital and work.

REFERENCES

EITZEN, D. STANLEY. 1985. *In Conflict and Order.* Boston: Allyn and Bacon.

LEVY, FRANK S. and RICHARD C. MICHEL. 1986. "An Economic Bust for the Baby Boom." *Challenge* (March–April): 33–39.

LUSTIG, R. JEFFREY. 1985. "The Politics of Shutdown: Community, Property, Corporatism." *Journal of Economic Issues* 19 (March): 123–152.

HILL, RICHARD CHILD and CYNTHIA NEGRY. 1985. "Deindustrialization and Racial Minorities in the Great Lakes Region, USA." Paper presented at the International Conference on "Racial Minorities, Economic Restructuring and Urban Decline." Coventry, England: University of Warwick (September).

WALSH, JOAN. 1987. "Is Homework The Answer?" *MS.* (July/August): 122–125, 206–209.

NEW POLITICS FOR NEW TIMES

24 The Politics of Shutdown: Community, Property, Corporation

R. JEFFREY LUSTIG

The shadows of plant closings fall across the nation's cities. Towns, regions, and individual working people find themselves threatened by the destruction of jobs, plants, and the nation's basic productive capacity. A special issue of *Business Week* announces the nation's "industrial decline." A *Newsweek* cover declares "The Collapse of Industrial America." And the costs of that collapse are explained by a laid-off San Francisco shipyard worker to a legislative committee: "In reality, I'm not just out of a job, I'm out of a city. My kids are out of school. My wife is out of her education, and I'm out of self-respect."

The wave of current closures reveals a larger pattern of deindustrialization, a process of "widespread, systematic disinvestment in the nation's basic productive potential." Emphasis here falls on the word "productive." Millions of dollars are invested, but on speculative expansion rather than new industrial capacity. U.S. Steel exemplified the pattern when it spent $6 billion acquiring Marathon Oil at the same time it was closing fourteen plants in eight states and firing 13,000 workers. Emphasis also falls on the word "national," for much of the current productive investment is overseas. General Electric expanded its overseas payroll by 5,000 during the 1970s while it was diminishing its U.S.

Source: R. Jeffrey Lustig, "The Politics of Shutdown: Community, Property, Corporation." *Journal of Economic Issues* Vol. 19 (March, 1985) pp. 123–152. Reprinted from the *Journal of Economic Issues* by special permission of the copyright holder, the Association for Evolutionary Economics.

workforce by 25,000. RCA increased its foreign workforce by 5,000 while it cut domestic employment by 14,000.

PRIVATE AND PUBLIC: THE EXPECTED EQUATION

Shutdowns, slowdowns, and relocations, milked depreciation reserves, outsourcing, and balked investment raise a number of important issues. The causes of disinvestment, effects of capital mobility, need for rights of advance notice, authority of collective bargaining agreements, and intricacies of bankruptcy law have all been discussed in newspaper articles, classrooms, campaign speeches, and even in the courts. Fundamental to these issues, however, is a deeper question about the relationship of large-scale enterpise to community. The question involves on one hand corporations employing a significant fraction of a community's workforce, and on the other the community in its physical character as location and in its active character as workforce. The relationship of this property to community is assumed by people in the United States to be one of private actors and public jurisdiction. Basic to this assumption in turn is a root, inchoate distinction between public and private, state and society.

In the liberal tradition and presumably as codified in Anglo-American property law, the private realm is the domain of personal choice—of personal achievement, distinction, and economic change. It is the realm of voluntarism, the arena of freedom. Fundamental to the liberal project was the attempt to expand this realm by limiting the arbitrary power of some private parties over others. Diffusing power and decision-making authority among a multitude of small property-holders was the chosen historical means for doing this. The domain of the state, on the other hand, is that of compulsion (or possible compulsion) in the service of universal categories, such as that of citizenship, that transcend the particularities of civil society. It is the realm of politics. In contrast to the Middle Ages, liberal society divorces private and public, taking legal cognizance of neither status or class. Position and obligation in liberal society are seen as products of contract rather than prescription, of personal choice rather than law. The power that remains is entrusted to a state presumed neutral, objective, and constrained by a general law.

That the social situation of the last half-century has been characterized by a "new social contract" between Big Business, Organized Labor, and Big Government has not changed the expectations based on this older distinction. Effective agreements in the modern regime are characterized as products of collective bargaining because they are negotiated by large collectives rather than by individuals, and through bargaining rather than market competition. Yet the realm of business and labor is still regarded as private, voluntarist, and guided by self-interest, and that of the state as separate from this, vaguely compulsory, and guided by an objective public interest. A corporate decision to close plants, an investor's decision to withdraw investment—the initiation, that is, of a capital strike—is therefore regarded as a private action, and the injuries imposed, private as well.

Historically in America, there have always been events which belied this conceptual distinction and clouded the rationale based upon it. In the years after Independence state governments "shaped decisively the contours of economic life." They promoted manufactures, chartered corporations, passed protective tariffs, and even vested power in private corporations to "condemn land for rights of way or other facilities," thus transferring to private agents "immunity against torts and injunction proceedings" and effectively delegating perogatives of sovereignty. By the early nineteenth century state courts intervened directly in the transformation of property, redrawing bounds of protected activity and shifting rewards to entrepreneurs engaged in developing the capital function of their holdings.

The distinction between private and public has also been clouded by private initiative. It was not uncommon during the nineteenth century for railroads, for example, to lever tax, land, or financial concessions from municipalities by threatening to route their tracks elsewhere. A string of California spite towns remains to this day, monuments to local intransigence in the face of this dragooning of public bodies by private parties. The current largest landholder in California achieved its original stature through federal railroad land grants. In the case of modern businesses founded in this fashion, curious germs of public authority continue to survive at the heart of "private" enterprise.

Current deindustrialization revives and confirms this confusion of realms. The essence of the problem is regarded as public dislocation arising from private action. But even the disinvesting giants can often be found to have attained their fulsome stature through early, thirsty drafts at the public trough. Many have also been recipients of local concessions, granted in the hope of retaining their favor. Indeed, the process of deindustrialization itself, as B. Bluestone and B. Harrison remark on the basis of their analysis of tax codes, is a virtual "artifact of the polical process.

To the extent that a corporate-government partnership proves pervasive and systematic, the contemporary view that large-scale industry must be permitted to seek its advantage because of its private status, whatever the community costs, fails to do justice to the empirical situation. Policy responses based on that assumption will therefore fail to deal with the problems actually posed. They will also fail, I will argue, to preserve democratic relations. The analysis that follows examines this partnership, explains why the private/public distinction as applied to this form of property demands reconsideration, and suggests what a more adequate and democratic understanding of its relationship to community would entail. . . .

Contemporary shutdowns, relocations, and dismantlings reveal not simply movements of capital and misfortunes of communities, but more importantly shifts in the relations of power between corporations on one hand, and communities and unions on the other. Ultimately they force a reconsideration of basic issues: the relationship of largescale property to community, and the adequacy of the historical form through which working people in America have chosen to organize themselves and protect their rights on the job. The basic story with plant closures, to put it differently, is a political one, and it raises serious questions for citizens even beyond the precincts of the affected communities.

CONTEMPORARY DEINDUSTRIALIZATION

The Shutdown Pattern

A standard pattern or model has been popularly perceived in the events surrounding the most publicized closedowns, like those in steel and auto. It was evident with interesting variations in the timber industry of California's north-coast, and also appears to be emerging in other industries like high-technology and Iowa meat-packing. This pattern must be considered in detail before we move on to the larger theoretical issues raised by plant shutdowns. Its steps are familiar.

First, a large industry employing a significant portion of a community's work force announces plans to close or partially retire existing facilities.

It complains, secondly, of the sudden impact of an exogenous, disruptive event that effectively deprives it of freedom of action. The closedown is presented, that is, as an unavoidable, unforeseen disaster, and as forcing the corporation out of the community in search of new profits. Restored productivity and improved consumption are presented as the public's long-term stake in this capital mobility.

Massive unemployment and community dislocations follow, with the impacts noted by the San Francisco shipyard worker: uprooted lives, disrupted families, broken dreams.

Municipal fiscal crises emerge as demands for services increase but revenues fall. Public works (roads, schooling, police) deteriorate, as do property values.

Finally, the federal government enters the picture through established compensation programs or, as with Chrysler, California timber, and probably numerous cases in the future, through special programs patterned along the general lines of disaster relief. In the vocabulary of the new social contract this is regarded as a socialization of risk. In light of the real pattern of closedown subsidies and compensation, however, it turns out to be a socialization of corporate *costs*. What we will discover with current deindustrialization, in fact, is a surfacing and strengthening of that business-government partnership that has undergirded much of American history. On its current scale and with its current effects this partnership amounts to nothing less than a corporatist policy—in a polity being shaped by the crises of disinvestment along increasingly corporatist lines. The familiar proposals for reindustrialization fulfill rather than forestall these corporatist tendencies.

The Pattern Reexamined

The popularly perceived pattern or model captures part of the picture of plant shutdowns, cut-backs, and dismantlings. But it misses a number of facts or incidents that have occurred with a marked regularity. Considered separately these incidents seem anomalous or as being of merely anecdotal interest. Taken together they tell a different story from the familiar one about what is actually occurring.

First, the popular model exaggerates the inevitability of recent plant closures. By restricting the relevant time span it neglects past corporate policies and presents a false picture of the options open to affected companies. American steel provided the first and most dramatic of recent closedowns in Youngstown, Ohio, between 1977 and 1980. It was not inevitable, however, that the industry would have declined for the preceding thirty years to modernize its facilities. U.S. Steel provides fully a quarter of the steel used in America every year, but it had not built a new integrated facility since 1953. "Investment in steel is profitable; indeed, it appears that the American steel industry may be the most profitable in the world." U.S. Steel chose, however, to use its sizeable profits to pay high dividends and, in the seventies, to diversify until 44 percent of its assets were committed outside steel. During a period in which Japanese steel increased its assets by more than 23 percent annually, this company announced it would run its Youngstown mills only as long as they were profitable without new investment. And Lykes Corporation, which initiated the events in Youngstown when it laid off 4,100 workers with three days notice, milked the depreciation reserves of Youngstown Sheet and Tube; which it had acquired in 1969, to finance diversification rather than to invest in new steel-producing technology.

The automobile industry revealed a similar story before the shutdowns of 1979-80, which closed twenty-five plants and idled 50,000 workers in an industry in which almost 200,000 out of a 780,000-person work force were already on short-term or long-term layoffs.

Long-term cycles of resource availability, international capability, and industrial structure play an obvious role in plant closures. Even when these cycles are relatively fixed, however, the social dislocations and distribution of burdens to which they give rise are not. The available evidence in any case does not suggest that current closures have been either inexorable in their evolution or inevitable in their effects. . . .

The first problem with the familiar closedown model, then, is that it exaggerates the vulnerability of affected corporations and the limits on their options. A second, closely related problem is that the model overlooks the degree to which affected companies have often been complicitous with presumably external forces. The model raises an artificial boundary between endogenous and exogenous variables.

Even while U.S. Steel attributed its shutdowns to Japanese competition, Japanese "dumping" of steel, and federal environmental regulations, for example, it was vastly extending its own foreign assets and even, according to the *Wall Street Journal,* buying Japanese steel for its own sale on the West Coast. Lykes Corporation invested part of its Youngstown Sheet and Tube windfall in Taiwan. The automobile industry, while complaining of the foreign competition that had captured a quarter of its domestic market, actually invested a full third of its budget abroad, with Ford hitting the 40 percent mark in 1981. . . .

The effect of this second problem in conjunction with the first is both to underrepresent the real responsibility of the closing corporations and to obscure the discretionary character of their decisions. The popular model thus complements the needs of the corporations at the same time it reaffirms the older liberal outlook, for it denies the political character of decisions made by what are clearly non-governmental bodies. In the most notable cases and in these basic production industries, closedowns have followed from the conscious creation of obsolescence—what might be called the engineering of obsolescence. What is the purpose of such engineering and of the social dislocations which follow?

According to market theory and to the familiar model, the purpose would be to move capital away from unprofitable industries and from less productive to more productive uses. A third problem with the standard model is that the available facts do not support this expectation. Numerous plants have been closed (U.S. Steel's two works at Youngstown, Kaiser at Fontana, Sperry Rand's New York furniture plant) not because they failed to be profitable, but because they failed to be as profitable as the artificially inflated needs of their owning conglomerates required. Nor has the flow of capital, as an empirical fact, generally been toward new jobs or higher productivity. The Lykes scenario saw a company with no experience in steel purchase a steel company six times its size in order to capture its assets and close its mills, idling 4,100, without creating new jobs in return. Oil too has diverted its record windfall from deregulation into what was, by 1982, $16.2 billion worth of acquisitions of other companies rather than into new explorations. These examples reveal a larger trend. The 1970s witnessed a dramatic shift among the Fortune 500 toward using assets to acquire existing productive capacity rather than to create new capacity.

The effects of this paper entrepreneurialism are predictable. Productivity growth, which averaged an annual rate of 3.2 percent at mid-century, averaged only .8 percent during the 1970s, and fell in 1980-81, directly after the major closedowns, to an annual increase of .5 percent. The product of this investment pattern is not higher productivity leading to improved consumption, but larger funds for gaining access to other companies' cash flow, depreciation reserves, pension funds, tax losses, or expected closedown write-offs. The process—exemplified best by the conspicuous solipsism of the leveraged buyout—is one in which existing assets are reshuffled, debt burdens increased and interest rates also increased, without a corresponding rise in (and with an increased burden on) capital formation.

The fourth problem with the familiar closedown model, suggested by this, is that the burdens imposed by plant shutdowns are not equally shared as the models of industrial collapse or natural disaster would imply. The impact on individual workers has been massive and sustained. When plants close or restrict production, men and women lose jobs, accumulated seniority, health insurance, pension rights, and often cars and mortgages as well. They become susceptible to a higher incidence of physical and mental illness, violence, and early death (for example, from heart disease). They receive established unemployment benefits and in a few industries (like auto) supplemental unemployment compensation from the companies or the Trade Readjustment Act.

The REPP benefits were meager, compared to any real price that could be put on a person's livelihood, family stability, or marital security, and meager compared to the personal capital—the productive wealth consisting in a person's skill, knowledge, and years of invested experience—expropriated in the course of the lay-off.

Local communities suffer similar losses. Their jobs had provided not only for individual sustenance but for their own "material existence . . . and continuity as communities. Losing the basis of that continuity means that schools must be closed, policemen and firemen furloughed, property taxes increased, and constitutive patterns of community interaction destroyed. (Nationally, the U.S. Bureau of Economic Analysis estimates that every percentage point of unemployment costs the larger community more than $68 million in foregone GNP, $20 billion in federal taxes, and $3.3 billion in forms of public aid).

A more important community cost must also be credited against closedowns. The capital of a productive facility, Thorstein Veblen long ago noted, consists not only of its palpable plant and equipment, but also of its community infrastructure (roads, electricity, sewage, schools), and thirdly, of the intangible capital which resides in the technological skills, experience, and knowledge of the workforce. These latter two elements, combined with the plant, make up modern industrial capital. In cases where a corporation employs a significant portion of a community's workforce, that corporation becomes "the *de facto* owner" of the composite capital; a closedown, therefore, constitutes the unilateral destruction of community capital.

The fundamental loss to a community from shutdowns of the magnitude witnessed recently, however, is political. The loss of jobs and tax bases and the threat to its continuity as a living community restricts its autonomy, self-determination, and authority as a governing body. Like the captive towns of the nineteenth century it finds that private corporations have acquired the power to issue birth certificates and death sentences to entire communities.

Communities have responded to the threat of closedown with an array of tactics extending from as yet rare proposals for community-owned enterprises, and worker-community run enterprises (attempted both in Youngstown and Humbolt County), to appeals for federal block grants to compensate for the effects of closure and offers of local concessions in the hope of retaining jobs. The loans necessary for starting worker-community run enterprises have not, however, generally been forthcoming. Federal grants have produced few jobs, and almost none to the populations affected by lay-offs. The most frequent response has, therefore, been concessions. Detroit provided the largest and starkest of these in its use of the power of eminent domain to raze an entire community, Poletown, in order to clear 400 acres (at a cost of $200 million). It destroyed over a thousand homes, 155 businesses and numerous churches, displaced 3,200 people, and extended a twelve-year, $64 million tax abatement to boot, in order to induce General Motors to build an assembly plant and retain roughly 3,000 jobs. As a result of this painful exchange of a neighborhood for a factory and homes for jobs, GM retained "its existing workforce, which is experienced and . . . accustomed to the discipline of assembly work"—that is, a major element of its community capital.

It is when we pursue this matter of impacts and compensation that the story gets interesting and the purposes of current deindustrialization clearer.

What do the companies receive for the closure of their plants or release of their holdings? First, the ability to *close* marginal operations, to *concentrate, modernize,* and *diversify.* In cases where other forces can be blamed, they have been able to do this without incurring great public hostility.

Next, the companies benefit from the assumption by the government of the tab for the personal, infrastructure and *social costs* entailed in the closedowns or relocations. Though the corporations may have spent profitable decades in the affected locales and even received generous subsidies, they have not been required to contribute to federal or local compensation programs.

Third, the companies receive a number of *direct economic subsidies.* Some of these consist of worker give-backs and community concessions. The vast majority are derived from the federal government and the bulk of these from the exemptions and incentives of the tax code. Totalling a massive $62 billion by 1980, the tax allowances "dwarf all other economic development programs" of all levels of government. Tax write-offs for plant and equipment losses, liberalized depreciation rules, investment tax credits, and preferential treatment for capital gains all throw disproportionate rewards in the direction of companies engaged in close-downs.

Government also offers direct subsidies in the form of subsidized loans or loan guarantees when a proposed closedown is judged too costly to be accepted. A Chrysler failure, for example, would have deprived the treasury of $280 million in annual taxes and produced a sixfold job-loss in auto-related industries. A Continental Illinois Bank failure would have brought down 2,100 smaller banks. Direct subsidies may, finally, take the form of specialized policies (as with tax exemptions for foreign income, federal export insurance, or crop control subsidies) or one-time distributions of largesse.

Taking advantage of these various subsidies U.S. Steel was able to reduce its tax burden in 1978 to $8 million on $250 million in profits, for an effective tax rate of 3.2 percent. The year before it achieved a negative rate and received credits on $135 million. The benefits it received from the Youngstown closures were $200 million—twice the amount unsuccessfully requested by Community Steel, Inc. to keep a mill open. In a year when it idled 20,000 employees, General Electric, the nation's largest conglomerate, was rewarded by having its $22 million tax liability entirely cancelled. . . .

As the result of these other benefits, direct and indirect, economic and political, the largest companies succeed in acquiring *power* they had not previously enjoyed. The closing of plants, rerouting of capital, modernization of facilities, all do more than shift assets and shuffle jobs. They confer augmented power over the now-isolated and demobilized workers, the now-shaken communities, the bankrupted competitors, and even the market itself.

This power is ultimately political. And it possesses an institutional character that promises long-term effects than simple statistics on subsidies and exemptions suggest. Concessions like those made by Detroit entail a long-term loss of governing authority. Closures and outsourcing threaten labor not simply with loss of wages and benefits but with loss of the union's presence, its very existence, on the shop floor. The market itself is affected as diversified giants, who have had the costs of their capital reduced by federal programs, further retool, lower prices, or simply sit back and watch the shake-out of smaller firms in the face of increased costs and difficulties.

This augmentation of power has been a historical corollary of the quest for profits or for a secure profit-making position. Alfred D. Chandler, the business historian, writes that it characterizes the country's first merger wave at the beginning of the century. It was clearly the goal of the advice tendered by Merrill Lynch and by the Argus Research Corporation to the steel industry in 1977 that it "contract . . . its capacity base" in order to be able to artifically restrict output and raise prices. It is also implied in Bluestone's and Harrison's observation that "in all of the examples we're coming up with, management is running away as much from union work rules as from union wages." In a corporate economy and on the scale discussed here, the quest for market power translates directly into institutionalized social power. It is not surprising that eventually "capital mobility as a strategy increases the power of the firm vis-a-vis the State."

What is surprising is that this increase of power occurs today not by augmenting but by curtailing productivity. The effect of current plant closures demonstrates the truth of Veblen's claim that the growth of business power actually derives from its threatened "withdrawal of efficiency" or "sabotage"—not from its ability to make things happen but to prevent things from happening. "Command [over] the individual forces of the technological situation" was centralized and augmented by such things as investment strikes, "retarding" industry, causing "advised idleness of plant" or restricting output.

Such "maneuvers of inhibition," Veblen asserted and Seymour Melman recently confirmed, satisfy the pecuniary thought not the productive interests. In these terms, current deindustrialization constitutes a maneuver of inhibition on a grand scale.

This augmentation of power is the real stake in current deindustrialization. It is the political product of engineered obsolescence. The standard model misses this. But the model itself turns out to be part of the story it means to explain. It is an old truth of politics that power is revealed not by those who have the ability to provide answers but by those who frame the original questions. And the familiar closedown model's confusion of responsibility, denial of agency, and misrepresentation of effects all reveal that the closing corporations have shaped the terms of public discourse and have done so in such a way as to enlarge their own freedom of movement. The real cost of accepting these terms, therefore, is not simply confusion or inaccuracy, just as in the past the real significance of ceding control over a region's timber resources was not primarily thwarted conservation or spoiled beauty. The real price of accepting these terms is future power over the public.

Increasingly through the 1970s, large conglomerates in the United States have initiated what amounted to an extensive capital strike—one to which the government refused to respond, as a strike called by labor would similarly have drawn no response. Control of productivity and differential federal protection have permitted these giants, through this withdrawal of efficiency, (1) to externalize the costs, and concentrate the benefits, of modernization, (2) externalize the social costs of a depleted resource base, and (3) join with other conglomerates in the breaking of the mid-century social contact. Having availed themselves for four decades of labor peace, union discipline, and passage of wage costs to consumers, they now resume an older posture vis-a-vis working people and communities.

The bias in federal compensation in all of this, far from being haphazard or the result of misinformation, has been systematic; far from being only a *post facto* occurrence its intervention has been fundamental and originating. Its incentives have constituted a policy that positively encourages plant closures by subsidizing bigness, promoting concentration, encouraging foreign investment, and because its rewards attach to abandonment, acquisition, and new investment, increasing the pace of capital mobility. It has played a major role in widening the breach between the making of goods and the making of money.

When the patterns of impact and compensation are studied, then, it becomes clear that the standard pattern of shutdowns mistakes the phenomenon. It is not a case of private vs. public, of industry vs. regulation at all. It is a case of intimate and systematic partnership between business and government, of industry using government "regulations" to its own advantage, and of politics in the "private" realm. It is the industrial selfgovernance of the California Forest Practices Act writ large. Evidence of this partnership exists in still other forms:

(1) in the government's liberalized merger policy which encourages closure of redundant capacity in new multiplant firms. An early example saw Attorney General Griffin Bell approve the merger of the nation's seventh and eighth largest steel producers (held by LTV and Lykes respectively), over the objection of his staff, which led directly to the closure of Youngstown's Brier Hill Works and the loss of 1,250 jobs; (2) in the friendly ministrations of government officials: John B. Crowell, Louisiana Pacific's general counsel during the Park controversy, now the Assistant Secretary of Agriculture in charge of the U.S. Forest Service, urged the gift of 70,000 additional acres of timberland to Louisiana Pacific for Park losses, promoted the exemption of timber companies from $2 billion of their contractual obligations incurred for speculative purposes during more inflationary times, and has increased the harvest of Forest Service aspen in Colorado from 2 MBF to 50 MBF "in response to Louisiana Pacific's request" for waferboard material, and (3) in the separate laws and zoning grants of state governments.

The central question raised by shutdowns is not whether plants should ever close or industry reorganize. The process of what J. Schumpeter called "creative destruction" is endemic in our modern economy. Longterm cycles and sudden reverses are inescapable parts of the business world. But these cycles and reverses are not, we have seen, natural disasters. Humanly-fashioned institutions determine how they play themselves out in the lives of men and women. "Behind [even] the trade cycle there is a structure of social relations." The real question is not whether closedowns will occur but who is empowered by social institutions to initiate them, why they are permitted to occur, and how society permits their burdens and benefits to be distributed.

What has occurred in these terms is not a socialization of risk but the subsidization of private modernization and growth. The government has helped create a situation in which:

> The community pays for the trends toward concentration, capital intensity, and depleted [production resources] through persistent high unemployment, lower per capita income, and a growing reliance on government assistance. . . . [T]he community becomes more and more disenfranchised.

Government has aided the corporations transferring the heavy costs of modernization and concentration from the companies incurring them to the local community, working people, and the nation at large—appropriating the energies of the general taxpayer and the abilities of voluntarist organizations for the enhancement of private power. Because the beneficiaries of its policies are the largest corporations, the government, furthermore, has effectively created a classic system of privilege—"private-law," endowing one class of citizens with special liberties, power over others, and responsibility for implementing a national industrial policy.

The Failure of Response

Appropriate reindustrialization policies would have sought economic diversification to establish new bases for employment, creation of secondary products in affected industries, retention of domestic processing, and rehabilitation of the resource or capital base. Instead, public policies and incentives have encouraged the same practices which created the dislocation, rewarded those responsible for it, and compensated the community, if at all, with facilities for dependent forms of commerce (for example, tourism rather than primary production). Workers and community have gotten "relief." They have not secured new bases for autonomy and independence.

Two basic attitudes have characterized the public responses to plant closedowns. The first defends the right of private business to seek profits where it will and frowns on major public aid to those adversely affected. The second encourages such aid, but within a situation similarly shaped by private actors; it too sees the public good as something arising primarily from private planning for private goals. The first, conservative approach trusts to private profit maximization to register social costs. It opposes public aid as a tax on the diligent to support the incompetent, and as working to place recipients in tutelage to the state—forcing a reversion in essence from liberal back to feudal relations. The more liberal alternative doubts that private market contracts are capable of registering social costs or establishing proper criteria for allocating resources. Given the rise of organizational capital and of a complex, interdependent society, it looks to government to ease the interdependence and monitor the powerful.

The facts adduced in this study undermine the authority of both of these approaches because they impugn the basic view of the relationship of property and community on which the approaches rest. The latter approach threatens to cast individuals and entire communities in the role of supplicants. In an era during which large corporations, having gathered communities' productive potential unto themselves (like the proverbial Octopus of Frank Norris's novel), threaten to move on in obedience to a distant logic of profit-taking, corporate liberalism would make people petitioners for charity rather than bearers of rights. Mono Lake basin in California can be protected under common-law doctrines of a public trust. The snail darter can be preserved by federal statute. But timber, auto, or steel workers lack any such protection for their jobs and everything depending on those jobs. Under the dispensations of modern liberalism, workingmen and women are to become appellants for relief, wards of the state, reactors rather than actors and, ultimately, subjects rather than citizens. This is an approach ultimately inconsistent with the requirements of democracy—indeed, with the voluntarist freedom the distinction between private and public sought to preserve in the first place.

The former approach, which regards decisions about plant location as purely private, has been disproven by the evidence. The pattern of government subsidy and support in steel, auto, and timber (to say nothing of oil, agriculture, and meatpacking) reveals that large elements of publicly delegated monies, authorities, and power attach to current disinvesting companies. The market in any case simply lacks mechanisms for registering large-scale social costs (as the destruction of timber resources). Indeed, plant closures appear as the cheapest alternative for disinvesting companies for this precise reason; the government has permitted them, using market methods, to dismiss the real costs of the closures from their shutdown equations.

What is the source of this extensive business-government partnership? Closer analysis reveals that it lies, deeper than contemporary subsidies, in a corporatist potential of the modern corporate form itself. To understand this and to begin to find more appropriate responses to plant shutdowns it is necessary to reconsider how the corporation in America came to be regarded as a private entity in the first place.

CORPORATIONS: THE NEW PROPERTY

It was only during the 1890s that railroad corporations (in the vanguard of corporate capital) acquired a Supreme Court which extended to them the protections of private property. Prior to this in Anglo-American law and tradition, the corporation had not been regarded as a form of property at all. It was a special association, chartered by the government for special purposes, given special powers, and saddled with special obligations. A product of political favor, it was looked upon as a branch of sovereignty in the hands of a subject and closely regulated for that reason. The legacy of the Middle Ages (and of earlier Roman precedent), furthermore, was for ownership rights to vest in the association as a collectivity or organic unit.

Corporations were feared in liberal political culture not so much because their powers created threats in the marketplace as because they threatened the *grounds* of market activity itself. The most important of these grounds was the right to labor. "The property which every man has in his own labor" explained no less an authority than Adam Smith,

> as, it is the original foundation of all other property, so it is the most sacred and inviolable. The patrimony of the poor man lies in the strength and dexterity of his hands: and to hinder him from employing this strength and dexterity . . . is a plain violation of this most sacred property.

Land monopolies which, for example, engrossed large tracts of land thus deprived others of their rights to this "sacred and inviolable" property. And government was therefore entrusted with the obligation to oversee corporate activity and protect the neutral grounds of market activity. A desire to enforce this obligation in the face of this institutional threat constituted the precise basis for the calls for government "regulation" that emerged in the populist insurgency in America.

Prior to the 1890s authoritative doctrine, as stated in the *Munn v. Ill.* cases, also held that private property itself when affected with a public interest "ceased to be *juris privati* only," and became regulable. When an owner used his property "in a manner to make it of public consequence and affect the community at large . . . he, in effect, grants to the public an interest in that use, and must submit to be controlled by the public for the common good." . . .

Current plant closures reveal a historical paradox which has long lain dormant. Corporate capital, the creature of community, has risen above the community and acquired the power to condemn it. Closures also reveal that the intervention of the State has been essential to this emancipation, in the 1980s as in the 1890s. But this indicates that the emancipation is really a matter of legal title rather than social fact, of legal fiction rather than social reality. It is a spurious emancipation, for its beneficiary mixes public and private inextricably. The proper term for such an entity is not simply "corporate," an economic term, but *corporatist,* a political category.

A corporatist order is one in which the largest, organized interests in society grasp public powers, exercise an effective franchise over the processes of representation and policy-implementation, and dwarf individual citizens in the enjoyment of constitutional rights. A fully corporatist regime (toward which corporatist institutions tend) has been defined as a system of

> interest intermediation in which the constituent units are organized into a limited number of singular, compulsory, noncompetitive, hierarchically ordered and functionally differentiated categories . . . licensed [if not created] by the state, and granted a deliberate representational monopoly . . . in exchange for certain controls on their . . . leaders . . . [and] demands.

The new property is an innately corporatist instituation not only because it is the recipient of particular subsidies and supports, and not only because it possesses special powers (which though no longer legally recognized, inhere in such things as institutional immortality, rights of reorganization, limited liability in financial and apparently criminal matters, concentrated control of large assets, and control of a region's job base). Most deeply the new property mixes public and private because it embodies a delegation of the intrinsic power of the sovereign, recognized by the common law, to protect the public from becoming vulnerable to "a common charge or burden" and from being compelled to resort or pay virtual tribute to any unregulated private party.

The empirical proof of this delegation is the significant power modern corporations enjoy today to become *de facto* makers of public policy in the communities in which they reside. Their private plans bear all the marks of such policy. They are (1) public in their consequences, (2) binding on all in their domain, and (3) authoritative in an implicit or often politically explicit way. They assert a control over the commons from which no appeal is possible. The fundamental question posed by this property is therefore political, as a dissenting judge in the Michigan case challenging Detroit's use of eminent domain for razing Poletown recognized. "When individual citizens are forced to suffer great social dislocation to permit private corporations to construct plants where they deem it most profitable," he wrote "one is left to wonder who the sovereign is."

As a result of the workings of this property the source of sovereignity has been confounded. The private realm for the vast majority of people has become one of compulsion rather than of voluntarism. Small businesses, working people's property, and the diffused powers and abilities expected to flow from property ownership, have all become subject to nullification and condemnation. The conclusion to which analysis leads and which the current wave of closures confirms is that *the extension to "practical monopolies" of the protections of private property is the denial to the vast public of the freedom that was originally intended to be secured by that property.*

Community and property today are fused. The prospect of separating them again or of returning large property to the private realm is not realistic; nor would it suffice to restore productivity to modern industry. Business and government admit as much when they declare a Chrysler or a Continental Illinois too important to be permitted to fail. The mix of public and private in deindustrializing policy reveals an accurate, though distorted; recognition of this fusion. The choice for municipalities and regions faced with shutdowns is either to acquiesce in the existence of this new property, and the politics and power it implies, or to create a new relationship by recognizing the community component of capital, and the communities' and employees' legitimate rights to participate in the control of these enterprises. Ultimately no alternative to these two options is possible. . . .

CONCLUSION

Behind the familiar arguments about plant closings, slowdowns, and relocations, are fundamental questions about the rights of community in large-scale enterprise upon which the freedom and security of its citizens, and its own integrity and continuity as a community, depend. They are questions that are fundamentally political and that affect our most basic political institutions. The impact of corporatist closure policies on the voluntarist associations of community and trade union have been noted. The affect on the general status of the citizenry must also be acknowledged. What is forgotten by those who justify shutdowns on the basis of increased productivity (besides empirical accuracy) is that the dislocations thus incurred are products of coercion not choice. The loss of "job . . . city . . . school[ing] . . . self-respect" is a forced loss, the emigration that follows, a forced emigration. And the arbitrariness of a utilitarian policy which sacrifices personal rights to larger social goals is in no way diminished by the fact it is implemented by corporate elites instead of liberal planners.

The institutions of the federal government are also transformed as tax subsidies, loan guarantees, and special supports grow, but the skewed character of the largesse militates against public deliberation and accountability. Externally, the powers of making and implementing a national industrial policy are farmed-out. Internally, substantive policies are first smuggled into tax policy, previously simply as a source of revenue, and then appropriated in an *ad hoc* manner by the Executive Branch as administrative agencies make an "end-run" around the legislature. The scope of intervention and the spectre of an executive State grows as the legitimacy of government paradoxically diminishes.

These are the political effects of current deindustrialization. Their gravity was discerned a century ago, and in terms almost identical to those of the judge in the Poletown dissent. It was in the wake of the Supreme Court ruling which first established the new corporate property that Judge Seymour Thompson of St. Louis demanded to know "whether the corporation is to rule the State, or the State the corporation." Is the corporation in the service of society, or society in the service of the corporation? Either the fiction of privacy will continue to clothe the reality of an emergent corporatism and produce an undemocratic power over our communities, or a revived recognition of the public interest and of property rights vesting in community will work to strengthen individual rights and democratic action by helping transform the manner in which the society provides for their foundations.

NEW POLITICS FOR NEW TIMES

25 *New Politics For New Collars*

FRANK RIESSMAN

An exciting new formulation in American politics has been proposed by Ralph Whitehead, professor of public policy at the University of Massachusetts. He describes a group called "new collars" consisting of about 25 million people who, he believes, may be the new American mainstream. They are the moderate-income workers of the post-industrial economy whose jobs fall into neither the traditional white-collar nor blue-collar category. They may be nurses, taxi drivers, computer keypunchers, salespeople, beauticians, fast-food restuarant managers, small contractors, or teachers.

In their political outlook, they are patriotic, but not jingoistic. They tend not to be much concerned about the ins and outs of day-to-day politics, and they are somewhat removed from and skeptical about the political process. However, they feel a deep sense of commitment to their families and their neighborhoods. They are liberal on women's rights, abortion, and school prayer. On foreign policy issues, they also tend to be fairly liberal and probably are a significant portion of the population that for the first time is willing to consider the reduction of military expenditures.

The new collars are concerned about their own empowerment and the expansion of their sense of control both for themselves individually and in terms of their like-interest group affiliations. They don't like politics as usual and are

Source: Frank Riessman, "New Politics for New Collars," *Social Policy* Vol. 16, (Winter, 1986) pp. 3–4. Published by Social Policy Corporation, New York, New York 10036.

only periodically active politically, as perhaps in the freeze movement. In many ways, the populist theme is potentially appealing to the new collars with its emphasis on traditional values, family, localism, empowerment, and its strong opposition to big bureaucracy, elitism, and greed.

BRINGING THE NEW COLLARS INTO THE DEMOCRATIC FOLD

Problems arise when considering the new collars in relation to traditional Democratic party constituencies and issues. How can the Democrats win this new group without losing their old supporters? For example, minorities and a large number of poor women and men want jobs and the expansion of social welfare benefits, programs not especially of interest to the new collars who are more concerned with deregulation (less government) and economic growth. Of course, economic growth could bring with it jobs and expanded benefits, particularly if it were accompanied by the reduction of huge military expenditures, which the new collars favor (although this may be less likely in light of the budget-balancing legislation).

So too, the labor movement, another traditional Democratic party constituency, is protectionistic of the old industries and concerned with social welfare benefits for the poor and minorities, as well as the use of the government to provide health and education support—issues not generally espoused by the new collars.

The differences with the new collars may not be so great, however, numerous polls report widespread mainstream support for New Deal programming in relation to health, education, and social welfare expenditures. Thus, exit polls in 1984 found overwhelming support for both New Deal and Great Society programs: 80 percent of voters expressed support for equal or more social expenditures, and only 15 percent said they favored cutting funds for these programs.

But in order to reach out effectively to the new collars, it will be important to develop a differentiated view of government. On the one hand, strong criticism of government inefficiency, waste, and corruption needs to be highlighted with particular reference to military waste and business greed. On the other hand, even with considerable economic growth, there is likely to be significant unemployment and underemployment as has been evidenced in the last three years in a period of rapid expansion. The employment issue will surface more seriously as we move into the recession phase of the business cycle, probably in late 1986 and surely in 1987. (Upturns generally last from 36 to 48 months, and the present one began at the beginning of 1983.) A government role will clearly be needed to deal with the endemic unemployment problem.

There is, in fact, broad popular support for government providing jobs for the unemployed when private industry cannot do so. (The Fingerhut/

Granados Opinion Research group found this to be true even for Reagan Democrats in the 1984 election, 83 percent of whom agreed that "government has a responsibility to see that every able-bodied person in the country has an opportunity to work.") Moreover, there is the well-known finding of the Congressional Budget Office; namely, that for every 1 million jobs added in the economy, government funds are increased by $30 billion as a result of the increased taxes paid by employed workers and the reduction of unemployment insurance and other benefits. This is an extremely important point for the new collars (as well as the budget balancers) who are disturbed by any increase in taxes to help "other people" at their expense. One of the old shibboleths regarding the expansion of employment has always been the fear of inflation. This seems far more unlikely in the present period in light of the "glut economy" that has contributed to the halt in inflation.

Many people, of course, decry direct government job creation as an old tired economic strategy, badly despoiled by various public employment experiences such as CETA. It will be necessary to reframe job development approaches in fresh ways. In Great Britain and France, unemployed workers are allowed to take their benefits in a lump sum for up to six months *if* they use them to pool with others to start a new business. In western Pennsylvania, community and labor leaders are pushing for the creation of a Steel Valley Authority to take over closing plants and then broker them to another party such as employees, in order to keep the facilities going. Companies hiring workers displaced by imports could get a tax credit for half of the new employees' wages (for up to three years) if the firm retrains them. Governor Cuomo has proposed a job guarantee for every high school graduate as a basic antidropout strategy, thereby giving the job guarantee a different kind of appeal.

The Democratic party's best chance of winning the new collars is through economic programs that produce growth, jobs, and economic security, together with a populist-like platform stressing issues such as toxic waste, high rents and interest rates, unfair taxes, crime, and the need for improved public services.

Perhaps equally important is the need for a new politics that centers on a positive ethos, a spirit that is currently reflected in the widespread support for anti-apartheid policies, the sanctuaries for the politically mistreated, the Catholic Bishops' position on poverty, the concern for the homeless, the farmers, and world hunger, the pop music of Springsteen et al., and the rejection of the rampant selfishness reflected in mergermania.

If the Democratic party can develop these themes, it may well capture a large portion of the 25 million new collars.

NEW POLITICS FOR NEW TIMES

26 *Fighting Back*

LANCE COMPA

Union workers had little trouble reaching a contract settlement with Black & Decker in April 1984 after the company took over General Electric's main housewares plant in Allentown, Pennsylvania. Negotiations were peaceful, and Black & Decker said it looked forward to a long partnership with Allentown workers. But in November, Black & Decker announced it would close the plant.

The shutdown, planned for early next year, would eliminate the only union operation in the Black & Decker chain. Members of United Electrical, Radio and Machine Workers (UE) Local 128 called the decision a "betrayal" and, like some other union workers faced with plant closings, they are fighting back.

The union has set its sights on a weak spot—the company's entry into the housewares market. Black & Decker recently bought out General Electric's housewares division and soon must replace the GE label with its own. This may be the biggest brand-name changeover in history, and the tool company is backing it with a $100 million marketing blitz.

Allentown workers, and their supporters in area unions, churches, and community groups, have responded by holding stormy protests at a Black & Decker stockholders meeting near Baltimore, at the National Housewares Show in Chicago, and outside the offices of a New York advertising agency that is

Source: Lance Compa, "Fighting Back," *The Progressive* (October, 1985) pp. 32–34.

handling the name change. Union members have talked to editors of housewares industry trade journals, arguing that shutdown of the profitable Allentown plant will undermine Black & Decker's entry into its new market. They are also planning to meet with Wall Street stock analysts. "We want to shake investor confidence in Black & Decker's ability to pull off the move into housewares," says Paul Kokolus, president of UE Local 128.

The Allentown campaign is one community's fight to prevent a plant closing in this time of shutdowns and runaway shops. In the past decade, millions of U. S. factory jobs have been lost—almost 250,000 during the first half of 1985 alone. Steel mills, rubber factories, auto plants, electrical equipment shops, oil refineries, and other industrial enterprises have shuttered operations in cities and towns around the country. Semiskilled workers have lost positions that once brought a decent living.

The American labor movement was largely unprepared for this onslaught. Most unions, reluctant to openly question management's right to close a facility, have adopted a strategy of orderly retreat, calling for advance notice of shutdowns, greater severance pay, extended insurance coverage, early retirement benefits, transfer rights, and retraining and relocation funds.

These are important demands, but as a strategy, they concede too much. The plants still close, even as workers learn that shutdowns often stem from management's poor planning, shortsightedness, and greed, not from well-informed company decisions guided by neutral market forces.

The United Electrical Workers union, at Allentown and elsewhere, has tried to develop tactics to block "final" plant-closing decisions. In their battles, UE members have challenged management-knows-best assumptions. Instead of offering concession packages, they have gone on the offensive. These efforts illustrate some possible strategies and tactics for labor in general.

The Westinghouse Air Brake Company (WABCO) announced in 1981 that it would close an outmoded, five-story rubber mill in Wilmerding, Pennsylvania, where gaskets and rubber parts for rail-car braking systems were manufactured. The company told UE Local 610 that is planned to build a new rubber mill in North or South Carolina.

The Wilmerding mill, with 300 workers, was part of a larger WABCO transportation division employing 4,000. Dan Marguriet, UE's business agent at the time, led the union effort against the closing. He organized a drive to gather political support in Pittsburgh and valley towns east of the city, and the union held plant-gate rallies and shop-floor protests in the rubber mill and throughout the WABCO complex.

After nine months of protests and growing political pressure, the company agreed to build a new rubber mill just ten miles from the old one and employ the same workers under the same union contract. "We had some leverage since they were only planning to move part of the operation," say Marguriet. "But it still took strong action to make the company stay."

Aside from saving 300 rubber mill jobs, the victory gave WABCO workers the confidence they needed for a successful seven-month strike to fight givebacks the following year, Marguriet says.

Members of UE Local 277 also won a strike against concessions in 1982 at Gulf & Western's Morse Cutting Tool plant in New Bedford, Massachusetts. In fighting the company, the union mobilized community support against the big conglomerate's practice of milking profits from Morse to finance acquisitions in non-manufacturing businesses instead of reinvesting money in plant improvements.

When Gulf & Western announced plans to close the Morse facility in mid-1984, the union and the community used the networks built during the 1982 strike to fight the closing. New Bedford's mayor said the city would exercise its power of eminent domain to take over the plant and keep it in operation until a suitable buyer could be found. Not since Harry Truman's shortlived seizure of steel plants had a factory been expropriated in the United States.

"Without our work in the strike and our educational job in the community, we could never have gotten the political support to make the eminent domain threat a credible one," says Local 277 President Rod Poineau.

Faced with the threat of seizure, Gulf & Western stepped up its own efforts to sell the plant. It found a buyer acceptable to the union and the city, and the plant remains in operation today. Labor relations are "excellent," Poineau says.

Even an unsuccessful fight can have a positive effect. In Charleston, South Carolina, a UE local took action in June 1984 when General Electric gave notice that it would close a facility that manufactured steam turbine generator parts.

GE said it was forced to consolidate steam turbine operations at the company's base in Schenectady, New York, because of a drop in construction of new power stations. At the time of the announcement, the productive, fifteen-year-old Charleston plant employed 450 workers, 330 of them represented by UE local 1202.

Under a national collective bargaining agreement between UE and General Electric, the company provided one year's advance notice of the shutdown and would give hourly wage earners severance pay ranging from $12,000 to $15,000, as well as twelve months' extended insurance coverage and $1,800 each for retraining courses.

But the workers did not want to go quietly. Local 1202 formed an Alternative Use Committee to explore product lines that could take advantage of the skills and equipment at the workplace. The committee's economic conversion plan concluded that the Charleston plant could manufacture parts for alternative energy systems, environmental protection devices, and tanks, containers, and piping for the treatment and transport of hazardous waste. The union presented its plan to General Electric in October 1984 and demanded that the company either convert the plant or find a buyer who would.

At the same time, UE launched a statewide campaign to build support among labor, religious, political, and community groups. Catholic Bishop Ernest L. Unterkoefler, who led a community forum on economic conversion, said that the UE plan reflected values expressed in the U.S. bishops' draft pastoral letter on the American economy. A Charleston legislator introduced a bill that would require economic conversion planning by any company contemplating a shutdown of operations in South Carolina.

"It was the biggest labor-community alliance in South Carolina since the '68 Charleston hospital strike," says State Senator Herbert Fielding, who played a key role in building the coalition that included participants ranging from Jesse Jackson, who led a plant-gate march and rally at the GE site, to Republican Senator Strom Thurmond.

General Electric went through with the Charleston plant closing in May, but UE members vowed to stay together and offer their economic conversion plan to potential employers. "We've got a ready-made force of skilled workers who can hit the ground running in a new operation," says Local President Carnell Gathers. "Sooner or later some company is going to want into this plant, and we'll be waiting for them."

UE's 1982 fight to save the Simpson Dura-Vent plant near Oakland also ended with a closing. Shortly after UE organized workers there, the union learned that management was planning to shut the facility down and move operations seventy-five miles north to rural Vacaville. Labor and community pressures delayed the closing for six months and succeeded in securing severance pay, extended insurance, and transfer rights for the workers.

What's more, the publicity and lobbying by Dura-Vent workers and supporters convinced the Vacaville City Council to adopt the nation's first municipal plant closing law. The ordinance barred public financing for companies seeking to run from other California locations to Vacaville, thus stripping Dura-Vent of its industrial revenue bond. It also established transfer rights for workers caught in a runaway shop and guaranteed resident status for employees transferring into Vacaville, making them eligible for low-income housing assistance. Similar legislation is now being considered in Santa Monica, Santa Cruz, Berkeley, and Oakland.

Workers in other unions have fought hard against plant closings, too. The United Steelworkers and their community allies in the Tri-State Conference on Steel have used twenty-four-hour vigils and mass rallies to rescue U.S. Steel facilities in Duquesne, Pennsylvania, from an appointment with the wrecking ball. A Bakery, Confectionery and Tobacco Workers local, in solidarity with locals at nine other plants around the country and backed by an aggressive community organizing effort, persuaded Nabisco in 1982 to reverse its decision to close a Pittsburgh facility. That same year, the United Auto Workers (UAW), with help from the University of Alabama, put together a plan for modernizing GM's parts plant in Tuscaloosa and convinced the automaker to keep the facility open. Chrysler, too, put aside plans to close a Detroit forge plant after UAW members there outlined ways to renovate and rebuild the facility without interrupting production.

Unfortunately, corporate executives still hold the trump card—the legal right to shut down facilities regardless of objections from workers, unions, or communities. Unions lost the combative influence of communists, socialists, and other radicals in their Cold War rush to conformity, and now the mainstream labor movement fights over workers' share of the economic pie without challenging the system that bakes it.

Perhaps that will change as unions begin to look carefully at each plant closing and fight back with tactics that take advantage of companies' vulnerabilities. The record surveyed here shows that victories are possible—even under current conditions—and that important political education can be carried out, win or lose. Labor resistance, political pressure, and public relations warfare hurt employers in the marketplace and force them to think twice about shutdowns. Such techniques can build momentum for a Federal law stripping companies of the absolute right to close plants, and help reestablish a spirit of militancy that can animate the labor movement.

THE CHANGING SCHOOLS

27 *Educational Requirements in a Changing Economy*

WILLIAM H. KOLBERG

Since 1900, Americans have experienced both economic boom and bust, enduring the hard times of the Great Depression, when as much as a third of the workforce could not find employment, and enjoying the boom years of the 1920's and 1960's, when a full-employment economy allowed workers to have their pick of jobs. Now, as we near the last decade of the 20th century, the U. S. is in danger of experiencing a unique phenomenon—both unfilled jobs and high unemployment.

It doesn't have to be that way. In fact, economic conditions predicted for the U. S. over the next 10 to 15 years offer the opportunity to cut unemployment significantly and, even more important, bring millions of Americans, mostly black and Hispanic, out of the trap of long-term poverty. However, if we don't respond to these conditions intelligently and effectively, we could be confronted with the development of a permanent underclass in America, alienated from society and unable to enter it.

The danger we face isn't a lack of jobs. Experts are predicting that the number of jobs will increase faster than the number of jobseekers. The danger is that too many Americans will lack the skills necessary to fill the jobs our economy is creating.

Source: William H. Kolberg, "Jobs Without People and People Without Jobs: The Coming Mismatch in the Information Society," *USA Today* Vol. 116 (July, 1978) pp. 18–20. Reprinted from USA Today Magazine, July 1987.

For years, people were afraid that computers, robots, and other technological innovations would eliminate millions of jobs. The reality has proved to be the exact opposite of the fear. Modern technology is generating the increases in productivity that make real economic growth possible. Yet, while we are eager to invest in new equipment, we don't seem to realize that a comparable investment in human beings is necessary if we are to take full advantage of the benefits of our new technology.

The American workplace is being revolutionized by computers and competition from abroad. Jobs are being created, transformed, and destroyed by the millions. While it's impossible to predict what the skill needs of the future will be, there's no doubt at all that low-skill positions are being replaced by jobs requiring more extensive education and training.

While the jobs are changing, the people who will fill the jobs are changing too. Our workforce is growing older; more and more women will move into the labor force, a powerful trend that has been developing for several decades; and the number of working women with children will continue to increase as well.

Perhaps most striking will be the actual decline in the number of young people coming into the labor force. At the same time, more and more of these young people will be black and Hispanic, and more of them will be poor. More of them will be coming from backgrounds that severely limit their chances for success.

LOOKING AT NUMBERS, LOOKING AT JOBS

Present figures indicate that, from 1985 to 1995, the number of jobs will jump by 15,600,000, making a total of almost 123,000,000 jobs. During the same time period, the number of people in the labor force—those either working or out looking for jobs—will climb by only 13,700,000, for a total of over 129,000,000. In effect, we will be gaining 2,000,000 jobs, enough to cut the unemployment rate by about two percentage points and bring it down to about five percent, generally regarded as the "full-employment" level.

Of course, such a welcome development isn't going to happen automatically. To obtain a net growth of 2,000,000 jobs, our economy will have to enjoy virtually continuous growth. Companies must be able to eliminate unneeded positions and fill new ones without delay. It's no secret that these new jobs will not be the jobs of the past, or even the present. The shift away from manufacturing and toward the service and information industries has been so pronounced that, in the last 15 years, only five percent of the new jobs created were in manufacturing, while 90% were in service and information.

We can be sure that the rate of technological change will not show in the next 15 years. Not only will new jobs be created, requiring skills that don't even exist at the present, but millions of traditional jobs will be restructured.

This doesn't mean that we will all be sitting at computers, though many of us are already, and more will be. It does mean that few of us wil be able to get along without adding to the skills we already have. Moreover, the pressure on those just coming into the labor force to develop new skills will be even greater. Economists are predicting that, in the near future, three out of four jobs will require education or training beyond the high school level. However, according to the best estimates, about 23,000,000 adults in the U.S. are functionally illiterate. One of the great dangers looming in the near future is that this problem could grow substantially worse.

The number of entry-level workers—our young people just entering the labor market—is expected to drop between now and the year 2000. At the same time, the number of high school dropouts, already running at about 1,000,000 a year, is likely to climb. In other words, while the demand for new workers will be increasing, fewer young people will be available, and even fewer will have the basic skills necessary to perform entry-level work. Businesses are already competing among themselves, and with the armed services, to attract and keep employable young workers. This competition is likely to grow fiercer and more costly in the future.

This projected increase in the number of functional illiterates in the workforce is only one aspect of what is potentially the most disruptive social development over the next 15 years—a dramatic increase in the number and percentage of minorities in the entry-level labor force. These young people want, need, and deserve jobs, but all too often, by the time they reach employment age, the effects of poverty, discrimination, and poor schooling have left them virtually unemployable.

Perhaps the best way to bring this problem into focus is to look at the number of children who are growing up in poverty in this country and to look at the effects of poverty on children. First of all, the number of children in poverty families is increasing, up from 16.8% in 1975 to 21% in 1985. For blacks and Hispanics, the rates are far higher—46% and 39%, respectively.

One of the most dramatic effects of poverty is its impact on educational attainment. Only 43% of black young people who come from poor homes ever graduate from high school. For whites, the figures aren't much better—a graduation rate of only 53%.

Looking at these statistics, some of the other numbers, however painful, aren't really surprising. When almost half of all black children grow up in poverty, and most of those don't graduate from high school, it isn't too surprising to learn that scarcely more than half of young black males aged 16 to 19 are even in the labor force, that less than a third of them are employed, and that one-quarter have never been employed.

Our rising drop-out rate goes hand in hand with another of this country's most disturbing new trends—the increase in teenage pregnancy. About 1,000,000 teenage girls become pregnant each year. Twelve percent of today's teenage girls are already mothers. Half of these young women will not graduate from high school. Half of them are raising their children as single parents, and three-quarters of all single mothers under 25 are living in poverty.

What we are seeing is the development of a potentially permanent underclass, composed of blacks, whites, and Hispanics. Each year, hundreds of thousands of young people do not make the transition from school to work. They reach working age without the basic knowledge they need to learn even the simple skills necessary for success in an entry-level job. Without education and without opportunity, these people are becoming dangerously alienated from a society with which they have little in common.

EVERYONE COUNTS

At the same time that these people are struggling—and failing—to enter the world of work, others are in danger of falling out. Millions of Americans now holding jobs can not have any confidence that the skills they possess will assure them continued employment.

For example, during the five-year period 1979 to 1984, 11,500,000 Americans lost their jobs as a result of plant shutdowns or relocations, technological improvements, or shrinking output. While the vast majority found new jobs, 1,300,000 dislocated workers remained unemployed, despite an expanding economy. Of all dislocated workers, about 20% require remedial education.

We need all these people. Our economy isn't going to grow the way we predict, and the way we want, unless we have a closer relationship between education, job training, and private-sector employers—a relationship that begins with kids in high school and doesn't end until retirement. We need to train our young people right to get their first job, and be able to retrain them when necessary, to save them from dead-end and disappearing jobs.

One of the things Americans have had to learn in these past 15 years is that there is no magic formula for productivity and prosperity, given to us alone. Other nations have found out what we have been up to and are applying these lessons with a vengeance. If we do not grasp this opportunity for economic expansion, if we do not utilize the ingenuity and energy of all our citizens, the rewards that should be ours will pass to others.

There is no magic bullet, no simple blueprint, that will assure America's economic future. Any attempt to set up a single, rigid economic policy for a country as large and varied as the U. S. would create infinitely more problems than it could ever solve. What we do need is an awareness of both the opportunities and dangers that lie ahead.

We need to be aware of the importance of doing a better job in the public schools to reach at-risk students, to keep them in school, and to see that they graduate with real, marketable skills. Business must acknowledge its stake in education and training for young people, realizing that it is in danger of not getting the workers it needs.

Training and retraining already cost private industry some $30-40,000,000,000 a year. We need to build on this foundation, to create public/private partnerships at the state and local level that will take advantage of the resources of business, labor, education, and government employment and training programs. A close integration of public and private training expertise and facilities would help assure workers that they will be able to acquire new skills and, if necessary, new careers. Such a program could greatly assist small business, which leads in the creation of new jobs, but at the same time has limited capacity for training.

The basis for these partnerships already exist, in the form of 50 state job training coordinating councils and more than 600 local private industry councils set up under the Federal Job Training Partnership Act. These councils are giving the public and private sectors the opportunity to address the employment problems that arise in their states and communities in an informed, coherent manner. Although the councils have only been in existence for a few years, communities already are using them to break down the barriers between government and business, education and employment, and welfare and job training.

While we work to train our labor force to meet the demands of our economy, we also must reshape our nation's work environment to meet the needs of our changing labor force. The steady rise of the number of women in the job market, women who often are the sole support of a family, means that more and better child care is a necessity, as are, in many cases, more flexible hours as well. Since fewer workers will be coming into the labor force, we must make effective use of the workers who are already there, through retraining and elimination of barriers to the use of older workers.

The last time Americans talked seriously about bringing the poor into the mainstream of American society was during the late 1960's, the heyday of the famous and often-criticized "War on Poverty." Much of the rhetoric of that era sounds outdated now, and many of its programs have been discarded, but the goals that motivated that effort—full employment and a satisfying, productive life for all Americans—are hard to fault.

If we give our free enterprise system the trained workers it needs, we can begin to approach those goals. It's a job that is much too big, much too varied, for government to handle on its own. That is a lesson that we've learned since the 1960's. We also have learned that good things don't just happen on their own. The problems of illiteracy, poverty, and alienation aren't going to go away, aren't even going to get better, unless an informed, educated public/private partnership makes up its mind to address them.

THE CHANGING SCHOOLS

28 *Computers and the Changing Schools*

JUDITH A. PERROLLE

The ideal function of education in an industrial democracy is to provide equal opportunity for all to develop their talents and prepare for occupational success. For society, the school's function is to develop its human resources and place people in appropriate jobs and roles. "Education is not only a moral renovator and a multiplier of intellectual power, but also the most prolific parent of material riches. It is not only the most honest and honorable, but the surest means of amassing property."

Critics of this view argue that education has another, less obvious function for society. They believe that a major function of education is to maintain the existing stratification system by convincing unsuccessful individuals that they have "failed." If there are more talented individuals than there are high-status job opportunities, then those who do not find rewarding jobs might see their position in society as a social rather than a personal problem. From this perspective, one purpose of education is to convince individuals to accept the status quo as a "given" and to define their success or failure in personal terms. Education in this way serves as an agent of social control, teaching students to conform to existing social structures. From either perspective, educational institutions experience a crisis when other social institutions are going through a period of rapid change.

Source: From Computers and Social Change by Judith A. Perrolle, © 1987 by Wadsworth, Inc. Reprinted by permission of the publisher, pp. 114–119.

THE CRISIS IN AMERICAN EDUCATION

The report of the National Commission on Excellence in Education presents an overwhelming indictment of U.S. public education and calls for sweeping reforms. It found that nearly 40 percent of seventeen-year-olds cannot draw inferences from written material and that only one-third can solve a math problem of several steps. Among the commission's recommendations is a required three years of science, three years of mathematics, and one-half year of computer science in the high schools. Unfortunately, there was already a great shortage of math and science teachers in 1983, and 50 percent of those teaching then were rated as "unqualified" by their principals.

One proposed solution to the shortage of science and math teachers is to retrain mid-career high-tech professionals. Yet teaching pays poorly compared to technical and professional work. With a 1982 national average salary of $17,000 for twelve-year veterans of teaching and entry-level salaries as low as Mississippi's $11,275, very few mid-career high-tech professionals are likely to volunteer to help save America's schools. Although President Reagan interpreted the commission report as calling for "an end to federal intrusion," the commission assigns the federal government "the primary responsibility for identifying, funding and supporting the national interest in education." This view is reaffirmed in the Twentieth Century Fund Task Force on Federal Elementary and Secondary Education Policy. Yet it is not clear how education is to be funded, especially since budget plans have included cutting federal support for education.

COMPUTER LITERACY AS A TECHNOLOGICAL FIX

The term *computer literacy* encompasses a vague concept of something our schools are failing to provide for the future citizens of the information age. Joseph Weizenbaum considers public concern with computer literacy another instance of the American habit of searching for a simple "technological fix" for complex and fundamental problems. Weizenbaum argues that the ability to work with a computer will be like the ability to read and write, do arithmetic, or drive a motor vehicle. Although these were once rare skills that assured placement in a job, they are now expected of most adults in industrial countries. Although computer literacy may, like driver education, be taught in the schools, a rudimentary understanding of computers is no solution at all to our basic educational shortcomings. Indeed, if we fail to

> require literacy in communication, sciences, and mathematics before applying computers to the tasks at hand, we may find that we have developed individuals who may be proficient in coding programs, who may even write some simple

ones, but individuals who cannot do much more than rely on pre-packaged subroutines to accomplish anything technical or scientific of any significance.

The College Board's definition of *computer competency* considerably broadens the concept of computer literacy. It includes basic knowledge of the workings and use of computers, including how to develop programs, how to evaluate computer results through "mental calculation and estimation," the knowledge of when and how computers can be used, and "some understanding of the problems and issues confronting individuals—and society generally—in the use of computers, including the social and economic effects of computers and the ethics involved in their use." University innovations are integrating the use of computers with more general education, as in new programs at MIT, Carnegie-Mellon University, Harvard Medical School, and Oberlin. Even in professional computer science programs, the goal is "not to generate technicians" but to graduate "educated experts" with substantive areas of noncomputer skills and a base in the humanities and social sciences.

The outlook in high school and secondary school curricula is less promising. Most teachers do not understand computer science, and existing in-service training programs provide only rudimentary skills. Most new lesson plans are little more than computerized versions of study guides and rote teaching. Some notable exceptions are LOGO, Sesame Street instruction systems for children, and PLATO computerbased science courses. In short, computer-aided instruction is a long way from being a solution to the crisis in American education.

Prescriptions for improvements in education generally assume that we know what learning and teaching are about. Many theorists believe that education involves human elements that cannot be programmed. Social interactions among students and teachers, facilities of intuition and imagery, and shared biologically and culturally based "nature" are all part of the traditional teaching process. They would dispute the claim by artificial intelligence expert Patrick Winston that "the methodology involved in making smart programs transfers to making smart people." Also, at an estimated 1983 cost of $125 per pupil per year, many school administrators are unable to afford a technological solution for their problems even if they believe it would improve the quality of education.

In the 1983 *World Yearbook of Education,* devoted to an examination of computers and education, editor Jacquetta Megarry argues, "the explosive growth of knowledge and the spectacular advances in microtechnology have underlined the importance of teaching students *how* to learn and *how* to retrieve information, rather than facts and even skills which may become rapidly obsolescent." This is a call for a return to basics in education—not better techniques for transmitting basic facts to students, but a better teaching of basic reasoning from information. If the wisdom of the past is presented to students as just more data, it is not likely to be of much help to them in the future.

THE EMERGING ROLE OF THE PRIVATE SECTOR

From a conflict perspective, schools are one of the arenas in which the battles of social change are waged. During the industrial revolution, manufacturers organized institutes to train craftsmen in science and technical skills. It took a hundred years before the English state came around to their point of view and began to support government-financed education for the industrial working class. Behind the furor over the current state of American education are strong pressures from industry and business, as spelled out quite clearly in the 1983 report of the Business—Higher Education Forum. The decline of American industrial preeminence in the world is, in this view, a business problem to be solved by restructuring education to serve corporate needs. Elizabeth Useem concludes that there is a need for corporate involvement in education, but indicates that conflicts exist between the goals of businesses and educational institutions, and "despite the calls for reform that dominate the political airwaves, cutbacks and decline still haunt the hallways of American schools."

Reductions in student loans and federal assistance to universities may be offset by new private sector educational activities. In areas of particularly rapid technological change, businesses complain that students are inappropriately prepared by the schools. They often sponsor on-the-job training programs for workers in new technologies and professional education and reeducation programs. Some industries are trying to improve the situation by supporting universities with gifts of equipment, the loan of employees for university teaching, and cooperative work-study programs for students. Corporations have also begun to compete with universities in providing technical education. By 1985, corporate enrollments and expenditures were almost as large as those of American colleges and universities.

There is also a growing number of small-scale entrepreneurs interested in educating others with the aid of information technology. Some of these companies design professional training programs modeled after academic conferences, but sometimes they look like little more than sales pitches for new hardware or software. Other companies provide training in programming and home computer usage in order to solve part of the "computer literacy" problem. These new private sector efforts appear to be concentrated in the marketing of new techniques rather than in the general transmission of cultural information. Although they are certainly a part of an information production and distribution economy, most have not yet tried to compete with public and nonprofit schools as socializing agents.

One limit on the expansion of education for profit is the society's need for educated citizens. Most businesses want more than technical specialization, especially in their upper-level employees. Companies seek general skills in planning, defining and solving problems, and communicating ideas. Interaction skills and general cultural knowledge are more important to career success than many preprofessional students realize. To the extent that this is the case, we should expect strong business support for what Education Secretary William Bennett calls the university's function of "preparation for life," even as the private sector takes over the function of preparing people for work.

Yet there is disagreement over what preparing people for work means. Experts who predict that new computer technology will enhance employee skills assume the existence of a highly competent work force. Likewise, predictions that the technology will be used to routinize work and control labor assume a strong management interest in taking decisions out of employee hands. Unless education provides graduates with decision-making competence and employment skills, it is not likely that management will seek technological innovations that give employees greater autonomy. But unless business management believes that employees should be well education for their new jobs, it is likely that they will confine their efforts to job training rather than improving the quality of general education.

EDUCATION AND SOCIAL EQUITY

Efforts to solve the crisis in American education with technological fixes and private sector job training may be threatening the access of minorities and the poor to education. Corporations need more than one kind of employee. The well-educated college graduates who would supply companies with competent managers and researchers form only part of the work force. Some jobs require minimal training and labor discipline, not education. These can be obtained from a poorly financed public educational system that teaches "computer literacy" in the narrowest sense of being able to use a computer terminal without understanding or thinking about it. A student who is "computer literate" in the sense that he or she can push buttons in response to machine cues (perhaps selecting a picture of a hamburger on a fast-food sales terminal) can also operate a complex computer system capable of handling sales, orders, delivery, and inventory.

From a functionalist perspective, universal educational opportunity is essential for maintaining social equity. The uneducated represent a waste of human resources and a potential source of social unrest. If, as conflict theorists argue, education perpetuates social inequality by discriminating against the disadvantaged, new patterns of discrimination can be expected from shifts in the educational system. The emergence of a two-class educational system, with high-quality instruction for the elite and narrow vocational training for the rest, would make the main task of the schools to select elite students for separate education and to convince those left in deteriorating educational institutions that it is their own fault. And, because business's needs for labor are global rather than national, reforms in education for the underdeveloped world may increasingly take the form of training inexpensive workers rather than providing the human capital needed for economic development.

THE FUTURE OF EDUCATION

It is impossible to predict the future of educational institutions with any accuracy, but computers probably will not solve their problems. A gradual improvement in educational software should make computer-aided instruction a more useful educational tool than it is at present. The capacity of computer networks to support communication at a distance may be used to replace many teachers, through the development of centralized electronic instruction that resembles television studios more than tranditional classrooms. Or computer-aided instruction may be used by students for information and practice and by teachers in their interactions with students. The outcome is not a technological issue. It depends on how we understand education and its purposes. As private sector education expands, bringing the cultural transmission of knowledge more into the marketplace, we will have to make some political choices about how much we want education and socialization to remain public processes.

THE CHANGING SCHOOLS

29 *The Quiet Revolution*

BROCK N. MEEKS

A Quiet Revolution is taking place. It is a revolution of phosphor, integrated circuits, and carrier tones, and it is occurring within the framework of American education. It is fueled by computer conferencing systems, modems, and on-line databases—the counterparts of yesteryear's chalkboards, lecture halls, and libraries. Its participants speak in code, using terms like "telelearning," "electronic university," and "virtual classroom." The ultimate goal of this revolution is a simple one: getting an education without setting foot inside a classroom.

Education outside the classroom is not a new concept. We've all seen the correspondence courses offered on matchbook covers. However, such courses suffer a high attrition rate (better known around correspondence schools as the body count). Academic studies show that 70 percent dropout rates are not uncommon. The major problem is a lack of communication with the instructor and other students. Without this vital motivation, only the most dedicated students complete a traditional correspondence course.

HALLOWED HALLS OF PHOSPHOR

An electronic university exists within the confines of a mainframe computer and is managed by either electronic mail, a computer conferencing system, or both.

Source: Brook N. Meeks, "The Quiet Revoltion," *BYTE* (February, 1987) p. 183–190.

A student attending an electronic university interacts with instructors and other members of the class by using a computer and a modem.

Electronic mail and computer conferencing allow an on-line course to duplicate many of the dynamics found in a traditional classroom. However, on-line courses carry a tremendous advantage: You don't have to be in the same place at the same time. The instructor and students can be in several different states and time zones and "attend" class at any time. International students can also participate in such courses because these electronic classrooms are hooked to packet-switched carrier networks, making them accessible from almost anywhere in the world.

ON-LINE MENTORS

TeleLearning Systems of San Francisco operates the Electronic University Network (EUN). During its four-year history, TeleLearning Systems has delivered college-level courses electronically, using telecommunications as the link between instructor and student. However, EUN recently underwent a radical restructuring of its on-line program.

In the old system, courses were delivered on disk, with associated hard-copy reading and study materials. The course disk contained a fixed set of assignments and projects for the student to complete and upload to the instructor. It was classic computer-assisted instruction, with an element of telecommunications thrown in to make the process more convenient. EUN's new structure, 180 degrees from the old system, is aptly called Protege. (Protege refers to both the new program and the software that drives it.)

"The essential idea of the program is to connect proteges with mentors," says James Milokovic, vice president of research and development at EUN and designer of the Protege software. The concept of the Protege program is analogous to a Copernican revolution, says Milokovic. "Instead of the student having to revolve around the instructor being in a certain building at a certian time on a certain day, now the entire program, including the instructor, revolves around the student."

The Protege software is the heart of EUN's new program. It's completely menu-driven and operates on one-key commands. All assignments are completed using the Protege disk. On disk are knowledge templates for performing all course requirements. For example, if you want to send your on-line mentor a question or pose a problem, you simply compose your query in the electronic mail template. To send the note to your instructor, you turn on the modem, hit one key, and your message is automatically delivered. You never enter another keystroke.

Course assignments are now uploaded to the student one at a time. This way the instructor stays in personal contact with the student. "It is of significant interest to the instructor to know [whether] you work for [a particular company] or are a minister," says Milokovic. This type of personal contact lets the instructor customize the course to the workaday world of the student. For example, besides teaching you the basics of marketing, the instructor may have you apply your textbook learning to how your company really designs its marketing strategy.

Currently, 200 colleges and universities participate in the program that forms EUN and provide EUN with more than 200 on-line courses. Each university or college develops its own course for inclusion in the EUN course catalog. Credit is granted from the institution that developed the course. If you've taken a course in economics from a professor at Harvard University, then Harvard grants you the credit, not EUN. EUN is simply the link between you and your mentor. If someone looks at your college transcript, there is no difference between a class taken on campus or over the modem.

Besides the individual classes, EUN also offers degree programs: three different M.B.A. degrees from City University of Bellevue, Washington, and four undergraduate degrees from Edison State College in New Jersey.

All courses offered by EUN meet or exceed standard College Level Examination Program (CLEP) requirements. This means your accumulated credit can be transferred to more than 1700 colleges and universities that accept CLEP-level courses. Costs begin at $215 for an average three-credit course and include connect time. The price goes up another $25 to $50 depending on your choice of books.

TELECONFERENCING

The New York Institute of Technology (NYIT) has established the fully accredited American Open University (AOU) on a VAX-11/730 using the Participate (PARTI) computer conferencing software. Currently, AOU offers 130 online courses and three B.S. degrees in business administration, general studies, and behavioral science. Each degree program requires 120 credits for completion, exactly the same as NYIT's on-campus degree programs require. "These aren't special programs for on-line students," says Ward Deutschman, dean of NYIT's Center for Adult and Professional Education. "Because of accrediting requirements they have to be the same courses offered on campus."

AOU emulates a traditional classroom using teleconferencing. When a course is in session, all interaction among the students and instructors takes place in a course conference. Here, students ask questions and discuss the assignments prepared by the instructor. They can also reply to the comments of other students.

Subconferences, or branches off the main course conference, are also possible. For instance, you might use a subconference when students are working in teams on a particular assignment. "Students often form 'study group' conferences as well," Deutschman said. These on-line study groups allow students to share notes or compare personal observations about the class.

When starting at AOU, you work with a special advisor to develop a background portfolio. This portfolio takes into account any previous college credit, military and corporate training, and life experience. "We don't want to take your time or money for things you've already learned," says Deutschman. After your portfolio is complete, your advisor will work out a degree map that allows you to chart a course of study to complete your degree. If you just want to bone up on a subject or two while accumulating some extra college credits, you can take individual courses.

When you sign up for a class, you receive a course learning package containing a textbook, an extensive (100- to 150-page) course outline, sample tests, quizzes, course objectives, and instructor background information. Once on-line (AOU is on the Telenet packet-switched network), you will find all the support typically available to on-campus students.

For example, AOU students can access the on-line librarian for reference materials. The AOU librarians will look up your request and send you a message explaining where, in your local area, that information is available. If you happen to live in a remote area where extensive library support is unavailable, the AOU librarians can arrange for an interlibrary loan. Students can also talk to counselors and advisors on-line and send private correspondence to the instructor or to other students.

Costs are $85 per credit—which is less than an on-campus course—and most courses are for three credits. "On-line students don't tax our overhead and don't use buildings or electricity," said Deutschman, "so we don't have to charge them as much." Your course fees include six hours of on-line time. If you happen to use more than six hours, you are billed on a prorata basis. "But for what it's worth, [so far] we've never charged anyone for extra time on-line," said Deutschman adding that AOU is more concerned with having students interact on-line—not watching the clock.

CLASSROOM FOR THE WORLD

"We want to be a university to the world," says Connected Education's director Paul Levinson. "Our philosophy is that education shouldn't be tied to a classroom or geographic region," a heady claim for a program that saw its first on-line semester open in September 1985. However, more than a year later, Connected Education has both the statistics and the enrollment to back up that claim. It has already drawn a fair contingent of international students. Last semester Connected Education had students enrolled from Japan, Singapore, and Columbia.

In conjunction with the New York-based New School of Social Research, Connected Education offers graduate and undergraduate courses in the New School's Media Studies program. Levinson says a Ph.D. program is well into the development stage.

Each course offered by Connected Education brings together an expert in a particular field of study and students needing graduate-level course work. All the courses are taught via the New Jersey Institute of Technology's Electronic Information Exchange System (EIES). All courses run on the semester system and are two months long, during which time the students and the instructor discuss various issues relevant to the course topic. An interested student can obtain a detailed course outline and a required reading list before committing to taking a specific course.

Unlike the structured courses taught at AOU, Connected Education courses are propelled by the give and take of daily on-line dialogue. Although instructors follow a prearranged course outline, they're flexible enough to change the direction of the course according to the needs of the class.

During the course the instructor assigns various projects and reading assignments that the students are required to complete and comment on. Daily participation in the course is encouraged. To obtain a passing grade, each student is required to input at least two messages per week. In reality, students often upload several messages a day—message traffic in Connected Education's courses runs extremely high. Students must also take a midterm and final exam. These exams are in essay form and are uploaded to the instructor.

Connected Education's first semester offered courses such as "Teleworld," an analysis of the social impacts of new media; "Artificial Intelligence and Real Life," which discussed topics such as "Is protein necessary for intelligence?"; and "Propaganda: The literary Science." Other courses offered range from "Telelaw" to "Politics of the Information Age."

Each course costs $795 and earns three graduate-level credit hours. All on-line connect charges are included, and students receive unlimited time on the conferencing system for class participation. Students have electronic mail and real-time conferencing capabilities as well. Levinson says transfer of Connected Education's credits to other institutions is possible but at the discretion of the receiving school.

ON-LINE BRAIN TRUST

Western Behavioral Sciences Institute (WBSI), a 27-year-old "think tank," offers a unique program designed to help executives and government officials further their management and communications skills in several different areas. "We've developed a program that alerts today's decision makers to the new requirements of leadership," says WBSI president, Dr. Richard Farson. The on-line structure of WBSI's program offers the opportunity for people in leadership to learn from each other, says Farson.

Participants in WBSI's School of Management and Strategic Studies (SMSS), as the program is called, include college presidents, corporate CEOs, national and foreign government officials, military and CIA personnel, best-selling authors, and former astronauts, among others. The program lasts for two years and is divided into four modules of six courses each. Each course is a month long.

Course instructors are drawn from the nation's top experts. For example, Harland Cleveland, dean of the Hubert H. Humphrey Institute of Public Affairs, teaches a course called "The Knowledge Dynamic in a Global Society." Another instructor is Herbert York, director of the Institute for the Study of Global Conflict and Cooperation and U.S. ambassador to the Comprehensive Test Ban negotiations from 1979 to 1981. Combine experts like these with some of the nation's most influential citizens and you have an on-line brain trust that has no equal.

The SMSS program operates much like the Connected Education program. Participants interact with the instructor and others in the course via the EIES conferencing system. The advantage of such a program is that it doesn't take the participant away from the work environment as do some of the high-powered management schools were participants are isolated from their jobs for up to 13 weeks.

Uncommon in the world of electronic learning, WBSI holds a week of face-to-face meetings before each module begins. This allows participants to meet each other and the instructors. This has the particular advantage of building up a strong on-line community that keeps participant motivation high. It also gives new participants a chance to get some individual hands-on training in the workings of the somewhat electric EIES conferencing system. This semiannual meeting is so successful at developing relationships among participants that alumni continue to participate in the courses and often act as freelance advisors to "undergraduates."

There is a small catch: the price tag. A two-year stint at SMSS runs $25,000. This is usually paid for by the participant's employer. Although the cost is high, it includes all a participant's on-line time for the life of the course. WBSI will even provide you with the computer setup of your choice if you don't already own one. To make class participation easier, WBSI has developed its own communications software specifically designed around the EIES system.

In a review of the nation's leading business management schools, done by the *Harvard Business Review,* SMSS placed in the top five. The review claimed that other management programs should take a close look at SMSS and follow the program, both in "form and content."

ON-LINE ADVANTAGE

On-line education holds a decided advantage over traditional on-campus education for several segments of the population. The disabled community benefits from the built-in ambiguity of the medium; when participating in an on-line class, it doesn't matter how long it takes to answer a question or type at a keyboard. Such a class meets on the disabled person's own timetable. Professionals who have little time to attend evening classes but seek to upgrade their education can dictate the time and place as well as learn from an acknowledged expert in the field.

But how credible is this type of education? While opinions differ, tests show that a higher degree of achievement and retention from distance learning courses is possible.

Is on-line education for everyone? AOU's Deutschman sums it up this way, "If a person asked me whether he or she should enroll as an on-campus student or go on-line, I wouldn't hesitate to recommend the on-campus environment. The people we are interested in addressing are those whose schedules, locations, or physical handicaps make it impossible to obtain a conventional college education."

While it is unlikely that this quiet revolution will replace its traditional counterpart, one thing is certain: Computer conferencing will have a penetrating effect on the established system and force it to take a closer look at the goals, methods, and purposes of on-line education.

THE DIVERSIFICATION OF FAMILY LIFE

30 An Economic Bust for the Baby Boom

FRANK S. LEVY AND RICHARD C. MICHEL

A careful look at the economic data of the past decade shows that while the middle class as a group is not disappearing, the young middle class (principally the baby boomers) has experienced a dramatic decline in its ability to pursue the conventional American dream: a home, financial security, and education for children. To understand how this has come about and what the prospects are for the future, we must expand the discussion to look not only at trends in income inequality but trends in income growth.

For the last 11 years, the American economy has been in a quiet depression in which neither wages nor family incomes have grown. All around us, there are signs that the middle class is in trouble. Single persons are postponing marriage. Families are postponing having children. Good jobs in manufacturing are being lost. Young people feel substantial economic pressure. But official statistics seem to show something different. The family income distribution has roughly the same shape today as it had in 1947 (and all the years in between), while consumption spending per person rose faster in the last 10 years than it did in the fondly-remembered 1950s. In the following paragraphs, we resolve this apparent conflict.

Reprinted with permission of Publisher, M.E. Sharpe, Inc. 80 Business Park Drive, Armonk, New York 10504 USA, from the March/April 1986 issue of *Challenge*, pp. 33–39.

ACHIEVING THE AMERICAN DREAM

The material American dream is by no means a recent concept. In the middle of the 19th century, Alexis deTocqueville noted the importance of material well-being in American life:

> "In America the passion for physical well-being is not always exclusive, but it is general; and if all do not feel it in the same manner, yet it is felt by all."

What was true in deTocqueville's observation is equally true today. A central theme of modern American life is the chance to enjoy a "middle-class standard of living." This standard is not officially defined anywhere, but it exists in many subtle forms throughout our society, from television advertisements to casual conversations. The middle-class American dream has come to include such material goods as a single-family home, one or two cars (including a new one), a washing machine and dryer, a dishwasher, a color TV, raising and educating children, providing for a lengthy period of retirement, and so on. When economic times are good and incomes are growing rapidly, the standard dream can be easily expanded to include a certain amount of vacation travel, dance lessons for the children, a home workshop, a personal computer, etc., but the dream is much slower to contract when economic times are bad.

A healthy economy changes constantly, and so even in the best of times a large number of people are trying to achieve the American dream (or to achieve it again). Included in the number are young workers just starting out, retired persons seeking to replace their wages, workers who have lost jobs in declining industries, and workers who have been displaced from successful industries through technical change.

When incomes are rising rapidly, many of these people will be successful. In the United States, this was true from 1947 to 1973—real (inflation-adjusted) wages were rising 2.5 to 3.0 percent per year. Two examples translate these numbers into human scale.

Consider a man passing from age 40 to age 50. In terms of career, the big promotions are behind him, but earnings can still rise if earnings are rising throughout society. Prior to 1973, this is exactly what happened, and the average man passing from 40 to 50 saw earnings rise by about 30 percent.

Consider next a younger man passing from 25 to 35. He is in the "fast track" portions of a career, and receives the benefits of promotions as well as general rises in living standards. Prior to 1973, the average man passing from 25 to 35 saw earnings rise by about 110 percent.

Family incomes traced a parallel pattern. During President Eisenhower's two terms in office, the economy experienced one mild and one quite serious recession, but the census measure of median family income (i.e., the income of the average family) increased by 30 percent after adjusting for inflation. During

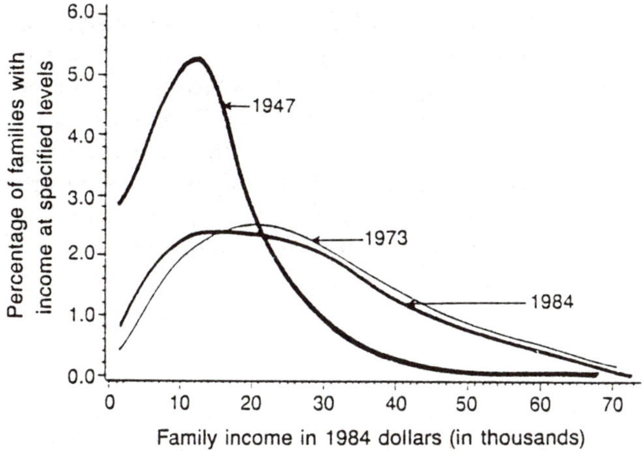

Figure 30–1 Smoothed histogram of the family income distribution 1947, 1973, 1984.

Note: This smoothed histogram was created for ease of graphic presentation. Technically, the height of a given curve at a given point represents the percentage of all families who have incomes in the $1,000 interval around that point.

Source (for all figures and table): Bureau of the Census data and authors' own calculations.

the eight years of Presidents Kennedy and Johnson, median family income increased by another 30 percent. And incomes kept growing through President Nixon's first term. Between 1947 and 1973, median family income had doubled from $14,000 to $28,000 (both figures are in 1984 dollars) and had never gone more than three years without setting a new record. During this time, measures of income inequality remained fairly constant, but the *whole income distribution* kept moving to higher and higher ground as most incomes rose (Figure 30-1).

But in some ways, 1973 was the last good year. After 1973, growth

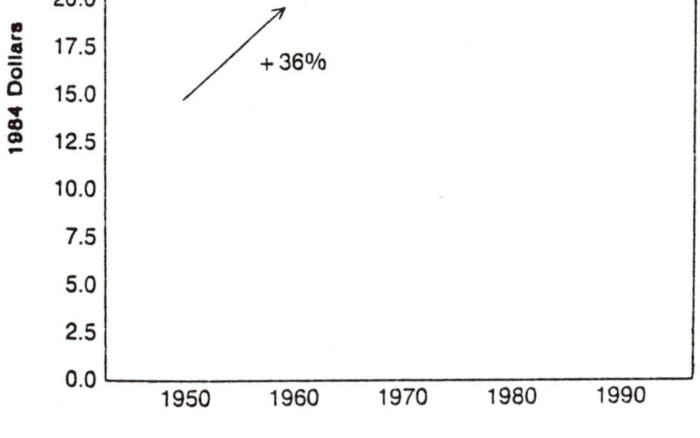

Figure 30–2 Average income gain for men pasing from age 40 to age 50.

largely stopped, and wages and incomes showed the effects. Previously, men passing from 40 to 50 had seen their earnings grow by 30 percent. But those who were 40 in 1973 saw their earnings over the next decade *decline* by 14 percent (Figure 30-2). And where younger men passing from 25 to 35 had seen their earnings more than double, those who were 25 in 1973 saw their earnings over the next decade grow by only 16 percent (Figure 30-3). Family incomes traced a similar path. From 1974–84, median family income (in 1984 dollars) remained below its 1973 high point ($28,167), and in 1984 it stood at $26,433.

WHAT WENT WRONG?

Like people in the middle of the Great Depression, we are not quite sure. The large OPEC oil price increases of 1973–74 and 1979–80 each reduced purchasing power by 5 percent. More important was the way in which worker productivity suddenly stopped growing after 1973. Rising productivity—rising output per man hour—is the ultimate source of rising living standards, and its sudden halt remains something of a mystery. But if experts do not fully understand the productivity slowdown, most attribute it to some combination of three generalized economic problems: rapid increases in energy prices (as distinct from high, stable energy prices), a labor force that was growing very fast, and a ten-year inflation that was very hard to break and which threw a wrench into the process of making sound economic projections.

In the 1973–84 period, as in earlier postwar years, the shape of income distribution (income inequality) remained relatively constant. But income distri-

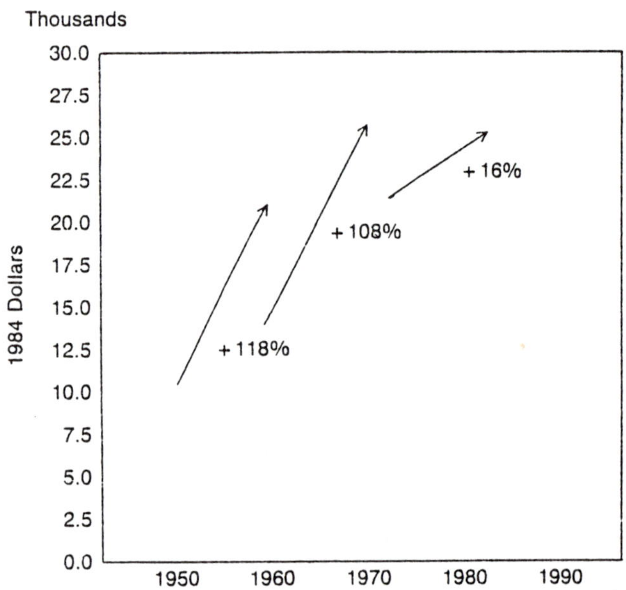

Figure 30–3 Average income gain for men passing from age 25 to 35.

bution stopped moving up. To the contrary, it slipped back a little (Figure 30-1), and this put the brakes on mobility. People who had already attained the middle-class dream found they had certain protections. One was job seniority (provided they were not working in a declining industry). Another was a fixed-payment mortgage, which kept their housing costs under control, particularly since mortgage payments stayed constant in the face of inflation. But for people who had not yet attained the middle-class standard (or who had lost the standard and were trying to regain it), it looked increasingly out of reach.

Stagnation's influence was widespread, as shown by two examples. The first depicts the growth of service-sector jobs relative to manufacturing jobs. Some commentators have argued that the explosion of low-wage service-sector jobs in the 1970s was doing us in. Implicit in this argument is the sense that service-sector jobs are something new. They are not: in the late 1940s, fully one-half of all employment (on a full-time equivalent basis) was already in the service sector, and this did nothing to stop 26 years of dramatic wage growth. During this time, service-sector jobs always paid 10-15 percent less than manufacturing jobs, but because wages were growing in *all* sectors, differences across sectors were not so important. In particular, a man could lose a good job and take a pay cut, but still have a sense that he could "grow back" to his old living standard in a few years. But from 1973 to 1984, when wages were stagnant, such thoughts of growing back were fanciful, and wage differences across sectors loomed much more importantly. A laid-off steel worker in 1984 had very little prospect of quickly restoring his former wages through promotions in a new service-sector job. His plight was thus far more serious than that of a laid-off steel worker in the 1960s.

The second example involves upward mobility across generations. In the decades prior to the 1970s, children expected early on to live better than their parents. Such is not now the case. A father-son example illustrates this dramatically. Suppose a young man of 18 or 19 is preparing to leave his parents' home. As he leaves, he sees what his father's salary would buy and he keeps the memory as a personal yardstick. In the 1950s or 1960s, the young man would have quickly measured up. By age 30, he already would have been earning one-third more than his father earned when the young man left home. But today, a 30-year-old man is earning about 10 percent less than his father earned when the young man left home. The fact that the young man's father owns a house with easy mortgage payments only sharpens the contrast in their economic status.

THE DEMOGRAPHIC SHIFT

If things have been so bad, how does one explain the fact that consumption per capita was rising? The answer largely involves demographics.

The most important demographic fact was the increasing proportion of the population who worked in the 1970s. During the decade, women of all ages entered the labor force in large numbers. Additionally, the baby-boom cohorts born in the 1950s and 1960s entered the labor force. The result was that the proportion of the population at work increased from 41 percent in 1970 to almost 50 percent today. Data on young families make this point even more dramatically. Among 25- to 34-year-old married couples, 47 percent of the wives worked in 1973. Today, two-thirds of all young wives are working. This greater proportion of workers helped prop up incomes.

In addition, many young persons in the 1970s postponed marriage or did not marry at all, a trend that continues today. Since 1973, the median age at first marriage has increased from 21 years to 22.8 years for women, and from 23.2 years to 25.4 years for men, the highest levels since the early 1900s.

Finally, the young families that were formed largely postponed having children of their own. Since 1973, the average number of children in young families (age 25 to 34) has dropped by a remarkable 27 percent. Among families with heads between 35 and 44, the decrease is only slightly less. This is a dramatic drop in the number of persons consuming goods in families, and it helped consumption per capita (i.e., per man, woman, and child) to grow faster than would have been predicted by wages alone.

On a national basis, then, consumption per head rose faster in the stagnant 1970s than it had in the booming 1950s, but the sources of increase differed between the decades. Per-capita consumption rose in the 1970s, because in aggregate terms the proportion of the population in the labor force rose from 41 percent to 48 percent. During the 1950s, the labor force remained constant at 40 percent of the population, and consumption per head rose because workers' wages rose.

The decade of the 1970s was thus the inverse of the 1950s. In the 1950s, wages were rising smartly, but per-capita income grew slowly because families were feeling sufficiently prosperous and optimistic to buy the most important consumption goods of all—children—in large quantities. In the 1970s, wage stagnation left individuals with two choices: decrease consumption or increase income through "quality of life" demographic accommodations. They chose the latter.

In sum, the 1973-84 period revealed just how much of postwar society was predicated on the assumption of rising real wages. When stagnation entered, it was clear that something was going wrong and that acquiring the middle-class dream was becoming more difficult. It was a short jump to the conclusion that income distribution was splitting apart into rich and poor. The issue, however, was not growing inequality of current income (the income distribution, after all, retained its shape) but a growing inequality of *prospects*—of the chance that one would enjoy a middle-class standard of living with a house, children, cars, retirement—the whole package. Those who already had it were largely able to retain it, while those who didn't have it saw their prospects dim.

THE ECONOMIC STATUS OF THE YOUNG MIDDLE CLASS

A family's economic situation depends on more than current income. Assets, savings, consumption patterns, debts, and prospects for the future are all important too. We can get a better sense of today's young middle class by examining these other variables and comparing them with the situation of young families in earlier decades. In this comparison, two variables stand out: one is wage growth; the other is the price of housing.

Begin by considering a young man who was 30 in 1949. On average he was earning $11,924 per year (in 1984 dollars), an apparently modest figure. But with the housing prices and interest rates of the time, he could carry the median-priced home for about 14 percent of his gross monthly pay. And his wages were far from their peak. Over the next ten years, his annual salary would rise by 63 percent in real terms (Table 30-1).

A 30-year-old man in the late 1950s faced similar circumstances. But by the early 1970s, the situation had worsened. For a 30-year-old man in 1973, carrying charges for the median-priced home would have absorbed 21 percent of monthly pay (up from 14 percent). More important, over the next decade, his annual salary (adjusted for inflation) did not rise at all (Table 30-1).

This man who was 30 in 1973 was born in 1943—too early to be an official member of the baby boom. How, then, did the baby boomers do? Too little time has passed to write their full story. But we know that to this point, they have faced stagnant wages, rapidly increasing housing prices, and very high interest rates. One figure is suggestive. Last year, a typical 30-year-old person who purchased a median-price home would have had to devote *44 percent* of gross monthly income to carrying charges. The future path of his or her earnings is, of course, unknown.

How did young families deal with these deteriorating circumstances?

Table 30-1 The Economic Situation of 30-Year-Old Men, 1949–83

Age 30 in	*Average Earnings at 30 (1984 Dollars)*	*Ratio of Monthly House Carrying Charges to Monthly Earnings at Age 30**	*Average Earnings at 40 (1984 Dollars)*	*Percent Change*
1949	11,924	0.14	19,475	+63
1959	17,188	0.16	25,627	+49
1973	23,580	0.21	23,395	− 1
1983	17,520	0.44	?	?

* Calculations based on median price of homes sold in that year and mortgage terms and interest rates prevalent in that year.

First there were the demographic adjustments we have already described: postponement of marriage, reliance on two earners, postponement of children. But other adjustments were more purely financial.

One was to get help from parents, particularly for housing. In 1980, almost one-third of all first-time home buyers got financial assistance from their relatives, compared to less than 10 percent in 1978. But this option has its limitations: not every family is comfortable enough financially to support younger members in this manner. The recent recession also affected families' abilities to transfer money to their baby-boom relatives. By 1982, the proportion of first-time buyers receiving financial assistance from relatives had dropped to eleven percent. While improving economic conditions are likely to make such assistance more feasible for many families, it appears that this strategy for buying a first home may have peaked.

Another adjustment was to buy smaller houses. Less than half of all new housing units today are single-family detached units, in contrast to more than 60 percent in the mid- to late 1970s. In the 1979 to 1983 period, the percentage of new units with garages and with two or more bathrooms declined, as did the median number of square feet per new unit.

More earners, fewer children, parental help, and smaller houses explain why home-ownership rates among young families declined only modestly. According to a recent study by the MIT-Harvard Joint Center for Housing Studies, home-ownership rates for households with heads under age 25 fell from 23.4 percent in 1973 to 19.4 percent in 1983. For households with heads in the 25-to-29 age group, the decline over this period was from 43.6 percent to 40.7 percent. (Actually, overall home-ownership rates as well as those for younger households peaked in the mid- to late 1970s, so the recent decline has been sharper than indicated here.)

ARE YUPPIES A MYTH?

When we turn from housing to other economic variables, a similar picture emerges. It contains relatively few Yuppies—those famous young couples who buy expensive imported automobiles and regularly eat at upscale restaurants. Yuppies are not a fiction. The baby-boom generation is very large, and even the small percentage who are well-to-do comprise a large enough number to make a strong impression on journalists, car dealers, and restauranteurs. But while Yuppies are not fiction, neither are they typical.

In 1984, the typical young American family (head, age 25–34) consisted of a husband and wife and a single child under age twelve. Less than half owned their own homes. Median pretax family income from all earners totalled $25,157, hardly enough to buy a BMW and eat out regularly. If this is the case, what are young families spending their money on? The answer comes as no surprise to those families: basic necessities.

Consumption expenditure studies by the Bureau of Labor Statistics (BLS) show that between 1973 and 1981 families whose heads were 25–34 experienced a 54-percent increase in their home fuel and utility costs, and a 65-percent increase in gasoline and motor-oil costs for their vehicles. In 1984 dollars, these increases amounted to almost $1,000 per year. Add to that an average annual increase of $231 in shelter costs (for higher rent and mortgages) and young families in 1981 were devoting almost 6 percent more of their disposable income to those three basic necessities than in 1973. These increases, moreover, took place as total consumption was shrinking. The same BLS studies show that between 1973 and 1981, the consumption of young families *declined* by 4 percent, or more than $1,000 per year in 1984 dollars.

With necessities increasing and total consumption declining, other items had to give. A young family in 1981 spent 14 percent less on furniture, 30 percent less on clothes, 15 percent less on personal care, and 38 percent less on charitable contributions than had a similar young family in 1973. And, contrary to popular belief, the average young family in 1981 spent only $47 more per year on food outside of the home (because there were so many young families, they could support a growing restaurant industury even though the average family did not dramatically increase its spending).

The final item to give was savings. In 1981, young families saved less than 1 percent of their after-tax income, contrasted with 4 percent for similar young families in 1973. A recent Federal Reserve Board survey indicated that the proportion of young families holding liquid assets in 1983 had declined from 93 percent to 87 percent relative to a separate survey taken in 1979. While this seems like a small change, it implies that almost one million more young families had no savings whatever to fall back on in the event of an emergency. And while savings diminished, debt increased. The same Federal Reserve Survey mentioned above showed that in 1983, 77 percent of families in the 25-34 age group had incurred some form of installment debt, compared to 67 percent in 1970. And recent monthly Federal Reserve Board data show that consumer debt is at an all-time high.

The savings and debt figures are not to be trivialized. The less a young family is able to save, the more it must postpone its entrance into the housing market. The greater its debt, the less its ability to qualify for mortgage loans, given the affordability guidelines used by financial institutions. Furthermore, nominal interest rates on consumer loans are 50 percent higher today than in the 1970s and, combined with low inflation rates, act to increase the length of repayments and relative value of outstanding debt for young families. In the 1970s, its was easier for young families to grow out of their debts, as inflation eroded the value of the dollars they paid back.

It is also worth remembering that older families had cushions of two kinds. One was homeownership. In the early 1970s, 80 percent of the population over 35 owned their own homes, and these homeowners were shielded from an important part of the decade's inflation. As prices went up rapidly and wages went up more slowly, homeowners' mortgage payments remained constant and more money was left for other things. A fixed mortgage did not fully compensate a family for the rising price of tomato soup or their children's college tuition. But anyone who doubts its importance need only ask a group of middle-aged families the following question: "Today, could you afford to buy the home you're now living in?"

The second cushion, for retired families, was social security. As late as 1965, a quarter of all elderly persons did not qualify for social-security benefits—they had not paid into the system long enough. But by 1975, more than 90 percent were covered, and the extension of these benefits substantially raised the incomes of the elderly. This impact was further magnified by 1972 legislation, which first increased social-security benefits by 20 percent and then indexed them to the Consumer Price Index. This legislation permitted social-security benefits to stay even (and in some cases exceed) the rate of inflation even as the wages of most workers were stagnant or declining.

Here, too, the cushion was not perfect. For example, social security could do nothing about the way in which inflation eroded most private pensions. Nevertheless, over the last 15 years, while income averaged across all families remained roughly constant at $25,000 (in 1984 dollars), the average income for families over 65 rose from $13,500 to $17,130. (Averages can of course be somewhat misleading. The elderly are a heterogeneous group, and about one out of every seven still have incomes below the poverty line. Poverty is particularly high among the "older elderly"—persons above 75—whose social-security coverage is low and whose private pensions have been eroded by inflation. But there is no question that the elderly as a group have fared better economically than other groups in the last decade and a half.)

PROSPECTS FOR THE FUTURE

People now in their seventies began their careers in the teeth of the Great Depression. But because they worked during the 1950s and 1960s, most went on to have comfortable lives. By recent standards, today's young middle class is not doing well. But the path of their lifetime incomes depends very little on what they can extract from other generations, and a great deal on how fast their wages will grow. If the economy were to return to anything like the steady wage growth of 1947 to 1973, the future of today's young middle class would be much brighter.

Faster wage growth, however, depends on increased worker productivity, a very difficult objective to achieve. Experience seems to show that government policy has some ability to control inflation and unemployment, but the restoration of growth in worker productivity is more problematic.

We have noted that the failure of wage growth in the 1970s seemed to come from three major factors: sharp increases in energy prices, a rapidly growing labor force, and serious inflation that became embedded in the economy. These factors seemed largely responsible for the slowdown in the growth of worker productivity which, in turn, slowed the growth of wages.

By 1984, each of these three factors had reversed. Energy prices were stable. Inflation was quite low. And the labor force, having absorbed the baby boom, was again growing slowly. But despite these favorable developments, recent worker productivity growth seems to be settling down to a disappointing one percent per year.

As commentators try to understand the current perplexing economic situation, some have pointed to the role that deficits and debt play in restraining wage growth. The federal budget deficit, for example, helps keep interest rates high, and this in turn contributes to the international trade deficit. From a pragmatic political and social point of view, the deficits are understandable. By spending more than we take in and by importing more than we export, we can, for a time, live beyond our means. In this way we avoid the limits of stagnant incomes, and avoid making difficult personal and political choices. But in the process, we are creating a set of conditions that, like conditions in the 1970s, may be antithetical to productivity growth: high interest rates (which make investments more expensive); imports subsidized by a strong dollar (which make domestic investment less profitable); and so on. We are trading off long-term growth for short-run gain. And if the current generation of middle-class Americans continues to avoid the problem, it will remain for baby boomers in the twenty-first century to bear the costs of resolving not only their personal debt but the national and international debts as well.

There is no painless cure for any generation in dealing with the nation's debt problems. Closing the federal deficit requires tax increases and budget cuts, each of which lowers the living standard of American families in the short run. Just closing the $200 billion deficit requires lowering living standards by an amount equivalent to the increase that comes from two good years of economic growth.

It is hard medicine to swallow, and yet that is one of the choices we face. The issue is not young versus old or even rich versus poor, but rather all of us versus the future. And there, to paraphrase the immortal Pogo, the enemy is ourselves.

THE DIVERSIFICATION OF FAMILY LIFE

31 Changes in Women's Economic and Family Roles in Post-World War II United States

HEIDI I. HARTMANN

In this essay I describe some of the changes that have occurred in the living arrangements of women and men since World War II. Together with changes in women's labor force participation and in their economic activity generally, these recent changes suggest fundamental alterations in gender relations. These changes have been dramatic and hold the potential for further, perhaps more sweeping, changes in the relations between women and men.

A great deal can be said about the effect on women of changes in the economic structure. The current structural transformation of the economy, sometimes viewed as a set of long-term trends in motion since World War II and sometimes viewed as a short-term economic crisis of recent origin, can be held responsible for the feminization of poverty, the high divorce rate, the increased labor force participation of women, and a host of other changes affecting how we live our intimate lives.

While many decry these changes, particularly for their supposed negative impact on women, children, and families, I argue that on the whole the economic changes of the past several decades have been positive for women. Women in advanced industrialized countries today have more access to economic resources independently of men than ever before in human history. They

Source: Heidi I. Hartmann, "Changes in Women's Economic and Family Roles in Post-World War II United States," from *Women, Households, and the Economy,* Lourdes Beneria and Catharine R. Stimpson (eds.). Copyright © 1987 by Rutgers, The State University. Reprinted with permission of Rutgers University Press, pp. 33–64.

have more control over the conditions of their lives, and probably have a higher standard of living relative to men than at any time previously. Of course, I do not deny that the recent economic downturns (in 1980 and 1982), coupled with cutbacks in social welfare programs, have affected women negatively. I argue only that they have not erased the progress of the past thirty to forty years. Confining my analysis to the United States, I will attempt to show how this general progress has occurred and demonstrate the potential for further, positive changes in women's status, as gender relations continue to evolve and to affect economic change as well.

RECENT ECONOMIC CHANGES

In order to describe the economic context for gender relations since World War II, let me digress for a moment to summarize some of the economic changes that have occurred during this period. In my view, the current economic situation has its roots in long-term economic and political changes within the United States as well as changes in the position of the United States relative to other countries in the postwar period. The current retrenchment of social welfare spending in the Reagan administration is at least partly a reflection of a U.S. economy that has been experiencing difficulty for some time. Most obvious among the long-term changes have been the decline in U.S. military hegemony since WW II; the deterioration of U.S. leadership in the international economy; increasing competition for world markets in manufactured goods from Western Europe, Japan, and other countries; increasing demands from third world providers of raw materials; and the movement of U.S. capital to more profitable production sites abroad. The competitive difficulties of U.S. manufacturing have contributed to the continuing decline in the proportion of manufacturing in the GNP and to the increase in output that originates in services. Many have argued that this shift to services is necessarily accompanied by a decline in average wages and a tendency toward a bifurcated income distribution (Bluestone and Harrison, 1983; Bluestone, Harrison, and Gorham, 1984; AFL—CIO, 1984).

These structural changes are reflected in a remarkable lack of economic progress for many people, at least on some indicators. Real incomes have not risen for U.S. families since 1970, although they rose substantially in the decade of the 1960s. Unemployment in 1982–84 was at its highest levels since the Great Depression.

This uneven economic performance, increases in military spending, and the tax cuts of Reagan's "supply-side" national fiscal and monetary policies have combined to produce an enormous deficit in the federal budget accounts. Though they are certainly exacerbated by the current administration's policies, the factors generating pressure for cuts in social welfare spending have a longer history. Federal social welfare outlays as a proportion of GNP actually peaked in

1976 at 12.1 percent, the year Carter was elected (Erie, Rein, and Wiget, 1983). Even with cuts in discretionary programs and slower growth in entitlements such as social security, the deficit looms large and is expected to remain so if tax policy or military spending policy is not reversed or moderated. Both continuing to live with a large deficit and taking drastic action to reduce it threaten prospects for sustained economic recovery. The long-term roots of the current fiscal crisis, then, lie in the generalized restructuring of the U.S. economy that has occurred over the past several decades. The future capacities of the U.S. economy are unclear, but the public expectation of substantial social welfare spending that has been generated by public programs developed over this same period is likely to create continued demand for social services, despite current pressures to reduce the deficit.

The real change in social welfare spending comes not from the recent Reagan cutbacks but from the previous, more gradual evolution that increased federal social welfare spending from 4 percent of GNP in 1950 to 11.5 percent in 1979 (Erie, Rein and Wiget, 1983). This long-term growth in social welfare spending has had significant effects on women's abilities to live independently of men outside of traditional families. The current cutbacks, I think, have to be seen not only as an attempt to discipline labor generally by reducing the social wage (an attempt made necessary by long-term economic difficulties), but also as an attempt to reassert patriarchal control and to turn back gains made by women in this same period. Reductions in social programs have the effect of forcing many back into traditional family forms for economic support, health care, child care, and so on. It must be recognized that the dynamics of state policy stem from gender struggle as well as from class conflict. Social structures and social policies have these dual determinants.

RECENT DEMOGRAPHIC CHANGES

Equally as striking as the postwar changes in U.S. economic structure and economic position vis-à-vis other countries are the postwar demographic changes that have occurred in fertility, family and household formation, and women's labor force participation in the United States. Briefly stated, since 1950 women are delaying marriage and childbearing, having fewer children over their lifetimes, and spending more time over their lifetimes living alone or as heads of their own households—living neither with parents or husbands. And, of course, women's participation in the labor force is at an all-time high. Let us look at some of these changes in more detail.

It continues to be true that by far the vast majority of women, and men, marry. For those between the ages of thirty-five and forty-four in 1980, over 90 percent had been or were currently married, a proportion that represents a modest increase since 1900 for both men and women. But compared to the immediate postwar baby-boom years, marriage is increasingly being delayed. For example, of those women between the ages of twenty and twenty-four, about 30 percent had never married in 1950; by 1980, fully 45 percent had not. Young women thus have more time before marriage to pursue higher education or work in the labor market than they had in 1950 and more time to live on their own (but they continue to have less such time than men, who marry on the average 2.5 years later) (Bianchi and Spain, 1984).

Divorce, as well as marriage, is now more prevalent than it was in 1950. Although the total rate of marital dissolution (death and divorce) has been fairly stable for the past one hundred years, divorce rates have generally increased as death rates have declined. Divorce rates were relatively constant in the postwar period until about 1965; since 1965 divorce rates have more than doubled. It is estimated that half of first marriages now occurring will end in divorce. And although remarriage is common, the proportion of adult women who were currently married declined from 67 percent in 1950 to 59 percent in 1980 (Bianchi and Spain, 1984). The decline has been greater for black women than for white women. Espenshade (1985) recently estimated that a black woman, on the average, can now expect to spend only 22 percent of her life in marriage and a white women only 43 percent.

Coupled with the fact that women live longer than men, and thus are likely to find themselves alone at the older end of the age range, these data regarding delayed marriage and increased divorce combine to generate a very large increase in the proportion of women living alone and/or not in families. The proportion of all households that were one-person households increased from 13 percent of the total in 1950 to 22 percent in 1980 (Kitagawa, 1981). As Table 31-1 shows, of 20 million single-person households in 1984, nearly two-thirds were women living alone (U.S. Bureau of the Census, 1985b). The increase in living alone is occurring at all ages but is particularly striking for the youngest and oldest women. Between 1940 and 1970, for young women eighteen to twenty-four years of age, the proportion living outside families (either alone or with unrelated individuals) increased from 11 to 17 percent of all women in that age group. For older women, those over sixty-five, the proportions living alone more than doubled, from 13.3 percent to 33.4 percent (Kobrin, 1976a and 1976b). Women in all unmarried statuses at all ages are increasingly heading their own households.

During the same period, women have also experienced a decrease in the time spent over their life cycles in childbearing and childrearing. More women are having children, but they are having them later and closer together and are having fewer of them. For example, although childlessness has declined for all women, women between twenty-five and twenty-nine had twice the childlessness rate in 1980 as in 1960 (Bianchi and Spain, 1984). Between 1950 and 1980 the total fertility rate fell from 3.1 children per woman to 1.8, well below the 2.1 rate necessary for population replacement. For most women, childbearing has become increasingly compressed within a span of only a few years' duration.

With the increase in nonmarriage, a significant and growing proportion

Table 31-1 Distribution of Households across Family Types, March 1984

	Number (thousands)	*Percent*
All households	85,407	100%
All families	61,997	72.4
All families with children less than 18 years old	31,046	36.3
Married couples with children	24,339	28.5
only husband in labor force	9,362	11.0
only wife in labor force	489	0.6
both husband and wife in labor force	13,889	16.3
neither husband nor wife in labor force	590	0.7
Female head with children less than 18 years old	5,907	6.9
Male head with children less than 18 years old	0.799	0.9
All families without children less than 18 years old	30,951	36.1
Married couples, no children	25,751	30.1
Female head, no children	3,971	4.6
Male head, no children	1,229	1.4
All nonfamilies	23,410	27.4
Female head	13,658	15.9
living alone	12,425	14.5
with nonfamily	1,233	1.5
Male head	9,752	11.4
living alone	7,529	8.8
with nonfamily	2,223	2.6
(All living alone)	(19,954)	(23.3)

SOURCE: U.S. Bureau of the Census, 1985b: tables 5, 18, 21; pp. 87–92, 171, 200.

NOTE: Of married couples with children, 38% have only the husband in the labor force, and in 57%, both husband and wife are in the labor force.

of women are bearing children outside marriage. Between 1940 and 1970, the fertility rate for unmarried women between the ages of fifteen and forty-four rose from seven births per 1,000 women to twenty-six per 1,000; the increase in extramarital fertility occurred at all ages. Since 1970 there has been some slowing in the rate of increase, and even in some years actual decreases in extramarital fertility rates for some demographic groups. But the number and proportion of childbirths outside marriage has continued to increase, as the number of unmarried women has also increased and fertility within marriage has decreased. Differences between black and white women are especially striking with respect to extramarital fertility. Between 1940 and 1976, the number of births per 1,000 unmarried women between the ages of fifteen and forty-four rose from four to thirteen for white women and from thirty-six to eighty-three for blacks. In 1976, 50 percent of all the births that occurred to black women occurred outside marriage; for whites the figure was 8 percent (Kitagawa, 1981). Although extramarital fertility has increased over this period for both black and white women, it is important to remember that fertility has decreased for black women just as it has for white women; in fact their fertility rates are converging (Hogan, 1983).

Recently, considerable media and public policy attention has been paid to teenage pregnancy. Actually, fertility has been falling among teenagers just as it has been for all women. But fertility outside marriage has increased for young white women (while, since 1970, it has decreased for blacks). Despite this convergence in extramarital birth rates among black and white teenagers, the levels are very much higher for blacks. In 1982 for black teenagers the rate was eighty-seven births per 1,000 unmarried females age fifteen to nineteen, while for whites it was eighteen per 1,000.

To some extent the greater fertility outside marriage and at younger ages of blacks compared to whites represents a different order of marriage and parenting. For first marriages between 1970 and 1974, for example, 38 percent of black women experienced childbirth before marriage compared to 6 percent of white women. As noted above, however, black women can nevertheless expect to spend less of their lifetimes in marriage and fewer of them will marry than white women. Recent estimates suggest that marriage over a lifetime will occur at a lower rate for black women than for white (77 percent versus 92 percent; Hogan, 1983).

The racial differences in fertility and marriage result in the by now familiar statistic that many more black children than white live with their mothers only. Among blacks, more children under eighteen now live with their mothers only than with both parents (Bianchi and Spain, 1984). Among all families with children under eighteen in 1984, 48.8 percent are single-parent-female among blacks, in contrast to 14.7 percent among whites (U.S. Bureau of the Census, 1985b). Because early childbirth has negative consequences for women's educational attainments and lifetime earnings, and because black women may face greater labor market barriers than white women, rates of poverty are approximately double among black single-parent-female families than white ones (in 1982, 56 percent versus 28 percent), and of course the incidence of poverty is higher for single-parent-female families (40 percent in 1982) than for husband-wife or single-parent-male (9 percent in 1982) (Ford Foundation, 1985).

As noted, however, the tendency for women to form and head households on their own has increased tremendously over the past thirty years in all age groups across all nonmarried statuses. And while heading families or households means poverty for some women, in general the tendency to form households is associated with the increases in economic well-being that make it possible, particularly the large increases in the labor force participation of women of all ages. Even the fact that young women with early first births establish their own households rather than live with their parents can be seen as an indicator of increased resources. And although a larger proportion of poor families are headed by women alone than heretofore (48 percent in 1979 versus

23 percent in 1959), actually a decreasing proportion of such households are poor. In 1959, 42.6 percent of families headed by women alone were poor; twenty years later the percent of such families who were poor had fallen to 30.4 (U.S. Bureau of the Census, 1985a: Table 1).

In sum, the experiences of marriage and childbearing have become more universal for women over the course of the twentieth century (and for childbearing, particularly since 1940), but as fertility has fallen and divorce has increased, living in families and raising children has become limited to a shorter period of the average woman's life. A substantial and increasing proportion of women raise children in households that do not include a husband or male partner. At any point in time, fewer people are living in marriages. Overall between 1950 and 1980, husband-wife couples decreased from 78 percent of all households to 61 percent (Bianchi and Spain, 1984). Single-person households have increased dramatically, and increasing numbers of couples are living in nonmarriage relationships. Thus, a general diversification of family and household types has occurred. Table 31-1 provides a general overview of family and household structure in the United States in 1984. (We cannot tell, of course, from the census data on which these figures are based to what extent people maintain important familial ties beyond households with either blood or fictive kin.)

Over this same period, as women's fertility has declined and household types have diversified, women's educational attainment and labor force participation have increased dramatically. Changes in educational attainment for women in the post—World War II period have been less dramatic than changes in labor force participation, but nonetheless substantial. For example, women have historically had lower college completion rates than men, but by 1981 college enrollment rates were approximately equal for men and women, and the subjects they study have also been converging (see Table 31-2). In 1980, nearly half the bachelor's and master's degrees conferred were awarded to women; by 1985, women earned over a third of the doctoral degrees—a substantial increase over 1950. For the population over age twenty-five, a statistic that reflects past as well as current trends, 20 percent of the men versus 13 percent of the women have completed college (these proportions have approximately tripled since 1950 for both sexes) (Bianchi and Spain, 1984).

Labor force participation rates, it is now well known, have increased dramatically since 1950 for every age group of women except those over age sixty-five. And increases have been greatest among women of childbearing age and among women who have children under six (their rate of labor force participation has nearly quadrupled). As Table 31-3 indicates, there is no longer a dip in labor force participation during the childbearing years. Average expected years of work at birth have increased from twelve years to twenty-nine years between 1940 and 1979-80 (Smith, 1982 and 1985). Women, collectively, are earning more money than ever before (even though their average earnings relative to men have not increased), simply because more women are in the labor

Table 31-2 Percentage of Bachelor's Degrees Awarded to Women to Discipline, 1966 and 1981, Selected Fields

Discipline	*1966 (%)*	*1981 (%)*		*1966 (%)*	*1981 (%)*
Agriculture	2.7	30.8	Health	76.9	83.5
Architecture	4.0	18.3	Home economics	97.5	95.0
Biological sciences	28.2	44.1	Mathematics	33.3	42.8
Computer and information sciences	13.0[a]	32.5	Physical sciences	13.6	24.6
Education	75.3	75.0	Psychology	41.0	65.0
Engineering	0.4	10.3	Social sciences	35.0	44.2
English and English literature	66.2	66.5	Economics	9.8	30.5
Foreign languages	70.7	75.6	History	34.6	37.9
			Sociology	59.6	69.6
			All fields	39.9	49.8

SOURCES: U.S. Department of Health, Education and Welfare, Office of Education, "Earned Degrees Conferred: 1965–66"; U.S. Department of Education, National Center for Education Statistics, "Earned Degrees Conferred, 1980–81."

[a] Data are for 1969, the earliest year available.

force for more of their lives than formerly. Women are thus contributing more financially to the support of themselves and their families; many are entirely self-supporting. Through increased labor force participation, women probably have access to more economic resources, independently from men, than ever before in human history.

INTERPRETATION OF CHANGE

One response to all these changes is to certify a family crisis and to bemoan the increased exploitation of women who must support households and children on their own or who bear the brunt of the speed-up that occurs when both adults must work outside the home. But another interpretation, one that I develop and support, is to see these changes as largely positive for women because they contribute to women's increased autonomy from men and their increased economic independence, whether or not they live with men. These changes probably raise women's own standards of living, since having their own sources of income probably allows them literally to spend more money on themselves. As Barbara Bergmann has pointed out, even though women earn less than men, their own earnings can bring them a standard of living comparable to that provided by men in the 1950s. Even with respect to total hours worked by women (housework plus wagework), women are probably better off now than they were just ten years ago. When women enter the labor market, total time spent working increases because wage work is added to housework: even though time spent on housework decreases substantially (from about fifty-five hours to

Table 31-3 Labor Force Participation Rate of Women

		Age group										
Year	*Birth cohort*	*20–24*	*25–29*	*30–34*	*35–39*	*40–44*	*45–49*	*50–54*	*55–59*	*60–64*	*65–69*	*70–74*
1955	1931–35	46.0	35.3	34.7	39.2	44.1	45.9	41.5	35.6	29.0	17.8	6.4
1960	1936–40	46.2	35.7	36.3	40.8	46.8	50.7	48.8	42.2	31.4	17.6	6.8
1965	1941–45	50.0	38.9	38.2	43.6	48.5	51.7	50.1	47.1	34.0	17.4	6.1
1970	1946–50	57.8	45.2	44.7	49.2	52.9	55.0	53.8	49.0	36.1	17.3	5.7
1975	1951–55	64.1	57.0	51.7	54.9	56.8	55.9	53.3	47.9	33.3	14.5	4.8
1980	1956–60	69.2	66.8	64.1	64.9	66.1	62.1	57.8	48.6	33.3	14.7	4.6
1985	1961–65	71.8	71.4	70.3	71.7	71.9	67.8	60.8	50.3	33.4	15.1	4.3

SOURCES: For 1955–1975: U.S. Department of Labor, Bureau of Labor Statistics. *Perspectives on working women: A Data Book.* Bulletin 2080. (Washington, D.C.: Government Printing Office, October 1980.) For 1980 and 1985: Employment and Earnings, January 1981 (Annual Data table 3). Employment and Earnings, January 1986 (Annual Data table 3).

NOTE: Labor Force Participation Rates of Women, 20 years and over, by age, annual averages, selected years, 1955–1979.

thirty-five hours per week), the combined total represents a work week of about sixty-five to seventy hours. In the past few years, however, several studies show that housework time for working women has fallen, not because husbands are picking up more, but because women are simply doing less (Hartmann, 1981b).

It is somewhat risky, of course, to infer a great deal about changes in gender relations from aggregate data such as these. My argument amounts to saying that patriarchy is weakening, and that women have benefited from these observed changes in living arrangments. But not all women have benefited equally. The fact that single black women who head households with children on their own are especially likely to be in poverty raises the issue of interpretation starkly. Does this greater autonomy from men represent choice or necessity? Is family break-up caused by poverty? Does it result from abandonment by men? These are important issues that the available research cannot yet resolve. Nevertheless, I believe my argument is sustained in its main points by what we now know.

Increased Autonomy for Women

The changes observed in the aggregate demographic data support the contention that changes in the degree, extent, and form of male domination have occurred. Three types of changes are considered: the trend toward nonmarriage and toward women maintaining households alone; the changes within marriage; and changes in responsibility for children.

First, as we have seen, the diversification of family forms has led to women having greater autonomy from men in their living situations; this is reflected in larger numbers of women living alone or heading families. The choice to maintain an independent household as a single woman has become common and socially acceptable. Only a minority of divorced women headed their own households in 1950, but most do so now. The possibility of "doubling-up" or living with other family members still exists, but it is no longer necessary. With the elderly, we have come to recognize and accept their desire to be independent, to live separately from adult children. Surely the average thirty-year-old divorced woman prefers to live on her own, too, rather than with her parents, for example. That more do live on their own, then, seems to be a superior choice, "revealed preferred," as an economist would say, chosen in preference to not doing so. Even among low-income women, maintaining a household on one's own generally became more possible over this period. A review of the literature on the impact of welfare policies on household formation shows that welfare programs do facilitate the establishment of a household once a new family has been formed, although they do not generally encourage births (Wilson and Neckerman, 1985). As further evidence of the impact of increased resources, the cutbacks in social welfare programs after 1981 were observed to coincide with a decrease in the rate of new household formation among young adults.

If it is relatively easy to conclude that maintaining one's own household once one becomes a single parent or is divorced is a preferred choice, it is more difficult to say that divorce itself, or parenthood without a male partner, is a result of choice, especially choice by women. Yet here too I would argue that evidence suggests it often is a woman's choice. First, divorce has increased over time as incomes have risen. People seem to want more divorce when they can afford it. The negative income tax experiments of the late 1960s indicate that guaranteed incomes did increase the rate of separation and divorce among poor families. In essence, the income supplement allowed many to escape from unsatisfactory marriages. More recently, Bane (1985) has found that for black women, undergoing a transition to a single-parent-female household is not associated with a transition to poverty status in a majority of cases; rather, poor black women heading their own households tended to be poor in their prior households as well. This finding suggest that it is not abandonment by males that makes these women poor. Rather, given their poverty, many apparently choose to live apart from men and head their own households. And as Petchesky (1984) notes, the increased tendency for unmarried teenage women who have become pregnant to raise their babies on their own (if they decide not to terminate their pregnancies) is an indicator of their increased freedom not be forced into unwanted early marriages. While some young women may want to catch a man and marry, many do not, nor do they choose abortions or adoption. Instead, they apparently believe that they have the capability to raise and support a child. Many older, single, professional women are choosing to have children without men (as noted, extramarital fertility is increasing for women of all ages). And women with the highest incomes and the best educations have the highest rates of divorce. In my opinion, these developments are on the whole indicators of the preferences of many women for an unmarried state. We also know that women living without men do less housework, with and without children; the presence of men in the household seems to create about eight hours of additional housework for women per week (Hartmann, 1981b). Is it not likely that faced with men's intransigence with respect to taking on more housework (housework by men, on average, has barely increased, despite women's increasing wage-work), and given that women have always been the primary child rearers anyway, that the increased propensity of all women to form their own house holds is in many ways a matter of choice?

A second set of observations about changes in the shape of male domination concerns changes within marriage. Women are achieving greater equality with men; through wage earning, they are changing family consumption patterns toward purchases that benefit them and toward greater personal consumption (Rainwater, 1979); they are doing less housework; and their greater willingness to leave marriages may be bringing about some improvements within marriages. As Spalter-Roth shows in a recent dissertation (1984), income per hour of labor (including both wage work and housework) has increased considerably for married women over the decade of the 1970s because their wage work has increased; their return per hour has increased relative to their husbands' as well. Further, an increasing proportion of women earn more than their husbands—12 percent in 1981 (Bianchi, 1983) versus 10 percent in 1976 (U.S. Bureau of the Census, 1978).

A third kind of evidence concerns children and here the picture is more mixed. Perhaps most important, parents' responsibility for children has probably declined because they have fewer of them, close together, and they themselves live longer; responsibility for minor children constitutes a smaller portion of adult lives than heretofore. Decreased fertility and increased control of fertility through contraception and abortion is especially liberating to women.[2] Intensive childcare represents a much more predictable and much shorter period of a woman's life. Beyond that, however, women have come to bear more of the responsibility for children alone, as single parenting has increased. On the one hand, this indicates their increased ability to do so. The right to have children without being dependent on men is no doubt important to many women. On the other hand, it indicates men's lessened responsibility for children, both in their daily care and financially. Few men pay child support faithfully after divorce or separation. This failure has clearly had a negative economic impact on women and children.

We also note that as children reach adolescence and young adulthood they have more opportunities to establish independence from their families of origin. Through college attendance, work, and/or delayed marriage, they often have a period of several years when they are living on their own. As Petchesky (1984) notes, some of the hysteria over pregnancy among teenagers has to do with the fact that pregnancy outside marriage is visible evidence of their independence.

We do not know enough about all of the ramifications for men's, women's, and children's lives, but taken together they do suggest that in general men exercise less authority over women and children in families than heretofore.

I would argue, then, that to the extent that there is a family crisis, it is by and large a healthy one, particularly for women. Even if it were not healthy, however, these changes are here to stay and are likely to continue in the same direction. Women are not going to go back home. The rapid changes we have been experiencing in family organization and in women's employment do cause hardship. Wage-working women with young children work long hours between work and family care. Many newly divorced women, especially older women who did not expect to support themselves, are left without labor market skills or experience. Some single women with children live in poverty, and/or bear the total burden of financial support and physical and emotional care of their children. And as we have noted, many men seem to have abrogated their responsibilities for children. Nevertheless, although some of the changes have negative implications, especially for particular groups of women, furthering these changes and reducing their negative implications will work to women's advantage.

Alternative Interpretations

A variety of alternative interpretations of these developments have been offered and will be discussed here briefly, some in more depth than others.

First, Ehrenreich has argued in *Hearts of Men* (1983) that many middle-class men may have left wives to lead more personally fulfilling, especially sexually fulfilling, lives. Her review of *Playboy* depicts a strident theme of ditching the wife for a sexy new woman. In her view, women's independence has been more the result than the cause of men's abandonment behavior. Tricked by a new sexual morality that worked against them, women, left by men, had to learn to fend for themselves. It is, of course, very difficult to say which came first (and perhaps it is unnecessary), but it should be remembered that marriage was not fulfilling for many women either, as early women's liberation literature described at length. Educated, middle-class women felt trapped, stifled—these were precisely the women with some ability to survive on their own; perhaps they often initiated leaving. And the sexual revolution made sexual exploration by women more acceptable; in particular, it increased the sexual experience of women before marriage.

Second, some have argued that the apparent gains for women, increased employment and increased wages, have come at the expense of men, or at least that at the same time men have lost jobs and earnings. In the past few years, for example, women's earnings have risen from 59 percent to 64 percent of men's earnings; but rather than representing a true increase for women, the improvement in the ratio could represent a deterioration for men. But there have definitely been some real labor market gains for women. They have entered professional, technical, and managerial jobs in large numbers, especially in the 1970s. During the 1970s, for the first time there was a substantial decrease in the index of sex segregation of occupations; having hovered around 65 since 1900, the index fell to about 60 in 1980, a proportionate decrease of 10 percent (Reskin and Hartmann, 1986). It should also be pointed out that more equality is not necessarily bad, even if it comes at the expense of men. If men on the average earn a dollar less per hour, and women a dollar more, the working class as a whole has not lost (assuming there are equal proportions of men and women in the labor force for this simple example).

Third, some have agrued that the dramatic changes observed in family life have been the result neither of women's nor men's choices, but have been forced choices necessitated by difficult economic circumstances (Currie, Dunn, and Fogarty, 1980). It is argued that the deepening capitalist crisis demands speed-up from workers and their families, and causes downward pressures on workers' wages to restore competitiveness. Women, especially married women, enter the labor market because two incomes are needed to support a family (due to inflation and wage erosion) or because their husbands have become unemployed (or reemployed at a low-wage job, having lost a high-wage manufacturing job). All must work harder simply to survive; neither men nor women progress toward autonomy.

This scenario probably does describe reality for some families (especially those of high-wage male blue-collar workers who have experienced substantial wage and job losses since 1980, but it is probably not an accurate picture for many who work in other sectors. In growth sectors (and in the economy overall, unemployment has decreased since 1984), workers may feel less pressed. Many progressives emphasize strategies to restore lost jobs and preserve those remaining through greater public control over investment decisions. In addition to preserving jobs in the industrial sector, however, it is also important to think about strategies for the new jobs; such strategies will be of greater benefit to women workers. If the newly available jobs pay less than those that have disappeared, one remedy is certainly organization—improving the conditions in the new jobs. In any case, women have made real gains in the labor market, and the increased contributions women make to family incomes have improved the prospects for gender equality at home.

The new arrangement for married couples, both men and women working for wages and women doing most of the housework, does amount to speed-up for women. The New Right is not entirely wrong in protesting this speed-up. Perhaps having one job, rather than two, was easier for women; having men to support them was a traditional and honorable means of economic survival for many women for many years. But it is a less secure route to economic survival than it once was. What is at issue, then, is how women subjectively experience these changes. Is the potential control a woman has over her economic future, when she works for wages (even at lower-paid women's jobs) a sufficient benefit to offset the hardship of the dual burden? I believe many women think so, though I cannot review evidence to support this claim here. In my view women are seeking wage work, not primarily because a deteriorating economy forces them to, but because they see it as a better strategy for personal economic security than the formerly available strategy of being dependent on a well-chosen man.

A fourth interpretation takes as given that many women have achieved some degree of autonomy from patriarchal control in the family (that the family has declined as a locus of men's control over women) but asks to what extent women may have traded one form of patriarchal control for another—a more diffused and generalized patriarchal control of the economy and public policy sometimes called public or collective partiarchy (Brown, 1981, and Spalter-Roth and Zeitz, 1981). Patriarchy has always been a social system, a system in which men's private power over women in families has been buttressed by such important social institutions as the church, the courts, the police, the legislatures, and so on. Those who stress the shift in patriarchal control from family to the institutions of the larger society help to broaden our understanding of patriarchy's social influence. Patriarchy does not operate only within the family (though it still operates there as well, despite some of the more optimistic observations made above), or only within the sex-gender system as narrowly construed to refer to biological and social reproduction. The sex-gender system, in this case the patriarchal sex-gender system, is everywhere.

In this view, women, having escaped from some family-centered patriarchal control, still face such control in their daily lives because economic and political systems are structured by gender: these structures include low-paid female job ghettos, lesser legal and civic rights, and public support programs (welfare, social security, unemployment insurance, health care) that treat women differently and less generously than men. Men, it can be argued, are less likely to be lords and masters at home (having determined that the costs of such a fiefdom are too high), but they are still more likely to exercise control in public domains—not only by controlling state policy, but also in the marketplace and in social life. In a way, the previous dominant form of patriarchy—male breadwinner, wife and kids at home—is being replaced with a new version—the swinging single male buying whatever services he needs and having sexual access to many women, on the one hand, and the single mother, working to support children or on welfare, on the other. The new version also includes the dual-career couple in which the wife is "super mom" and suffers from the double day, while the husband still maintains some patriarchal privileges (like not doing housework).

In this new scenario, the costs of family breakdown (the costs of the new male freedom) have been shifted to the state, much as O'Connor (1973) argued that the capitalist class was able throughout the post—World War II period to transfer many of the costs of doing business (education, training, research and development, etc.) onto the state, thereby generating a fiscal crisis of the state. The growth of the welfare state can be seen as a similar shift, transferring the costs of individual patriarchy to the state. The primary cost of family-centred patriarchy has been the financial support of dependent wives and children. These private costs become social when the system of private patriarchy breaks down: women are widowed or deserted, women have children outside marriage, and so on. Over time, additional costs, such as those of taking care of the elderly, or of educating college-age children, have also been shifted to the state.

But further shifts of patriarchal costs (and control) to the public sector seem unlikely for several reasons. First, some men may feel they have little access to patriarchal benefits through the marketplace (young men and lower-paid male workers) and they may resist the shift of patriarchal authority to policy makers. Second, the recent exacerbation of the fiscal crisis, which has resulted in the Reagan administration's attempt to relocate some of the costs of dependent care back in the family, may prevent further cost shifting, as much as some men might prefer it at least in the short run. Third, women's increasing political power is likely to influence the role of the public sector. Costs may be shifted to the state, but to the extent that women can exercise influence, patriarchal control may not follow. The state may come increasingly to serve women's interests as well as, or instead of, men's. The development of the state (as well as the development of the economy) must be understood as a result not only of capitalist relations and class struggle but also of patriarchal relations and gender struggle.

Obviously, we cannot know with certainty which interpretation of the significant changes that have been observed in family structure and women's economic participation is most accurate. Equally obviously, one or another interpretation will prove most convincing according to the particular circumstances under consideration. Women have experienced these changes differently from one another depending on their race, class, and individual histories. Clearly some have lost and some have gained in the transformations in economic and gender relations that have taken place. Continuing change, too, will produce further winners and losers. But on the whole, I believe these changes have been progressive for women and are likely to continue to prove so.

ROOTS OF CHANGE, SEEDS OF FURTHER CHANGE

It may be instructive to consider one of the changes discussed above, women's increased labor force participation, in order better to understand the sources of change. The increase in women's labor force participation can be seen as a result of contradictions inherent in an earlier accommodation of gender and economic relations. Specifically, the previous, pre—World War II arrangement of patriarchal capitalism (or capitalist patriarchy) contained the seeds of the current changes, and the current situation will lead to further changes. I have argued elsewhere (Hartmann, 1981a) that in the twentieth century a partnership between capitalism and patriarchy emerged in which the interests of both were served by allowing women a secondary place in the wage labor market, which kept them dependent on men and thus assured men of their continued services.[3] Women's place in the labor market has been characterized by the sex segregation of jobs and by low wages. Women's place in the labor market should be seen primarily as the result of patriarchal gender relations (operating in this case in a capitalist production environment).

Prior to World War II, the majority of married women did not work in the labor market; female-dominated jobs were populated by single women, young and old, who had to support themselves.[4] The labor of married women, though it often contributed to cash income via various forms of homework, was largely family-centered in childcare and housework. Married women generally relied on their husbands' incomes. When they worked, they were viewed as secondary earners; their positions as wives in some sense justified their lower wages. Thus, by ensuring a pool of low-wage workers (wives and by extension, all women), patriarchy created an attractive pool of additional workers who could be drawn into capitalist relations on favorable terms for the capitalists. As women are drawn into the labor market, the work they formerly did at home also became increasingly available on the market. The service sector grows because the availability of cheap female labor provides the supply and because the use of women in the labor market rather than at home also provides the demand for replacement services (fast-food replacing home cooking, for example).

In this way, the enormous growth of the service sector, an important aspect of the structural transformation discussed here, can be seen as a direct result not only of capitalist development but also of patriarchal relations. Without the availability of large numbers of low-wage women workers, it is not clear that capitalist expansion would have taken this direction so strongly. The growth of much of the business service sector is required by a shift toward large-scale, multinational business with its consequent need for more communication, management, record keeping, and so forth. Women's labor allows this shift to occur. And the shift toward the commercialization of personal services is required by women's increased labor force participation. Changes in economic structure, then, can be seen as the result of changes in gender relations as well as in class behavior.

Hence, capital may be gaining in the centuries-old struggle between capital and men over the deployment of women's labor power (and men may be losing). More of women's labor power is being deployed outside the home, where it is not under the direct control of fathers and husbands. Moreover, I would argue, women, too, are gaining in this struggle.[5]

The increased labor force participation of women also sets the stage for further transformation. As women work more and work more continuously over their lifetimes, they have come to challenge their low-wage, secondary positions. The women's movement has contributed to this general process of consciousness-raising concerning the value of women's work. Two examples of this challenge are the development of the comparable worth strategy and the beginnings of large-scale clerical worker organizing.

The comparable worth strategy challenges the relative pay levels of men's and women's jobs—nurses and plumbers, librarians and engineers, secretaries and sales representatives. Comparable worth is a direct assault on the economic, social, and cultural system that has undervalued women's work relative to men's. Such claims are usually initiated when groups of women workers begin to investigate the relative wage levels of men's and women's jobs at their workplace, the qualifications required for the jobs, and the methods of setting pay. Arguments are then couched in terms of entire groups of women workers being underpaid. (Treiman and Hartmann, 1981). The most significant part of this type of campaign may be its ideological and consciousness-raising elements. Such studies convince some women who might not have thought about it or articulated it that their jobs are underpaid, that they do deserve more for the work they do. Moreover, by bringing the bases for wage differentials into the realm of public discussion, comparable worth campaigns politicize the wage determination process. In both these ways, such campaigns set the stage for the unionization of women.

But the more revolutionary aspect of the comparable worth strategy arises because it creates the possibility that women will be able to support themselves financially on equal terms with men. Such an eventuality would revolutionize gender relations and create the possibility of true autonomy for women. Moreover, by raising issues about how women's work is valued, obvious parallels are drawn to the undervalued work women do in the home. Whether in a capitalist wage labor market or a socialist one, issues of the value of the work performed traditionally by men and by women must be raised. Such questions directly challenge patriarchal norms and patriarchal power bases.[6]

While capitalists would undoubtedly have continued to profit from women's secondary position in the labor market, women's own challenge to their secondary status creates a new opportunity for working-class unity between men and women. Men are being presented with a second chance to incorporate women into the labor market as equals. This progressive response to the changes we have observed can improve women's and children's lives more than a return to greater economic dependence on men. Men, and society in general, should have greater responsibility for children, and women should have higher incomes. While men have something to lose from this approach, in the long run they also stand to gain.

CONCLUSION

To conclude, let me outline briefly the progressive social policies that would enhance women's ability to be self-sufficient, clarify what I am not arguing in this essay, and comment on the importance of understanding social change as the outcome of both gender and class struggle.

Employment-related policies that would be important to women's advancement toward autonomy include increased collective bargaining and unionization, continued emphasis on equal employment opportunities and affirmative action, and, of course, comparable worth. For these policies to be effective, consistent full employment must be achieved. This in turn will no doubt require greater public control over what have generally been up to now private investment decisions. Improving working conditions on the job will also require a greater degree of employee involvement in work decisions. Hence, advancing women's specific interests as workers will also lead to the enhancement of workers' rights and economic prospects in general. Moreover, that women have greater responsibility than men for children needs to be recognized, and social programs to ameliorate the effects of this difference must be developed. Policies to provide subsidized child care, parental leave, and so forth will at least initially benefit women more than men, but they obviously benefit male parents as well.

A central demand that should be raised and supported by feminist and progressive groups now is for universally available free child care. Because childbearing has become more universal and is limited to a shorter period of adult life, achieving consensus about supporting families for this limited period should become increasingly possible. Social security provides a useful analogy; it supports people for a limited (though growing) period of their lives, and everyone pays for it through payroll taxes. Universal, employment-related parental insurance financed via a payroll tax could provide parental leaves as well as funds for child care. Child care itself could be available free (or at modest cost in addition to the normal payroll tax) to the actual current users.

In general, I believe most benefits should be tied to employment or participation in training programs. As working for wages increasingly becomes the norm for all women, the fact that poor, young, minority women are "stockpiled" on welfare programs increasingly disadvantages them. They, like all women, need to learn labor market skills and progress toward self-sufficiency. Of course, not everyone is able to work, and social programs that provide a decent standard of living for those unable to work are needed as well.

In my view, enhancing economic autonomy for women so that they need not be dependent on individual men should be the central goal of the women's movement; greater social responsibility for children is a critical component of that goal. One would also hope that individual men would adjust to the new realities and take on their share of child care and housework.

I am *not* arguing that women, who would be able to be truly economically autonomous with these new arrangements, would not choose to live with or even marry men. They might. They might also choose to live in groups or with other women or alone. But whatever their choices they would be less coerced by economic considerations than they are presently. Marriage would no longer be the central path to economic survival that it now is. I am also not arguing that the transition to this new arrangement has been painless. Some women have been negatively affected. But I am arguing that social policies such as those described above can do much to ameliorate the negative effects. In the long run we will gain more than we lose.

I want to close by explicitly arguing against the tendency to believe that everything that happens in capitalist societies is the result of capitalist development and only capitalist development. Even structural transformations that are most often thought of in economic terms—foreign competition, industry shifts, declining productivity—are in fact shaped by both gender and class relations. The reindustrialization issue itself can be seen as an attempt to preserve and reestablish the male sector of the economy. Lane Kirkland's recent unabashed, enthusiastic support for highly capital-intensive pursuits like the shale oil projects in the mountain states can be seen as a rather backward-looking approach—if the sectors where unions have been strong can grow, he seems to be thinking, then unions can grow. Kirkland apparently believes it would be easier to reindustrialize America than to learn how to unionize new kinds of workers.[7] It seems inevitable that fewer and fewer men will have "macho" jobs in the economy—those associated with hard physical labor and rugged working conditions. With technological change and sectoral shifts, more and more men will, like women, work in the service sector. The recent emphasis on sex role research, the new androgynous personality, and so forth, might be seen as an effort to retrain men for "women's" jobs, or at least to make these jobs increasingly ideologically acceptable to men. Although men will increasingly enter the service sector, I would not expect sex segregation in the labor market to eliminated entirely over the next few decades; that will require many, many years. But the lines between blue-collar and white-collar jobs and professional and support jobs will most likely blur. These developments hold the promise of progressive change if we can develop new strategies that build upon them. Just as the increased mobility of capital noted by Bluestone and Kuhn calls forth new solutions, so might these sectoral and structural shifts. The new solutions will undoubtedly require a new consciousness on the part of men as well as of women.

Not only is there a tendency to attribute everything to capitalism in thinking about economic change, but there is an accompanying tendency to view capitalist crises as necessarily negative. There can be no question that budget cuts, unemployment, and poverty make people's lives worse and that women are particularly severely affected by them; inasmuch as these are the short-run effects of the current crisis, they are serious and unhappy results. But the longer structural transformation that is also under investigation here has had and will continue to have many positive effects, particularly for women, inasmuch as it entails fundamental changes in gender relations as well. Placing gender relations at the center of the analysis causes us to have a very different view of the transformation. There is potential now for continued truly revolutionary changes in our intimate lives and in women's ability to be economically autonomous (perhaps for the first time in human history). These are important, progressive changes. The only question is whether we as a society are ready for them.

NOTES

I would like to express my appreciation to Lourdes Beneria and Catharine Stimpson for insisting that I do this paper; members of my study group in Washington, D.C., especially Phyllis Palmer and Roberta Spalter-Roth, for trenchant discussions of many of the issues raised in this paper; several colleagues, Micaela di Leonardo, Judith Stacey, Lourdes Beneria, and Catharine Stimpson, for particularly close readings of the first draft and many helpful suggestions; and several groups, in addition to the attendees at the Conference on Structural Transformation, whose spirited argument and probing questions have helped clarify the argument: the Women's Studies Forum at the University of Pennsylvania and groups at the University of California at Los Angeles, University of California at Berkeley, Evergreen State College, Bucknell University, and Harvard University. I would also like to thank Lucile DiGirolamo of the National Research Council staff, and Gillian Marcelle, an intern with the council, for assistance in completing this paper. The views expressed, however, are mine, and not those of the National Academy of Sciences.

1. Elaine McCrate (1984) has recently formalized these notions in a feminist model of a non-clearing marriage market.

2. It is important to keep in mind that the legal right to abortion is only slightly more than a decade old; and the right to contraceptives is only about twenty years old.

3. During the periods of the emergence of capitalism and the industrial revolution, wage labor became the dominant mode of economic survival, a mode that challenged earlier family-centered production systems and created a challenge to men's continued control of women's labor power and to women's ability to continue to contribute to the economic support of their families. That challenge was resolved by the establishment of family wages (at least as the ideal), which allowed men to support their wives and children at home, retaining the advantages of the deployment of women's labor power at home (Hartmann, 1976; 1981a).

4. Many poor women and minority women did work in the labor market, even when they were married, and many married women earned money at home. In general, the "patriarchal accord" was less relevant for poorer and minority men, since, in the patriarchal hierarchy, they enjoyed fewer patriarchal privileges than majority men. They often did not have wives at home. But even in these groups it was believed that men should be the primary earners.

5. Contrast this feminist interpretation with that of Karl Kautsky (1971:26), an important socialist of his day: "The capitalist system of production does not in most cases destroy the single household of the workingman, but robs it of all but its unpleasant features. The activity of woman today in industrial pursuits . . . means an increase of her former burden by a new one. *But one cannot serve two masters.* The household of the working-man suffers whenever his wife must help to earn the daily bread." Apparently one type of slave these socialists were not interested in liberating was women.

6. The issue of comparable worth also fundamentally alters our understanding of the way the labor market operates. In earlier work on job segregation, for example, I essentially assumed that some jobs have low wages, and, through a process of struggle between men and women, those are the ones allocated to women. But another way to look at the same result—women in low-paid jobs—is to see it as the result of an attempt to devalue whatever it is that women are doing and to pay it less. In this view, the pay levels of jobs themselves are seen to be the result of gender struggle on the job. Just as Marxists have increasingly begun to look at class struggle as it occurs directly on the shop floor as well as in the larger society, feminists have to look at gender struggle in the workplace—how new jobs get allocated to each sex, how their pay levels are determined, and so on. This perspective can lead to a transformation of accepted Marxist theories. As Philips and Taylor (1980) point out, Harry Braverman's book *Labor and Monopoly Capital* (1974), emphasizes only class struggle: it is always in the interest of capital to divide workers and to deskill as many components of the job as possible. Taylor and Philips argue, however, that the process of deskilling is sometimes only apparent: the jobs may not actually require less skill, rather they are labeled unskilled and devalued. Male workers

often play a large role in the process of devaluing and labeling women's work and in determining relative wages rates. Wages are gendered for both women and men.

7. Rothschild (1981) has criticized this view fairly thoroughly; growth rates in the industrial sector would have to be very great over the next several years in order to bring about any reversal of the secular trend toward the service economy.

REFERENCE LIST

AFL-CIO. 1984. *Deindustrialization and the two tier society: Challenges for an industrial policy.* (Washington, D.C.:) Industrial Union Department, AFL-CIO.

BANE, MARY JO. 1985. The politics and policies of the feminization of poverty. Paper prepared for the Colloquium on The Changing Situation of Black Americans and Women: Roots and Reverberations in U.S. Social Politics since the 1960s, University of Chicago, April 26–27.

BIANCHI, SUZANNE M. 1983. *Wives who earn more than their husbands.* Special Demographic analyses CDS-80-9. Washington, D.C.: GPO.

BIANCHI, SUZANNE M., AND DAPHNE SPAIN. 1984. *American women: Three decades of change.* Special Demographic Analysis CDS-80-8. Washington, D.C.: GPO.

BLAUSTEIN, ARTHUR I. 1982. *The American promise:* Equal justice and economic opportunity. New Brunswick, N.J.: Transaction Books.

BLUESTONE, BARRY, AND BENNETT HARRISON. 1983. *The deindustrialization of America.* New York: Basic Books.

BLUESTONE, BARRY, BENNETT HARRISON, AND LUCY GORHAM. 1984. *Stormclouds over the horizon: Labor market crisis and industrial policy.* Brookline, Mass.: Economic Education Project.

BLUESTONE, BARRY, AND SARAH KUHN. 1983. Paper presented at the Conference on Women and Structural Transformation: The Crisis of Work and Family Life, Rutgers University, New Brunswick, November 18–19.

BRAVERMAN, HARRY. 1974. *Labor and monopoly capital: The degradation of work in the twentieth century.* New York: Monthly Review Press.

BROWN, CAROL. 1981. Mothers, fathers and children: from private to public patriarchy. In *Women and revolution,* edited by Lydia Sargent, 239–67. Boston: South End Press or London: Pluto Press.

CURRIE, ELLIOTT, ROBERT DUNN, AND DAVID FOGARTY. 1980. The new immiseration: stagflation, inequality, and the working class. *Socialist Review* 54 (November–December):7–31.

EHRENREICH, BARBARA. 1983. *The hearts of men.* Garden City, N.Y.: Anchor Press/Doubleday.

ERIE, STEVEN E., MARTIN REIN, AND BARBARA WIGET. 1983. Women and the Reagan revolution: thermidor for the social welfare economy. In *Families, politics, and public policy: A feminist dialogue on women and the state,* edited by Irene Diamond, 94–119. New York: Longman.

ESPENSHADE, THOMAS J. 1985. Marriage trends in America: estimates, implications, and underlying causes. *Population and Development Review* 11(2):193–245.

Ford Foundation. 1985. Women, children, and poverty in America. Working Paper New York NY (January).

HARTMANN, HEIDI I. 1976. Capitalism, patriarchy, and job segregation by sex. *Signs* 1, 2 supplement (Spring):137–169.

HARTMANN, HEIDI I. 1981a. The unhappy marriage of Marxism and Feminism. In *Women and revolution,* edited by Lydia Sargent, 1–41. Boston: South End Press or London: Pluto Press.

HARTMANN, HEIDI I. 1981b. The family as the locus of gender, class and political struggle: the example of housework. *Signs* 6(Spring):366–394.

HOGAN, DENNIS P. 1983. Demographic trends in human fertility and parenting across the life-span. Paper prepared for the Conference on Biosocial Life-Span Approaches to Parental and Offspring Development, sponsored by the Social Science Research Council, Elkridge, Maryland, May.

KAUTSKY, KARL. 1971. *The class struggle.* New York: W. W. Norton.

KITAGAWA, EVELYN M. 1981. New life-styles: marriage patterns, living arrangements, and fertility outside of marriage. *Annals of the American Academy of Political and Social Sciences* 453 (January):1–27.

KOBRIN, FRANCES E. 1976a. The fall in household size and the rise of the primary individual in the United States. *Demography* 13 (February):127–38.

KOBRIN, FRANCES E. 1976b. The primary individual and the family: changes in living arrangements in the United States since 1940. *Journal of Marriage and Family* 38 (May):233–39.

MCCRATE, ELAINE. 1984. The Growth of Nonmarriage among U. S. Women: An unanswered question for the new family economics, and an alternative. Department of Economics, University of Massachusetts, Amherst (October).

MOORE, KRISTIN A. 1985. Teenage pregnancy: The dimensions of the problem. *New Perspectives* 17 (Summer):11–15.

O'CONNOR, JAMES. 1973. *The fiscal crisis of the state.* New York: St. Martin's Press.

PETCHESKY, ROSALIND POLLACK. 1984. *Abortion and women's choice: The state, sexuality, and reproductive freedom.* New York: Longman.

PHILLIPS, ANNE AND BARBARA TAYLOR. 1980. Sex and skill: notes toward a feminist economics. *Feminist Review* 6:79–88.

RAINWATER, LEE. 1979. Mothers' contribution to the family money economy in Europe and the United States. *Journal of Family History* 4(2):198–211.

RESKIN, BARBARA F. AND HEIDI I. HARTMANN, EDS. 1986. *Women's work, men's work: sex segregation on the job.* Washington, D.C.: National Academy Press.

ROTHSCHILD, EMMA. 1981. Reagan and the real economy. *The New York Review of Books* 28 (February 5):12–18.

SMITH, SHIRLEY J. 1982. New worklife estimates reflect changing profile of labor force. *Monthly Labor Review* 105 (March):15–20.

SMITH, SHIRLEY J. 1985. Revised worklife tables refect 1979–80 experience. *Monthly Labor Review* 108 (August):23–30.

SPALTER-ROTH, ROBERTA. 1984. A Comparison of the living standards between husbands and wives in dual-earner couples, 1968–1979: A feminist analysis. Ph.D. dissertation. The American University, Washington, D.C.

SPALTER-ROTH, ROBERTA M. AND EILEEN ZEITA. 1981. Production and reproduction of everyday life. In *Dynamics of world development,* edited by Richard Rubinson, 193–209. Beverly Hills, Calif.: Sage Publications.

TREIMAN, DONALD J. AND HEIDI I. HARTMANN, EDS. 1981. *Women, work, and wages: Equal pay for jobs of equal value.* Washington, D.C.: National Academy Press.

WILSON, WILLAM J. AND KATHRYN M. NECKERMAN. 1985. Poverty and family structure: the widening gap between evidence and public policy issues. In *Fighting poverty: What works and what doesn't,* edited by S. Danziger and D. Weinberg, 232–282. Cambridge, Mass.: Harvard University Press.

U.S. Department of Commerce. Bureau of the Census. 1985a. *Characteristics of the population below the poverty level: 1983.* Current Population Reports. Series P-60. No. 147 (February). Washington, D.C.: GPO.

U.S. Department of Commerce, Bureau of the Census. 1985b. *Household and family characteristics, March 1984.* Current Population Reports, Series P-20, No. 398. Washington, D.C.: GPO.

U.S. Department of Commerce, Bureau of the Census. 1978. *Money income in 1976 of families and persons in the United States.* Current Population Reports. Series P-60, Number 114. Washington, D.C.: GPO.

THE DIVERSIFICATION OF FAMILY LIFE

32 *Home Computer Sweatshops*

PHILIP MATTERA

At 6 A.M., Ann Blackwell slips out of bed in her home in Elgin, South Carolina, and walks to the den, where she activates a Texas Instruments 770 terminal. The green screen flickers as the machine begins transmitting Blackwell's work from the day before to a computer located in the offices of her employer, Blue Cross-Blue Shield of South Carolina. A few hours later, having seen her children off to school and cleared the breakfast dishes, she returns to the terminal and gets to work processing physicians' claims. With breaks for household chores and lunch, she remains at the keyboard until it is time to prepare dinner.

For the past five years Blackwell has been working at home this way as a participant in what Blue Cross calls its "cottage keyers" program. Blackwell puts in an average of fifty hours a week at the terminal, mostly while her two children, aged 7 and 13, are at school; but when she is behind quota she works at night as well. She is paid 16 cents per claim, each of which requires about ninety seconds to process. By completing about 2,000 claims a week (the company requires a minimum of 1,200), she nets about $100—after deductions for taxes and equipment rental charges paid to Blue Cross. That is for a fifty-hour week, with no paid vacation time, no paid sick leave and no fringe benefits.

Blackwell admits that the work is boring and lonely, but she likes not having to get dressed up and drive to an office. She also likes the flexibility but

SOURCE: Philip Mattera, "Home Computer Sweatshops," *The Nation* (April 2, 1983), pp. 390–392.

dispels any notion that her job is a cushy one. "You've got to discipline yourself," she says.

Like Ann Blackwell, several hundred clerical and professional workers across the country are participating in home work projects, which, if they prove successful, could radically alter the conditions of employment for millions of people. These typists, claims processors, data entry clerks, computer programmers and other "information workers" have agreed to abandon the office and perform their tasks at home, on computer terminals electronically linked to a company's main office. They are part of a new trend toward what is called, in management jargon, "telecommuting," "remote work" or "flexiplace."

The revival, in high-tech form, of the home workplace, which had largely vanished under capitalism, was heralded several years ago by futurologist Alvin Toffler in his book *The Third Wave.* Toffler described the "electronic cottage" in glowing terms and predicted that its advent would alter "our entire economy, our cities, our ecology, our family structure, our values, and even our politics . . . almost beyond our recognition."

Hyperbole aside, there is no doubt that a wide expansion of electronic home work will have serious social consequences. But what Toffler and other propagators of the idea ignore are the negative effects that the shift of work to the home could have on the worker. As Dennis Chamot of the A.F.L.-C.I.O. warns, "The potential for exploitation and abuse is tremendous."

Electronic home work is a curious amalgam of late twentieth-century technology and what in some instances amounts to early nineteenth-century working conditions. Ann Blackwell's situation at Blue Cross is not necessarily typical of the conditions faced by all the people who have participated in home work programs sponsored in recent years by companies such as Control Data Corporation, Aetna Life & Casualty Company and Continental Illinois Bank. Yet the use of home work, particularly by clerical workers, is being promoted in management literature in part as a way to cut costs. The message seems to be that home workers, predominantly women, can legitimately be denied the pay levels and benefits enjoyed by their colleagues in the office. Managers are also enticed by the prospect of a work force for which absenteeism and tardiness are virtually eliminated.

If for the boss electronic home work means potentially lower labor costs, for workers the arrangement means isolation and an end to the socializing and informal cooperation that are often the only things making a job bearable. In an advertisement for Lanier Business Products' Telestaff System, the company praises the virtues of a physically dispersed labor force: "The Telestaff Station. It brings work to your office. Not people."

Companies such as Control Data that pride themselves on being socially responsible emphasize the humanitarian aspects of home work, portraying it as a boon to the handicapped and to those who cannot travel to an office. Addressing a communications conference in 1981, Control Data vice president Gail Bergsven described her firm's experiment as part of a "holistic" management approach through which "the corporation can help itself by helping its employees."

The workers apparently are not encouraged to help themselves. None of the current home workers belong to unions, and union leaders charge that companies are so enthusiastic about shifting clerical work to the home because they see it as a way of keeping their office workers unorganized. "It's a union-busting tactic," states June McMahon, research director of the Service Employees International Union. Because of the difficulty organizing home workers and preventing employer abuses, the executive board of the S.E.I.U. approved a resolution last year calling for a Federal ban on electronic home work.

Thus, as it has so often in the past, new technology has transformed the role of labor as well as the techniques of production. The growing corporate interest in the electronic cottage reflects other forces at work—notably the structural changes now taking place in the economy, which have been accelerated by the current recession. Information processing is gradually replacing heavy industry as the driving force in the American economy, and with the advent of microcomputers, some of this work can be farmed out.

Home work could also feed into another trend—business decentralization. Many employers have decided that gathering a large work force under one roof is not the most profitable way of doing business. Advances in flexible automation are making it possible both to reduce the payroll and to disperse the remaining employees throughout a constellation of small facilities. Unionized companies have been making greater use of subcontracting, parceling out work to lowcost nonunion suppliers. Last year unions representing 8,000 workers at General Electric's plants in Schenectady, New York, struck for two days to protest G.E.'s transfer of work to such companies.

At its worst, decentralization has caused the resurgence of the sweatshop and an exploitative underground economy. Home work in a more traditional form—garment making—has burgeoned among undocumented workers in New York City and Los Angeles, as well as in small towns in the Northeast. The Reagan Administration, which favors any measure that will lower labor costs, has sought to repeal Federal rules, issued during the 1940s, which ban home work in various kinds of light manufacturing.

Home work, in both its traditional and electronic forms, represents decentralization carried to its logical extreme: shrinking the production unit down to a single worker in his or her home. Since so few office workers are unionized, there is a heightened risk that some service businesses will become new runaway shops, cutting back on office staff and giving their jobs to low-paid workers toiling in their homes. Unable to bargain for higher wages, these computer pieceworkers might be forced to supplement their earnings by putting family members to work. In view of the growing computer sophistication of young people, this could give rise to a new form of child labor.

The danger of a domestic electronic sweatshop is obscured by the depiction of the electronic cottage as a form of entrepreneurship. Promoters like Toffler assume that home workers cease to be mere employees and are transformed into independent contractors. But it is just that status that makes it easier for business to deny them fringe benefits and hourly wages. Karen Nussbaum, executive director of 9 to 5, the National Association of Working Women, argues that the idea of entrepreneurship "does not bear much relevance to clericals. What's happening is that the employer is shifting overhead to the worker and evading demands for decent pay and benefits." For clerical workers, working at home could be "an unmitigated disaster," according to Nussbaum.

Some feminists take a more tolerant attitude toward electronic home work, arguing that it provides homebound women with their only means of earning an independent income. Margrethe Olson, who is doing research on the phenomenon at the New York University School of Business Administration, acknowledges the potential for abuse, but views home work as an opportunity for women with young children to work part time on a flexible schedule. (Olson insists, however, that it should not be viewed as a substitute for day-care facilities.) Supporters also point out that working at home enables people to avoid the daily commute, the sometimes stifling atmosphere of the office and the need to get dressed up to go to work.

Some of the positive attitudes toward home work stem from a selective definition of the kind of work involved. Actually, the skill levels needed vary. First there are the low-paid clerical workers doing routine tasks requiring little experience or expertise. They have the least bargaining power and are most vulnerable to exploitation. Then there are better-paid professionals and executives who may bring work home to do on their personal computers. There are also limited numbers of professionals who are participating full time in home work experiments, as well as an even smaller number of computer specialists who can sell their services by means of their home terminals. For some of the latter, entrepreneurship is often genuine—and quite lucrative. Computer magazines are full of tales about clever software producers setting up shop on their kitchen tables and becoming overnight millionaires.

It is usually the upscale professionals and entrepreneurs that futurologists like Toffler have in mind when they proclaim that electronic home work is the harbinger of a glorious new age in which work and social life will be decentralized. The state of complete separation of home and workplace, which was brought about by the Industrial Revolution, will end, they argue, and people will become more involved in their communities. The "home-centered society"—currently being promoted through electronic shopping and banking—is a key to Toffler's "third wave."

Yet with their exclusive focus on technology, these futurologists have ignored crucial considerations of class and gender. Nussbaum of 9 to 5 points out that the "homecentered society" would affect people differently, depending on their income levels. The upbeat view of it, she says, "is a projection of the ideals of the upper-middle class. It has a completely different effect for the working class. Being at home without money is like living in a prison."

The political problems associated with home work are all the more serious because most of the people who will be engaged in it, especially at the clerical level, are women. The prospect of being "home-centered" may not be so appealing to a generation of women who have been abandoning the home and entering the labor market in record numbers. Some feminists express concern that the rise of electronic home work will push employed women into dead-end jobs and make it more difficult for them to escape the housewife role. The simple fact that workers are isolated from one another can make home work a form of disempowerment, depriving women of the measure of social status that working with others provides.

Such fears are indirectly confirmed by the eagerness with which the New Right has embraced the concept of home work, which they believe would strengthen the nuclear family. Conservative Republican Newt Gingrich of Georgia has introduced what he calls the Family Opportunity Act, which would give tax credits to families that purchase home computers for business or educational purposes. Last year in a press release, Gingrich said the legislation would "strengthen neighborhoods and allow working mothers with preschool children to earn a living while staying at home." At a forum on the family co-sponsored by the Moral Majority last summer, home work was praised as a way for women to earn some money without neglecting their families.

If the future is not as bright as the futurologists paint it, neither is it entirely black. For one thing, some business executives have mixed feelings about the arrangement. Many managers are uneasy about losing the ability to oversee directly the work of their underlings. Even with the strictest controls, they assume that anyone working at home is goofing off. Moreover, some analysts argue that separation from the supposedly stimulating office environment tends to lower productivity. And many managers surely sense that employees will resist being sent home to work.

Still, even if the electronic home work force does not reach the 5 million to 10 million level that some corporate think tanks are predicting for the 1990s, those workers, particularly women, who are compelled or pressured to accept it will face new problems. The labor and women's movements also face a difficult choice: Should they seek a ban on home work or try to organize home workers? Both groups are only beginning to address the problem, yet business is far along in its experiment. The challenge for activists is to catch up with the transformation of work that is taking place, especially now that the changes are hitting home.

PART V
Spatial Reorganization: Cities, Communities, and Regions

Terms such as *rust belt, urban decline, sun belt,* and *boomtown* are well integrated in the vernacular of many Americans. The national and global reorganization of cities, communities, and regions are visible and dramatic outcomes of the four interrelated forces that are transforming this society. Economic changes are remaking the urban and regional geography of America. These industrially determined changes in spatial structure are intertwined with new patterns in job access, housing, and transporation. They are also responsible, in large part, for new forms of structural unemployment and urban poverty.

Two overarching trends are responsible for the reorganization of America's cities, communities, and regions. The first is the movement of industries, resources, jobs, and population from traditional manufacturing centers of the northeast and northcentral states to the South and West. The second is the movement throughout the country, and especially in the old industrial cities, to outer suburban and rural areas (Oxford Analytica, 1986:211–212).

The movement of manufacturing, retail trade, and consumer services to the suburbs is among the most significant transformations in progress. This transfer is not merely a mass migration of business from one location to another. On the contrary, it signifies the fundamental structural shift from an economy based on the manufacturing of goods to one based on information and services. As this shift accelerates, it contributes to the simultaneous transformation and decline of the employment and residential bases of the city. High levels of structural unemployment are the result of a gap between new urban job structures and the skill levels of urban residents. A chief outcome of the movement of jobs from the central cities to the suburbs has been the formation of new patterns of spatial isolation that are making inner city minority residents a superfluous people in the service-information society.

Urban decay is closely linked with the development of new spatial forms that encompass work and living for millions of Americans. The old metropolitan dominance of cities is giving way to new collections of interdependent villages which are business, retail, housing, and entertainment focal points. Los Angeles is the prototype of the urban village, but virtually every American city is affected by suburban sprawl. Many contradictions are emerging within the new suburbs. Here, too, residential patterns and job opportunities are often mismatched. Many service workers cannot afford to live in the modern urban settings where they are employed.

As the business of America is being transferred to new locations, many communities are paying a high price. American cities are robbed of home-grown business by mergers, acquisitions, and move-outs. The movement of capital creates urban decay in some communities, while others undergo phenomenal growth. Both kinds of communities, however, experience traumatic dislocations. The decaying communities cannot provide basic services due to lost tax bases. Boom communities cannot meet the demands for new roads, sewage treatment, schools, hospitals, recreation facilities, and housing. Silicon Valley is the prototype of regional growth with abounding problems and strains. Furthermore, boom communities are not immune to busts, plant closings, and job loss due to capital flight.

Capital in the global economy is extraordinarily mobile. This mobility is reorganizing entire regions of the country. Increasingly, different parts of the country are in competition for highly mobile businesses and plants. The combination of new technology and capital flight will affect individual cities in different ways:

> Some cities like Boston and Pittsburgh have been able to take advantage of their investment in higher education and use that base to shift into areas such as electronics and robots. But other cities will not be as fortunate. It should always be remembered that while new technology creates new jobs, only competitveness guarantees that those jobs will be in the United States rather than elsewhere. (Oxford Analytica, 1986:353)

Regional economic growth is not only the result of industrial development but of investment choices that involve the movement of capital for the maximization of profit. Corporate relocation could well foster new kinds of regional sectionalism as different regions compete for business. Will communities keep labor costs and local taxes down? Will they invest in local education and other forms of specialization? How will cities and communities exploit the growing service and high-tech sectors of the economy? These will be decisive issues for the future.

Competitive pressures force more firms to abandon locally owned businesses and build corporate empires with offices in distant cities. The result is that "cities that for generations proudly referred to themselves as the 'headquarters' location of one or more nationally known corporations are finding themselves in the category of 'branch-plant' town" (Peirce, 1987:6C). The alteration of our urban and suburban landscapes by the repositioning of firms in the international marketplace reveals the enormous power of capitalism in shaping society. Cities, and communities expand and decline in response to the needs of capitalism. Yet, within these developments the contradictions that are endemic to capitalism are also exposed. While the expansion of the microelectronics industry has generated rapid regional growth, it has also transformed the class structure and imposed new patterns of residential segregation throughout Silicon Valley. The opposing forces of demise and development have set the stage for new forms of inequality and polarization by region, urban location, class, and race. These are the themes that emerge from the readings included in this section.

REFERENCES

Oxford Analytica. 1986. *America in Perspective: The Social, Economic, Political, Fiscal, and Psychological Trends that Will Shape American Society for the Next Ten Years and Beyond.* Boston: Houghton Mifflin.

Peirce, Neal. 1987. "Firms are Seeking Greener Pastures." *Denver Post* (Sept. 22):6C.

33 *General Motors and Flint, Michigan*

MICHAEL MOORE

The official unemployment rate had hit 27 percent, and it seemed that only an act of God could save Flint, Michigan. Unable to communicate with higher beings, the Mayor of Flint did the next best thing: he paid $20,000 for TV evangelist Rev. Robert Schuller to come to Flint to help rid the city of its unemployment plague. Four thousand Flint citizens packed the municipal auditorium, while thousands at home watched the event live over Flint's public television station. Under a forty-foot banner than declared, "Tough Times Don't Last—Tough People Do!" jobless auto workers who had survived by selling Amway products and Mary Kay cosmetics made their testimonials. They were the warm-up for Schuller, the "preacher of positive thinking," who finally bounded across the stage and launched into an attack on the "negative Neds and Nells" in the audience who couldn't find a job.

"Don't fix the blame; fix the problem!" Schuller shouted from the podium. "You can't hate the company for laying you off! That's not going to get you rehired! They're not giving jobs to angry people!" The assembled, many of whom had been given tickets to the event at the local unemployment office, were admonished: "Never look at what you've lost; look at what you have left! Don't exaggerate your unemployment problem! What you have is an idea problem. . . The only difference between you and the supersuccessful is that they

SOURCE: Michael Moore, "In Flint, Tough Times Last," *The Nation* (June 6, 1987), pp. 753–756.

have bigger dreams!" It was the white rip-off of "I *am* Somebody," and it played well with the General Motors and United Automobile Workers officials who joined Schuller on the dais. The preacher concluded with a simple piece of advice: "If you are unemployed, you should look at this as a great opportunity to travel." With that, Schuller was out the door.

That was five years ago. Since then, not a single new auto job as been created by G.M. Tough times, unlike TV evangelists, do last, and Flint, Michigan, the city with more G.M. factories and workers per capita than any other city in the world, is heading toward economic collapse. Last December, G.M. closed its Flint Chevy V-6 engine plant. At the end of May it shut down the Flint Truck and Bus line. In the fall it will mothball one of its oldest Flint facilities, the Fisher Body plant. At least 9,000 people will lose their jobs, and an estimated 30,000 more will be laid off in local businesses as a result of the lost income and tax revenue. Just nine years ago G.M. employed nearly 80,000 people in Flint; by the end of this year only 48,000 will still have jobs.

General Motors was founded in Flint on September 16, 1908. Nearly thirty years later the fledgling United Automobile Workers of America organized a sit-down strike in two of Flint's G.M. factories, including the Fisher Body plant, an unprecedented takeover that was to last forty-four days. The strike sparked numerous other actions in plants across the country, and the modern U.S. labor movement was born.

Over the years, Flint became the quintessential company town. Most people worked for G.M. Your house was built by G.M.'s construction company, and G.M. held the mortgage. You bought your car from G.M., and the company held that mortgage, too. General Motors, the saying went, would take care of your every need, from womb to tomb. Then, in the 1960s, G.M. discovered that it could make cheaper cars in the nonunion South. By the 1970s it decided that the Third World was even better. Buick engines would no longer be built in Flint; a town near Juarez, Mexico, would produce them. By the mid-1980s almost a third of the parts in the average G.M. car were made outside the country.

In 1982, G.M. claimed it was going broke, so the U.A.W. agreed to slash workers' wages and benefits, saving management more than $2.1 billion. But G.M. wasn't broke, and the money it saved from those concessions helped it buy Hughes Aircraft, Electronic Data Systems and several hightech firms. Close to 250,000 G.M. employees had permanently lost their jobs by 1984, the year G.M. posted a record profit of $4.5 billion. By the end of last year G.M. had made another $6.8 billion, and had announced that, over a three-year period, it would cut 25 percent of its white-collar work force and close at least eleven plants in the United States. Before the year was out it would also open twelve factories in Mexico.

What's good for General Motors has become what's bad for America, and even worse for Flint. Record rates of suicide, spouse abuse, alcoholism and crime—Flint surpassed Miami and Detroit last year with the highest rate of violent crime—have accompanied the dramatic rise in unemployment. Nearly 30,000 people have left the Flint area since 1982 and moved elsewhere in search of work. Those who remain live in a city with more than half its downtown demolished or boarded up. Thousands of families line up at a single location every other month to pick up Federal surplus cheese and butter. Nearly 70 percent of the city's young black men are unemployed.

The city's elected leadership, instead of seeking solutions to the crisis perpetrated by General Motors, has helped to pave the way for G.M.'s departure. Between 1976 and 1986 the Flint City Council voted to grant G.M. a 50 percent reduction in taxes on $1.3 billion worth of property. With each new tax abatement it received, G.M. promised, as required by the state law, to create new jobs. Instead, as with the concessions it got from the U.A.W., the company eliminated 18,000 jobs and used the savings to facilitate its move from Flint. Tom Kay, a G.M. lobbyist in Flint, sees it another way. "These tax breaks are perfectly legal," he said. "G.M. is no more greedy than a senior citizen receiving Social Security or anyone else who receives assistance from the government." Frances Cleaves, a Flint auto worker, disagrees: "It's extortion, plain and simple. They took the money and ran." Neither the City Council nor the state Attorney General's office have asked G.M. to return the funds.

In a series of moves that recall images of the dance band that was ordered to keep playing as the Titanic sank, the Mott Foundation, established in the 1920s to financially support Flint's educational and recreational facilities, has sought to shift the community's attention away from the G.M. layoffs. In 1982 the foundation, set up by G.M.'s largest stockholder and three-time Flint Mayor, Charles Stewart Mott, announced that it would make the company and the automobile a tourist attraction. It convinced the Federal, state and local governments to contribute $25 million to build a theme park called AutoWorld. The all-enclosed amusement center, the largest of its kind in the world, offered two rides through the "humorous history of automobility," a movie about car commercials, a giant car engine, and an assembly line complete with robots and an "auto worker" singing a tender ode to them called, "Me and My Buddy." The ironic highlight was a life-size recreation of what Flint's main street looked like before the plywood and the wrecking ball took over. The price of admission was $8.95, and one market-research firm projected that almost a million tourists a year would descend on Flint. Predictably, few came, and AutoWorld was closed in 1985 after seven months of operation.

AutoWorld was not the first scheme that the Mott Foundation had presented to the city as a means to create low-paying jobs. In 1981 the foundation and the city prodded the U. S. Department of Housing and Urban Development into spending $13 million to build a luxury Hyatt Regency Hotel in the heart of downtown Flint. (The Hyatt corporation, wisely, refused to put any of its own money into the project.) The hotel was seldom more than half full, and last year it went into foreclosure. The latest project being pushed on the city is Water Street Pavilion, a James Rouse endeavor similar to his other enterprises, including the South Street Seaport in New York City and Faneuil Hall in Boston. Open now for sixteen months, Water Street Pavilion is becoming a symbol of the"new Flint" that the foundation had not intended: more than a half-dozen of its stores have already closed.

When G.M. chair Roger Smith appeared before the television cameras last November 6 and announced that 30,000 more G.M. employees would soon be out of work—one-third of them from Flint—many area residents finally realized that G.M. was abandoning them. Wage concession, tax breaks, even improved-quality audits on the assembly lines, were not enough to keep the company in its hometown. "If we took a 90 percent pay cut and worked on Christmas Day," said union activist Mike Westfall, "we still wouldn't be able to compete with the slave-labor wages G.M. is using to exploit Mexican workers."

Westfall, Cleaves and others in Flint have decided that their only recourse now is to fight back. "General Motors knows that if it can get away with this in Flint, the birthplace of the U.A.W., it can pull this off anywhere," said Cleaves. A National Coalition to Stop Plant Closings has been formed in the city and is organizing other communities to press for the passage of a law that would bring a halt to factory closures across the country. Three rallies, two in Flint and one in Detroit, have been held in the past two months to protest the layoffs. Overflow crowds have heard speakers such as Jesse Jackson and U.A.W. pioneer Victor Reuther condemn General Motors for bringing about a depression in Flint at a time when the company is still number one in the Fortune 500.

Perhaps the most significant developments in the coming months will stem from the U.A.W.'s contract negotiations with the Big Three auto makers. The current contract expires in September. Union observers have noted that the rank and file is displaying more intense militancy than it has for quite some time. Many locals are pressuring the International to include language in the next contract that would prohibit the exportation, and exploitation, of jobs in Third World countries. All factions in the union agree that there will be no more concessions.

But Cleaves and the members of her caucus at a Flint G.M. parts plant are suspicious of the same union leadership that five years ago was so willing to give G.M. what it wanted. "The only thing union leaders in this town are interested in it 'Let's make a deal,' " said Cleaves. "They're more apt to tell you management's position before management does, as if they're just another arm of G.M."

The anger over what many feel has been the International's weak response to the impending layoffs, came to a head in Chicago in April at a U.A.W. convention called to formulate strategy for the contract talks this summer. Hundreds of rank-and-file members showed up to picket outside the McCormick Place convention center and demanded, unsuccessfully, to be heard on the convention floor. "The union's leaders are trying to distract us with all this talk about trade legislation," Cleaves said. "The problem isn't the Japanese; it's General Motors."

The talk of a strike—either at Ford or, more likely, at G.M.—looms large, and everyone seems to be preparing for it. Flint area merchants complain that they have seen at least a 20 percent drop in sales in recent months, a sure sign that people are hoarding their savings in anticipation of a long and bitter strike. Westfall, though, believes that G.M. and the U.A.W. will agree on an "illusionary contract," which will provide for an attractive retirement plan—full benefits, say, at the end of twenty or twenty-five years of service—but will have the net effect of helping G.M. reduce its overall work force.

People like Westfall and Cleaves believe that this year, this contract, this series of layoffs, could be the last chance for the citizens of Flint to halt their city's descent into a permanent state of poverty. This isn't Youngstown, Ohio, they point out, where the steel industry hit the skids, or Peoria, Illinois, where International Harvester went bankrupt. This is Flint, Michigan, home of the richest company in the world, a corporation so wealthy it can pay former board member H. Ross Perot three-quarters of a billion dollars to stop him from criticizing Roger Smith in public.

Meanwhile, in Flint, insult continues to be heaped onto injury. A national demographics magazine has rated Flint "the second-worst place to live in the country" (Pine Bluff, Arkansas, was first). There was no Christmas parade last year; instead, the week before the holiday, a convoy of tractor trailers appeared and hauled away the entire assembly line of the just-closed Chevy engine plant, transporting it to its new destination, in Mexico. When ABC's *Nightline* decided to focus on Flint's woes a few months ago, an unemployed Flint resident hopped into the network's truck parked in front of City Hall five minutes before air time and drove off, thus canceling Flint's fifteen minutes of fame. And AutoWorld, that domed shrine to the industry that built and then destroyed Flint, may soon be gutted and turned into a giant church.

Perhaps they could invite the Rev. Robert Schuller back for a little more of that positive thinking.

34 *Corporate Location Decisions*

JOE R. FEAGIN

CAPITAL FLIGHT

GE CLOSES PLANT IN LOS ANGELES. THOUSANDS OF WORKERS AFFECTED. FORD PLANT CLOSING, THOUSANDS UNEMPLOYED. AUTOMAKER BLAMES HIGH COSTS, MOVING TO MEXICO. AREA ECONOMY DEPRESSED BY TEXTILE COMPANY LAYOFFS. NEW ENGLAND TOOLMAKING COMPANY MOVES TO SOUTH CAROLINA. STATE GOVERNMENT TO REDUCE TAXES TO KEEP INDUSTRIES FROM MOVING SOUTH.

Headlines such as these are regularly repeated in newspapers across the nation. All regions and most cities have been affected by a company closing down a facility. According to recent estimates, between 1970 and 1980 alone, 32 million jobs were lost because of plant, store, and office closings; another six million jobs were lost because of permanent cutbacks short of closure. Many areas have experienced shutdowns and lost jobs. This is part of a disinvestment process being aggressively carried out by the leaders of American capitalism. Some areas are actually losing plants, called runaway plants, to other areas. But in many cases plants, offices, and stores are closed in one area and the investment capital is moved elsewhere—to the Sunbelt or Third World countries. In the 1970s twelve states lost more jobs from closings than were added by new businesses starting up. Capital flight from one city to another or from the United States to overseas can be part of a broad corporate plan to run from taxes, or to find cheaper land, or to find cheaper and less militant workers. The withdrawal of capital from an area has often been the work of large firms. In a New England

SOURCE: Joe R. Feagin, *The Urban Real Estate Game* (Englewood Cliffs, N.J.: Prentice-Hall, 1983), pp. 20–37.

study, Barry Bluestone, Bennett Harrison, and Lawrence Baker found that large companies were responsible for half of the jobs lost in that area by plant closings between 1969 and 1976. Such a withdrawal has disrupted communities, cost workers their homes, increased health problems, and remade the face of the industrial heartland of America.[1]

This scenario recurs in the United States. In the fall of 1979, U.S. Steel closed down operations in Youngstown, Ohio—with 3,500 jobs lost. In the spring of 1980, Ford closed its Pico Rivera plant in Los Angeles—with 2,000 jobs lost. In the winter of 1982, General Electric laid off hundreds of workers at its Ontario, California, metal iron plant. That corporation's capital is being reportedly invested in plants in Singapore, Brazil, and Mexico—areas with much lower wages. The Ontario plant was profitable, but in the corporate view not profitable enough. In the spring of 1982 it was announed that the famous Seth Thomas clock factory, resident in Thomaston, Connecticut, for more than 160 years, would be moving its operations to Atlanta, Georgia. The president of General Time, Inc., the parent company that owns the factory, was quoted to the effect that relocation to Georgia would mean lower operating costs. As a result, most of the plant's 175 workers will lose their jobs.[2]

Milking Plants

A major example of corporate relocation was the closing of a large steel plant in Campbell, Ohio, In 1969 this profitable Youngstown Sheet and Tube Company plant was bought by Lykes Corporation, a large conglomerate. Before its sale to Lykes, the plant had received millions in annual investment, but Lykes reportedly dropped the investment substantially, allowing the quality of the facilities to deteriorate. Apparently Lykes used the profits from the Ohio plant to pay off other debts and to buy nonsteel companies. After eight years most of the Campbell facility was closed as unprofitable. Many conglomerates by up profitable companies and slowly "milk" them for capital to be used for other investments. Local workers and communities pay a heavy price for this disinvestment. In the Campbell case the local community lost millions of dollars in tax revenues and even had to borrow money to run its schools. In addition, 4,900 jobs were lost. The total social cost of the death of this plant has been estimated at $72 million over the first few years in terms of increased welfare and food stamp costs, related governmental expenditures, and lost taxes to government. Ripple effects, including the loss of thousands of jobs in other local businesses, could be seen across the area.[3]

Human and Social Costs

A conventional business perspective on plant or capital relocation from one area of a city to another or from northern cities to southern cities suggests that unregulated corporate mobility is necessary for the health of the U.S.

economy. Capital must be free to move. Corporate executives do not wish governments to intervene, except to facilitate capital mobility with tax incentives and other subsidies. They see no harm coming to working people from unrestrained capital mobility. Some even suggest that working people should simply follow corporations wherever they or their capital go. New jobs may, or may not, be created in abandoned areas.

The United States is the only advanced industrial nation that allows this mobility for corporate enterprises. Western European nations, for example, limit the ability to corporations to move plants and offices; governmental permission is usually required.

Corporate flight from one U.S. city to another is not just a matter of money. It involves serious social costs: broken lives for working people, who lose jobs and livelihoods as well as homes and their consumer goods. Whole cities, such as Youngstown, Ohio, become economicaly depressed when companies leave. There are no longer enough firms and people to pay taxes for good schools and city services. People who are permanently laid off, particularly older workers, can have great difficulty in finding new jobs. Corporate flight and plant closings have ripple effects across the faces of cities. Taxes decrease because companies leave and unemployed workers pay fewer taxes. Money for urban services is reduced. Police officers and fire fighters are laid off. Local small businesses, grocery stores, taverns, gas stations go bankrupt. Downtown shopping areas become ghost towns, with many stores boarded up. Cities are left to die.[4] . . .

THE WORLD AS A MONOPOLY BOARD: LOCATING FIRMS FOR PROFIT

Conventional Views

There is an extensive economics and business literature on how companies make location decisions. This literature is highly technical, but its assumptions are simple. Location decisions hinge on profitability. One of the founding fathers of industrial "location theory," August Lösch suggested that industries should locate at the point of maximum profits, where the total income exceeds total costs by the greatest amount. Cutting costs is a critical aspect of location decisions.[6]

More recently, Melvin Greenhut has traced out the conventional business view of location decisions and listed major factors that influence decisions.

1. Cost factors at the location: transporation costs, extent of unionization, wage costs, taxation
2. Demand factors at the location: extent of competition by other firms
3. Cost-reducing factors: cities that specialize in one type of industry (presumably including governmental subsidies)

4. Revenue-increasing factors: areas that especially need the product made
5. Personal interaction factors: advantages from interaction with other companies and executives[7]

Conventional location analysts see a hugh market of smaller autonomous companies competing with one another for space. Their model is one of smaller companies, no one of which has coercive or dominant market power. But this is the major error in the assumptions of these analyses. The cost and profit factors they identify are generally accurate, but the theories ignore the rise of large companies (*monopoly capitalism*) and, most dramatically, neglect the social and human costs of corporate location decisions.

Big Capital and Little Capital

The rise of big corporations has changed corporate location and capital mobility decisions significantly. Big companies, such as a manufacturing corporation, can more easily abandon a community and reduce wage or materials costs by moving than smaller companies.

Before 1880 most capitalist companies were small. By the early 1900s mergers of companies were creating giant corporations, first in the oil, steel, and grain industries, later in the auto industry. In 1901 U.S. Steel, a company formed from mergers, already controlled $1.4 billion in the form of hundreds of industrial plants. Decentralized capitalism was on the decline. Industrial concentration was the trend. And the trend continues today. Every week newspapers detail mergers of corporations, which create ever greater concentration and centralization of wealth in many industries. Today only 500 companies control 80 percent of all the manufacturing plants and related assets in the United States. This means economic and political dominance for larger companies.[8]

Location decision making is shaped by the size of a corporation and its access to capital. Many large companies have the internal profit-generated funds to expand and relocate without borrowing a great deal; large corporations can use their credit worthiness to secure capital from lenders for new development projects. Large size gives corporations the ability to plan and build long-term projects; and large size enables companies to separate their operations spatially with greater ease because such companies have the resources to reintegrate operations using transportation and communication networks, even across wide areas of the United States or the world. Many firms have put part of all of their administrative operations in office towers in downtown areas of cities such as New York, Chicago, San Francisco, and Houston; they have placed research and development facilities in suburban rings or regional cities; they have located their manufacturing operations in despersed areas, including Sunbelt cities, suburban or rural areas, or towns and cities in Latin America or Asia.

Large-scale building projects in U.S. cities reflect a dispersion of corporate functions. Industrial parks house the manufacturing facilities of large corporations and related smaller companies. Office towers contain managerial and information (for instance, computerized accounting) processing; they have thus become the administrative centers for corporations. Shopping centers are filled with retail corporations, which sell the products of the manufacturing firms. Downtown office tower development exploded only when corporations became large enough to separate their administrative and production operations. Sometimes downtown development is explained in terms of the need of businesspeople to have much face-to-face contact. But this business need existed well *before* downtown skyscrapers began to increase dramatically in numbers. It was not until the 1910–1925 period that many national corporations could afford to separate administrative headquarters and production facilities.[9]

Suburban developments are often built by development and construction corporations as are apartment buildings, to house those who work for large corporations and the smaller businesses dependent on them. Today's cities, in part, reflect the different functional components of national and multinational corporations in the geographical layout and location of development projects, including office towers, industrial buildings and parks, and shopping centers.

And then there is the impact of urban development and corporate location on individual consumers and workers. Large-scale investments generally do not follow people, but in fact lead and shape the movements of consumers and workers. These are what Michael Smith calls "mobility-inducing investments." The movement of industry and shopping facilities to suburbia or Sunbelt cities has a significant effect on workers, consumers, and small businesses. Often they must move. And downtown business and residential areas in older cities may decline as a result.[10]

New location decisions by large companies shape the decisions of smaller firms dependent on them. For example, the location of major oil and gas companies in Sunbelt cities such as Houston has brought thousands of smaller companies, oil tool companies and smaller oil companies, law firms and accounting firms, to locations nearby. And many retail (for instance, restaurant) facilities have located there to serve the thousands of workers and executives.

Central Cities: No Corporate Philanthropy?

In recent decades many companies have decided to locate their manufacturing and other operations in the "better business climate" of outlying suburbs or in Sunbelt areas. They cite strong unions and expensive (in tax dollars) city services as problems for doing business profitably in central cities. This view may not be shared by ordinary working people, who see worker organizations such as unions as a way to protect themselves from greater employer exploitation.

Traditional social scientists view the suburbanization of industries as both normal and good.

> Directed by the sixth sense of profit, entrepreneurs of blue-collar industries have sought alternative, more cost-effective sites in the suburbs, exurbs, and nonme-

> tropolitan areas which are far better suited to today's advanced transportation and production technologies and which avoid what have become serious negative externalities of inner-city locations.[11]

Here is a sign of the corporation-biased assumptions that underlie much contemporary urban analysis. Profit is described as a unique human sense, a "sixth sense," which drives enterepreneurs to avoid the"negative externalities" (that is, the human problems) of inner-city locations. What is really being said is that owners and top executives follow the profit logic of a capitalistic system; this system's rules of profitability dictate that the needs of working people in cities for such things as decent-paying jobs be more or less ignored in the never-ending quest for expansion of the profit rate.

A survey by the Fantus Company, a firm providing major location assistance to industrial corporations, found that *only* 112 of 700 companies would even consider building a plant in or near poverty areas in cities. These firms would not build in such locations without the following conditions:

1. A large docile labor pool
2. Lack of well-organized opposition from unions and community groups
3. Ample land and adequate utilities and roads
4. Lower taxes
5. Absence of massive housing deterioration
6. Government-subsidized training programs and services

The Fantus report summed it up: "These companies are interested in good publicity and profits. None of them indicated any interest in philanthropy."[12]

Do Regulations Hurt?

Government red tape and regulations, including environmental standards, are not major considerations in most company decisions about plant and office locations inside and outside cities. Even though corporate officials complain about regulations, they generally do *not* cite these as the significant factors affecting their siting of new plants and offices. Moreover, many government regulations reflect a compromise between the interests of businesses and local citizens' pressures.

Unless citizens mobilize and organize to protest, the routine behavior of most government officials is to enforce local regulations to the degree that business leaders will accept. Variations from zoning laws protecting residential areas from commercial development are granted; environmental reports are filed and ignored; development permits are granted without the studies required by law. Unless local citizens protest, property development interests in cities get much of what they want. For example, in the late 1970s Dow Chemical Company dropped its plan to build a half-billion-dollar petrochemical plant on the metropolitan fringe of San Francisco because a coalition of community groups, a labor union, and environmental groups threatened lawsuits and forced officials to enforce local environmental laws, which Dow decided would be too costly to obey. Still, this corporate retreat may be an exception, because of the lack of citizen protest in other cities. For the most part, government regulations seem to have a minor effect on corporate location decisions.[13]

USING UP CITIES AND MOVIN' ON: UNEVEN URBAN DEVELOPMENT

Cities grow and die largely in response to the needs of capitalism. Cities are old inventions, but their character and existence have always reflected the choices of powerful decision makers. Over the last century those decision makers have been corporate owners and top executives. Regional "flights" are only one type of corporate location decision. Where corporate executives decide to locate their production and distribution facilities, directly or indirectly, affects the lives of most Americans, whether the choices involve locating a new facility in a suburb, deciding on a downtown location for an office tower, or moving investments from northern cities to the Sunbelt.

Waves of Urban Growth

Large corporations have sought out the most favorable places for their operations—the central cities of large metropolitan areas. By the 1920s metropolitan cities dominated most states, even though a majority of the population lived outside them. Virtually all industrial production was located in factories in the top twenty-five metropolitan areas; New York City, Chicago, Philadelphia, Cincinnati, Pittsburgh, and Milwaukee were the dominant manufacturing areas. Many cities were specialized centers. Albany made shirts; Troy, shirt collars; New Bedford and Fall River, textiles; Waltham, watches; Bridgeport, furniture; Dayton, cash registers; Detroit, automobiles. These centers in turn attracted skilled and unskilled American workers.[14]

Typically, a few key industries are at the heart of each successive epoch of industrial growth. Usually the location decision of any one company does not give a city its characteristic industrial structure—rather, the set of decisions made by several, dozens, or hundreds of companies are involved. In each distinctive period of U.S. industrial growth older industries, such as steel or automobile production, do not necessarily die out. They are just less dominant. And each new wave, such as today's computer and electronics industries, is reflected in distinctive urban development.[15]

However, there can be inertia in the urban environment. Iron and steel industries, for example, have been highly localized; in the 1920–1960 period most steel was made in the areas around Pittsburgh, Cleveland, Detroit, Youngstown, and Chicago. There has been a slow shift westward in steelmaking, reflecting western demand for steel and new sources of ore and coal. But this shift has been gradual. The high cost of fixed capital equipment, of land and buildings, can shape locational decisions and enforce stability for a long period of time.[16]

New industries frequently locate in new cities, such as electronics in Sunbelt cities. Older cities may continue as before or they may stagnate and die, as Detroit seems to be doing. Some older cities may capture part of the newer industries; yet others may actually lose their industries, as when New England towns lost their textile factories to southern cities. Older cities tend to give way to the dominance of newer cities.

Designing Cities for Worker Control

David Gordon has noted that U.S. cities have been shaped by the changing character of capitalism in three periods: commercial capitalism (1750–1850), industrial-competitive capitalism (1850–1900), and monopoly capitalism (1900-). The cities that grew dramatically in each of these periods tended to be of different kinds. In the commercial period merchants' activities generated and influenced cities; ports became major cities as trade expanded. New York became the dominant commercial metropolis. Central cities were diversified in terms of income, but poorer workers generally lived outside the central areas. Merchants engaged in land speculation; the lot-and-block grid system across many cities was first implemented by these early land speculators.[17]

Later, from 1850 to 1900 industrial capitalism and the factory system dominated city structure. Factories required large numbers of wage laborers, so it is not surprising that the crowded central cities became centers of industrial capitalism. Traditional discussions of capitalistic location have emphasized that transportation and communication efficiency and access to other companies are among the most important factors in city location. But these industrial cities grew well beyond the point where transportation and access were efficient. Cities became greatly congested. Gordon suggests that a primary concern of capitalists was labor control, that is, keeping cheap workers available in large numbers. The impersonality of cities as large as New York and Chicago, in the beginning at least, made it more difficult for workers to organize successfully. During this period factories were located in downtown areas; segregated working-class residential areas—tenements—sprang up nearby. Higher income urbanites tended to leave the downtown centers, moving farther out to suburbs as streetcar lines expanded. Gordon argues that these industrial cities became clogged not because of technological necessity, but because the dominant capitalists there required a concentrated mass of relatively impoverished workers to make their enterprises profitable.

The twentieth-century city, in contrast, grew up under a monopoly capitalism dominated by a few hundred huge corporations; it is characterized by significant decentralization of manufacturing, the movement out of central city areas. Modern technology is usually offered as the reason for this change: trucks and telephones made manufacturing in outlying areas possible. But Gordon notes that this decentralization had already begun in 1900, well before the truck was an important factor. Manufacturing decentralization was to a substantial degree a reaction to the militant labor unions and growing class conflict. Companies able to finance plants outside the central cities moved out to find cheaper labor and to reduce the ability or workers to organize. Large-scale union organizations and rebellion in major central cities took place in the 1870–1915 period, and this contributed to the decentralization of much industry to outlying city areas.[18]

Shift to the Sunbelt

The 1980 census revealed major population losses for major northern cities; St. Louis, Cleveland, Detroit, and Pittsburgh suffered dramatic losses of more than one-fifth of their populations over the 1960s and 1970s. Chicago, Boston, Newark, and New York also lost significant proportions. Substantial manufacturing and other employment have shifted from the northern states to the South in the last two decades. This Sunbelt shift began in the 1960s considerably before the well-publicized fiscal crisis of northern cities such as New York. These crises are substantially the result of that shift, not its cause. Between 1960 and 1980 "capital stock" (new plant and equipment) grew at twice the pace in the South as in the North. In roughly the same period the North lost a total of 800,000 manufacturing jobs, even though the nation gained 2.3 million jobs. Moreover, the shift of jobs from the North has been faster than the shift of population.[19]

Why the tilt to the Sunbelt? Why have many companies chosen to invest in new operations or expand existing operations in this area? Several explanations have been offered. Some have argued that capital movements to the South have followed workers seeking more attractive life-styles, an argument contradicted by data on jobs moving faster than populations. Even more often, business and allied political leaders talk about "fiscal responsibility," "balanced budgets," and "good business climate" in describing what are for them "good" and "bad" cities. What does the phrase"a good business climate" mean? It is a code for lower wages, moderate living costs, fewer unions, lower taxes, and a cooperative (in other words, conservative) governmental environment, as outlined by a handbook of the Urban Land Institute, a major development industry think tank.[20]

There are other reasons for Sunbelt growth. From the beginning Sunbelt cities have not had as many physical barriers to new industrial development. Most were initially smaller, less built up, with open spaces. They did not have the older industrial structures of Frostbelt cities.

How does the shift to the Sunbelt work? Some multilocation corporations disinvest by delaying maintenance or ignoring machinery replacement needs in a northern plant, store, or office facility. Then they invest the savings in new and different operations in the Sunbelt. Or sometimes facilities may be moved from plants being phased out to expanding plants already in the Sunbelt. By various means the profits from northern operations can be rechanneled. An estimated 2 percent to 5 percent of manufacturing changes result from the actual physical relocation of the same plants. John Rees's study of location decisions by manufacturing firms in the Sunbelt cities of Dallas, Fort Worth, and Houston found that nearly 6,000 such plant decisions were made in the 1960–1975 period. Of these, four in ten were plant expansions, and just over one-third were new plants. Only one-seventh of these decisions were plant relocations from outside the area; the rest were acquisitions. The majority of plant location decisions did not involve relocating existing plants. Major causes of Sunbelt growth have been the expansion of existing companies and the creation of new companies. However, an important component of this process is the shifting of operations by a large multilocational company from one region to another, expanding employment in one region—the Sunbelt—and contracting employment in another—the Frostbelt.[21]

Manufacturing and other business growth stem mostly from capital shifts into businesses distinctive in the Sunbelt. The six pillars of this Southern development, according to Kirkpatrick Sale, are defense, oil, advanced technology, real estate and construction, tourism, and agriculture—all industries that are relatively new that have seen a dramatic new infusion of capital since World War II. The Sunbelt reflects a new round of corporate profit making in cities.[22]

Most Sunbelt cities have good business climates in part because the federal government has channeled financial aid to them and away from northern cities. This began in the 1930s under the New Deal and accelerated with major military expenditures in World War II. In the last few decades substantial federal expenditures have gone to support military—industrial complexes (e.g., Dallas-Ft. Worth), as well as to subsidize highway, sewerage, electric, and other services, providing the support structure necessary for expanded corporate development in the South. In the last decade or two federal resources (military and civilian) have been dispersed disproportionately to cities in the South and West, to the disadvantage of cities in the North. These expenditures make it clear that the rise of the Sunbelt cities is closely tied to government action. Some even speak of this regional conflict over federal aid as the "second war between the states."[23]

The Government's Urban America Report

A 1980 report sponsored by the Carter administration suggests that the federal government should go further still in aiding corporate growth in Sunbelt cities, to the point of letting northern cities stagnate or die. This report, prepared by the President's special Commission for a National Agenda for the Eighties, calls on government to stay away from the declining cities of the North such as New York and Cleveland and explicitly notes that "contrary to public wisdom, cities are not permanent." A basic idea is that the federal government should not try to create jobs in the declining central cities of the North but rather should keep "hands off," or at most help people migrate to the Sunbelt, where corporations are generating new jobs.[24]

The *Urban America* report views the Sunbelt as an area of "economic viability, benefitting large numbers of poor southerners." But the business climate is not necessarily good for poorer working people in the Sunbelt. Robert Firestine notes that demographic data show that "despite the dramatic income expansion in the Sunbelt in recent years, this economic growth has apparently done little to alleviate the relative extremes in income distribution which have marked the South, at least for decades." [25] Census data indicate that extraordinarily large proportions of inner-city residents in Sunbelt cities remain very poor in spite of the region's heralded prosperity. Incomes are almost as unequally distributed in the South today as in the past, particularly for blacks and Chicanos. The South is on the whole much less generous than the North in educational, health, welfare, and other social service expenditures. Part of the price workers and their families pay for business prosperity in the South is less desirable working conditions and lower quality of services, not only for the poor, but for middle-income workers as well.

Nor is business development in the southern cities necessarily permanent. Surprisingly, the South is losing certain manufacturing jobs. Recently for every 100 new manufacturing jobs created in the South, 80 manufacturing jobs were lost. Yet in the North for every 100 jobs created, 110 were lost, so the net gain in the South was clearly better. The balance is favorable for the South, but why does it also see plant closings? The *Urban America* report is quite candid on this point: "Industrial activity that was attracted to the South since the 1950s by lower wages, greater labor control, and lower energy costs is often found relocating outside the United States to achieve even lower production costs.[26] " Capital will flow from Sunbelt cities when profitability dictates that it do so, without regard to the human costs.

Shifts within the Sunbelt

Some movement out of Sunbelt cities is already observable. To take one recent example, advertisements in the Houston area have boldly proclaimed the advantages of good schools, good jobs, less pollution, and fewer traffic jams *elsewhere*. These ads are not intended to seduce workers to Houston from other cities. Rather, they are intended to attract firms and people out of Houston to outlying cities within a hundred-mile radius. Even southern business leaders and publications, such as the *Houston Business Journal*, have come to recognize that "a city, like a macrocosm of businesses that create it, can reach a point of diminishing returns."[27] Houston's traffic, pollution, and poverty-related crime—all creatures of unbridled Sunbelt capitalism—have escalated business costs. Thus companies such as Brown Oil Tools, Inc., an old Houston firm, are moving out from the "shining new" metropolis to small towns such as Huntsville. The views of a former top company executive were recently summarized by a business reporter:

> He doesn't hedge when he gives reasons for moving Brown to Huntsville: traffic, equality of life and the availability of high quality labor. . . . The time element in commuting is become an important consideration for workers, and he contends that Houston is making too little progress in its public transportation system and in the congestion of its freeways.[28]

Even the industrial heartland of the North may come back into fashion as a place for profitable investment. By the early 1980s some real estate investors were saying that big money was to be made in northern cities and their suburbs. One investment consultant noted that moderate growth areas outside the Sunbelt do not have the increasing Sunbelt problem of supply (e.g., new development projects) exceeding the demand. Developers in the Sunbelt are too optimistic and are overbuilding and outpacing demand, "so that they're not really reaping the benefits of growth—because the supply is so ample."[29] This investor recommends urban areas where developers are not building so aggressively. He suggests investing in the suburban areas of Boston, Columbus, and even troubled Detroit.

This uneven oscillation of investment shows clearly that corporations have no commitment to urban communities anywhere. They do not exclusively prefer rural or urban areas, or the Sunbelt or Frostbelt as places to make profits; rather they will move freely from one place to another, even encouraging competition between local communities bidding for their jobs with tax incentives and other subsidies, with little concern for human costs.[30]

GOVERNMENT SUBSIDIZES BUSINESS

The *Urban America* report put the federal government's seal of approval on corporate mobility whatever its social costs. But government do a lot more than create commissions to make life easy for corporations.

Welfare Assistance for Capital

Government officials in many cities and states have competed vigorously to entice companies to their areas. Thousands of local agencies and officials compete, with government grants and subsidies, for corporations seeking to move their investments. City and state governments play a type of "Russian roulette" to attract businesses; businesses pit one government against another for the best deal. The result is a broad array of tax and other subsidies that are in fact unnecessary, making taxpayers shoulder more and more business risks.

In a recent speech a senior vice-president of a major industrial location firm emphasized his company's success in winning $30 million of subsidies in Canada for one of his client corporations, including "a free, fully serviced site with heavy utilities in place, reduced local property tax levies, a federal grant of over $10 million, [and] a provincial training grant." [31] Similar packages have been worked out for companies in many areas of the United States.

Some state governments, especially in the Sunbelt, have restricted unionization with right-to-work laws. Texas, Virginia, and Oklahoma have advertised that "labor-management relations are excellent" and that "wage rates are considerably below those found in major manufacturing areas." Northern states such as New Hampshire have also advertised this lack of unionization. North Carolina has proclaimed in a promotion brochure that its right-to-work law "*guarantees* employee freedom" and "*preserves* right to manage." [32] And a recent advertisement for plant location in Florida ran as follows:

> *Got Labor Pains? Central Florida Delivers!* Florida's "right-to-work" laws contribute to a low incidence of labor un-rest and work stoppages. We are ranked 43rd among the fifty states, with less than 13 percent of non-agricultural labor unionized.[33]

This ad was placed by a county development group in cooperation with the Florida State Department of Commerce.

Enterprise Zones: Company Towns?

Federal government subsidies for business have been quite substantial. They have included major programs for urban renewal in central cities and other public investments that cleared land for new business developments. Here we can note one major proposal to redo central cities that was developed in the early 1980s. In 1982 the Reagan administration presented its first serious program for dealing with cities, an "urban enterprise zones" proposal to deal with poverty-stricken central-city areas. The basic ideas was to entice companies back into cities by providing an incredible array of giveaways. For companies setting up shop in central-city areas there was to be a sharp reduction in taxes, including investment tax credits, payroll employment tax reductions, income tax and capital gains tax reductions, and numerous others. Also proposed was a significant reduction in governmental regulations in the areas where companies relocate. Eligible areas, in the final Reagan proposal, would be chosen on the basic of poverty and unemployment levels, and local governments would apply to the federal Department of Housing and Urban Development for designation as urban enterprise zones. Seventy-five zones in the first few years were suggested. (An estimated 350 cities have areas that could meet the criteria.)[34]

The Reagan administration's 1982 Enterprise Zone Tax Act was presented to Congress in a form substantially different from the original proposal, which was far more extreme in its attempt to do away with minimum wage, health, environmental, and civil rights regulations for firms locating in the enterprise zones. The original proposal was called by some "frontier" capitalism because of its throw-the-workers-to-the-wolves philosophy. It was an attempt to assist companies in making a profit in central-city areas. The proposal would attract some companies seeking to make superprofits, although judging from similar past programs the corporate commitments would be temporary. These proposals are in line with the Fantus survey of corporations mentioned earlier. They include many items on the shopping list of concessions corporate executives want.

An official of a public employees' union called the Reagan urban enterprise zone proposal "an excuse for the creation of company towns." Such proposals do reflect the profit needs of corporations in seeking out locations. Jobs for ordinary working people are thereby created, but at the heavy social cost of letting corporations dictate the terms. In this sense, then, urban enterprise zones would be company towns built with federal assistance.

CONCLUSION

U.S. cities are changing, as industrial and commercial corporations make decisions to abandon one area and move to another. This constant threat of job loss has made many American working people fearful. Their community, their city may be next. Bluestone and Harrison have articulated the dynamics succinctly:

> Working people's feelings about this are not without founcation. Deindustrialization is occurring—on a surprisingly massive scale. It can be seen from North to South and East to West. It is happening as the largest, most powerful corporations in the nation shut down older plants in the industrial heartland (*and* in the Old South) to move to new industrial zones in the New South and overseas . . . in an effort to maximize their profits and increase their span of control over the global economic system.[35]

NOTES

1. Barry Bluestone, Bennett-Harrison, and Lawrence Baker, *Corporate Flight* (Washington, D.C.: Progressive Alliance, 1981), pp. 12–13; "Capital Moves: Who's Left Behind," *Dollars and Sense,* April, 1981, pp. 7–8; Barry Bluestone and Bennett Harrison, *The Deindustrialization of America* (New York: Basic Books, 1982), pp. 9–30.
2. "Capital Moves," p. 7; "Seth Thomas," *Austin American Statesman,* April, 1982, p. A7.
3. "Capital Moves," p. 8; David Smith, *The Public Balance Sheet: A New Tool for Evaluating Economic Choices* (Washington, D.C.: Conference on Alternative State and Local Policies, 1979), pp. 3–4.
4. Bluestone, Harrison, and Baker, *Corporate Flight,* pp. 24–35.
5. Karl Jenkins, "Don't Let Them Close This Plant," quoted in "Big Three' in Mexico," *Labor Today* 19 (June, 1980), p. 6.

6. August Lösch, *The Economics of Location* (New Haven: Yale University Press, 1954), pp. 28–30.
7. Melvin Greenhut, *Plant Location in Theory and Practice* (Chapel Hill, N.C.: University of North Carolina Press, 1956), pp. 175–285. I am drawing on the summary in David M. Smith, *Industrial Location,* first edition (New York: John Wiley and Sons, 1971), pp. 144–146.
8. E. K. Hunt and Howard J. Sherman, *Economics,* fourth edition (New York: Harper and Row, 1981), pp. 112, 575–579.
9. M. Storper, R. Walker, and E. Widess, "Performance Regulation and Industrial Location: A Case Study," *Environment and Planning* 13 (1981), pp. 321—338; David M. Gordon, "Capitalism and the Roots of Urban Crisis," in *The Fiscal Crisis of American Cities* (New York: Vintage Books, 1977), pp. 104–105.
10. Michael P. Smith, *The City and Social Theory* (New York: St. Martin's Press, 1979), p. 240.
11. John D. Kasarda, "The Implications of Contemporary Redistribution Trends for National Urban Policy," *Social Science Quarterly* 61 (December, 1980), p. 385.
12. The report is quoted in Richard A. Walker, "A Theory of Suburbanization: Capitalism and Construction of Urban Space in the United States," in *Urbanization and Urban Planning in Capitalist Society,* edited by Michael Dear and Allen J. Scott (London: Methuen, 1981), p. 401.
13. Storper, Walker, and Widess, "Performance Regulation and Industrial Location, pp. 325–327.
14. Diana Klebanow, Franklin L. Jonas, and Ira M. Leonard, *Urban Legacy* (New York: Mentor Books, 1977), pp. 157–158.
15. Alfred J. Watkins, *The Practice of Urban Economics* (Beverly Hills, Calif.: Sage Publications, 1980), pp. 130, 154.
16. Smith, *Industrial Location,* pp. 350–360.
17. Gordon, "Capitalism and the Roots of Urban Crisis," pp. 93–94.
18. Ibid., pp. 94–102.
19. President's Commission for a National Agenda for the Eighties, Panel on Policies and Prospects, *Urban America in the Eighties: Perspectives and Prospects* (Washington, D.C.: U.S. Government Printing Office, 1980), pp. 18, 27, 42.
20. Urban Land Institute, *Industrial Development Handbook* (Washington, D.C.: Urban Land Institute, 1975), p. 229.
21. John Rees, "Regional Shifts in the U.S. and the Internal Generation of Manufacturing Growth in the Southwest," paper presented to Conference on the Committee on Urban Public Economics, Baltimore, Johns Hopkins University, May, 1978, pp. 11–13; John Rees, "Manufacturing Change, Internal Control and Government Spending in the Growth Region of the United States," Paper presented to the 24th North American Meetings of the Regional Science Association, Philadelphia, November, 1977, pp. 1–7.
22. President's Commission, *Urban America in the Eighties,* pp. 41–42; Kirkpatrick Sale, *Power Shift* (New York: Random House, 1975).
23. President's Commission, *Urban America in the Eighties,* p. 78; Felix Rohatyn, "The Older America: Can It Survive?" *New York Review of Books 27* (January 22, 1981), pp. 13–16.
24. President's Commission, *Urban America in the Eighties,* pp. 18–42; see also "Burning Up the Snowbelt," *Time,* January 12, 1981, p. 19.
25. Robert E. Firestine, "Economic Growth and Inequality, Demographic Change and the Public Sector Response," in *The Rise of the Sunbelt Cities* (Beverly Hills: Sage, 1977), p. 199; c.f. also Peter A. Lupsha and William J. Siembieda, "The Poverty of Public Services: An Analysis and Interpretation," in *The Rise of the Sunbelt Cities,* p. 174.
26. President's Commission, *Urban America in the Eighties,* p. 42.
27. "Corporate Flight to the Suburbs," *Houston Business Journal,* January 12, 1981, p. 1.
28. Ibid.
29. Steven Lewis, "Institutions Will Dominate Development in 1981, Predicts White: Money Market to Remain Volatile," *National Real Estate Investor* 23 (February, 1981), p. 60.
30. Robert Goodman, *The Last Entrepreneurs* (New York: Simon and Schuster, 1979), p. 49.
31. Robert M. Ady, "Shifting Factors in Plant Location," *Industrial Development,* November–December, 1981, pp. 16–17.
32. Quoted in Goodman, *The Last Entrepreneurs,* p. 37.
33. Advertisement in Conway Publications, *Site Selection Handbook/78 23* (February, 1978), p. 70.
34. John Herbers, "New Urban Program," *New York Times,* March 28, 1982.
35. Bluestone and Harrison, *The Deindustrialization of America,* p. 47.

35 *The Transformation of Cities and Suburbs: How Business Is Creating Urban Villages*

CHRISTOPHER B. LEINBERGER AND CHARLES LOCKWOOD

Since the end of the second World War urban planners and academicians have warned Americans that suburban sprawl is an evil to be fought with every ounce of our national energy. They have condemned cars as the cause of city-killing growth. They have idealized mass transit. Yet, other than business and political interests in city downtowns, few have paid attention to their cries. We Americans have allowed sprawl to happen. In fact, we have reveled in our low-density suburban housing, our automobiles, and our decentralized living.

Now it appears that the much-reviled postwar suburban sprawl, with its sea of split-level houses surrounding retail businesses and apartment complexes strung randomly along its highways, was merely a transitional phase between the tranditional compact pre-war city and today's metropolitan area. Our cities are becoming groups of interdependent "urban villages," which are business, retail, housing, and entertainment focal points amid a low-density cityscape. Each urban village has its core—a kind of new downtwon—where the buildings are tallest, the daytime population largest, and the traffic congestion most severe. And each urban village has its outlying districts, which may stretch as far as ten miles from the core.

SOURCE: Christopher B. Leinberger and Charles Lockwood, "How Business Is Reshaping America," *The Atlantic Monthly* 258 (October 1986), pp. 43–52.

Urban villages represent a dramatic restructuring of America's cities and suburbs—one that is already affecting how millions of Americans live and work. Almost every city is swept up in the urban-village phenomenon—not only fast-growing Sunbelt cities like Atlanta and Phoenix but also slow-growing older ones like St. Louis and Kansas City, and archetypal cities like New York and Baltimore.

THE URBAN VILLAGES OF LOS ANGELES

Los Angeles is perhaps the most evolved example of the urban-village phenomenon. Although downtown Los Angeles has recently experienced an unprecedented boom in office-building, the metropolitan area has simultaneously given rise to sixteen smaller urban-village cores, among them Century City, Costa Mesa/Irvine/Newport Beach, Encino, Glendale, the Airport, Warner Center, Ontario, Pasadena, Universal City/Burbank, and Westwood.

One of the easiest ways to gauge the growth of Los Angeles's urban villages is to look at the downtown's market share of the metropolitan region's office space over the past decades. (Office space is used as a standard measure of changing patterns of urban development in this article, because the office is the "factory of the future." In parts of some metropolitan areas 40 to 60 percent of the people hired for newly created jobs go to work in office buildings—not only white-collar workers but also support staff like janitors and food-service workers and security guards.) Since 1960 downtown Los Angeles's share of the metropolitan office market has declined from 60 to 34 percent, as expansion has shifted to the faster-growing urban-village cores in the suburbs.

The Costa Mesa/Irvine/Newport Beach complex, in Orange County, provides a good example of how the urban-village phenomenon is reshaping greater Los Angeles. Until the early 1960s a series of small towns in a county whose total population was 868,000, the complex today is California's third largest downtown, as measured by office and business-park space. The first phase in the creation of the urban village's core was the construction of a regional shopping mall, as it has been in many other locations. In this case there were two—South Coast Plaza, in Costa Mesa, and Fashion Island, in Newport Beach, both of which opened in the early 1970s. Next developers completed modest two-story office buildings nearby for local professional firms, and light industrial facilities for aerospace and high-tech outfits.

In 1980 the population of Orange County reached 1,932,708, and jobs there numbered 1,670,100. The boom transformed the emerging Costa Mesa/Irvine/Newport Beach subcenter. Office space there now totals 21.1 million square feet—well behind downtown Los Angeles's 36.6 million square feet, but gaining fast on downtown San Francisco's 26.8 million. Both the architecture and the tenants of the new high-rises are often the equal of those of office buildings in Los Angeles and San Francisco. The urban village's new hotels include a Westin, a Meridien, and a Four Seasons. South Coast Plaza has more sales than any other shopping center in the country. It is projected that once an addition is completed later this year, the plaza's annual retail sales will surpass those for both downtown San Francisco and the "Golden Triangle" of Beverly Hills. South Coast Plaza has most of the exclusive shops found in those locations.

Many of Los Angeles's established urban villages are gaining their own identities. Los Angeles's aerospace industry is increasingly concentrated at the Airport and in Torrance. The entertainment industry, which has traditionally been located in Hollywood, is now moving over the Hollywood Hills to Universal City/Burbank. Los Angeles's insurance companies, traditionally found in the mid-Wilshire Boulevard district two miles west of downtown, have recently selected Pasadena as an insurance subcenter, because it is closer than the mid-Wilshire location to neighborhoods from which they can expect to attract clerical workers and is also close to an executive-housing neighborhood.

The insurance and financial industries are transforming Pasadena, a city of 128,500 residents eight miles northeast of downtown Los Angeles, into an urban-village core. Perhaps the best-known city of its size in America because of the annual Tournament of Roses Parade and Rose Bowl, Pasadena's history and image have long been conservative and "old money." However, by 1970 downtown Pasadena was marred by empty storefronts, fading turn-of-the-century resort hotels, and half-empty parking lots. Low-income, primarily black and Latino northwest Pasadena was troubled by crime, gangs, drugs, and welfare dependency.

In the early 1970s the I-210 freeway was completed, connecting the city to the rest of the San Gabriel Valley and eventually resulting in a spurt of office-building. Pasadena became a back-office employment center for insurance and banking offices that wanted to hire middle-class Angelenos living in the rapidly growing San Gabriel Valley suburbs. Since 1980 twenty-eight mid- and high-rise office buildings have gone up.

Because of all this office construction, the character of downtown Pasadena has changed radically. A shopping mall several blocks from City Hall, which opened in 1980, is among the most successful in southern California. Entire blocks of handsome 1920s and 1930s two-story commerical buildings have been restored, and restaurants, espresso bars, and shops are now moving into ground-floor retail space that had gone begging for many years.

On the edge of downtown, within walking distance of the new offices and shops, badly run-down single-family housing has been replaced by new condominiums and townhouses. Thousands of houses in nearby residential neighborhoods have been renovated as young families have discovered the city's turn-of-the-century architecture and tree-shaded streets.

Change has not come easily to Pasadena. Not all of the city is being renovated; residents of the still decaying northwest section fear that new construction will squeeze them out of the neighborhood. And the many new jobs have created traffic problems. Today about 60,000 people commute to work in Pasadena, whereas approximately 25,000 leave the city for jobs elsewhere.

Pasadena residents worry about the long-range impact of the downtown construction on the overall community. Why has Pasadena attracted back-office operations instead of professional and corporate tenants, which would be more in keeping with the city's lingering blue-blood image? Does Pasadena want more high-rise office towers of any kind? How will the growing traffic affect downtown Pasadena and the nearby reviving neighborhoods?

WHY SUBURBS ARE BECOMING URBAN VILLAGES

This debate over growth is being repeated in neighborhood associations and city councils across the nation, in such widely dispersed locations as Walnut Creek, California; Princeton, New Jersey; Schaumburg, Illinois; and Bethesda, Maryland. In fact, attempts to control growth and traffic congestion are among the hottest issues in local politics today. And the reason is that the urban-village phenomenon is reshaping our cities. However, most people do not see the larger picture; they simply see the consequences. The phenomenon has occurred simultaneously in all kinds of cities across the nation, mostly for five reasons. Four of these have helped to create or further the postwar pattern of sprawl. When the fifth reason comes into play, it is easier to see why our metropolitan regions are coalescing into these sub-centers.

One reason for the growth of urban villages is that the nation's economy is shifting from a manufacturing to a service and knowledge base. The number of industrial jobs has declined from one third of all jobs in 1920 to one sixth today. The figure may drop to less than one tenth by the year 2000, with service and knowledge employment making up the difference, much as agricultural employment fell and industrial employment rose during the transition to an industrial economy in the late nineteenth and early twentieth centuries. From 1973 to 1985, while five million blue-collar industrial jobs were lost, the service and knowledge fields—which are a diverse as computer programming, the professions, retail sales, and fast food—accounted for all of the nation's employment growth (from 82 million to 110 million jobs). The Bureau of Labor Statistics reports that the number of service and knowledge jobs can be expected to grow by nine million in the next ten years. The shift to a service and knowledge economy has greatly accelerated the restructuring of our metropolitan areas by creating a need for much more office space. In eight years, from the beginning of 1978 through 1985, 1.1 billion square feet of office space was constructed in the United States—the equivalent of 220 World Trade Center towers.

People are willing to live near the office now, whereas they were reluctant to live near factories that were dirty, noisy, and visually unattractive. Traditionally the poor and the working class have lived near factory districts. Across the nation office buildings and high-tech business parks have appeared in middle-class and even exclusive suburbs, such as the house country around Valley Forge, outside Philadelphia; Bellevue, near Seattle; and the Buckhead district, in the north Atlanta suburbs. In Newport Beach, California, south of Los Angeles, a new subdivision of homes costing from $300,000 to $2 million abuts a Ford aerospace facility doing research and small-component-part assembly for military contracts.

A second reason for the emergence of urban villages has been changes in transportation patterns. Business shipments have shifted from rail to truck and Americans have decided that they prefer automobile commuting to mass transit. With fixed-rail transportation, all shipments must go from central terminal to central terminal before delivery. With trucks, shipments can travel from door to door. Similarly, with automobiles, people can go from door to door whenever they choose, without having to pass through central stations and terminals at fixed times. By 1960, the first year that the U. S. Bureau of the Census kept track, 69.5 percent of all commuting trips were made by car. By 1980 the figure had increased to 86 percent.

A third reason has been recent telecommunications advances, notably the dramatic drop in long-distance telephone costs. More and more day-to-day work is being done over the telephone. Not only cheaper long-distance telephone rates but also overnight mail, telecopiers, Zap-Mail, and computer modems allow communication without physical proximity.

Fourth, it is cheaper for businesses to operate in urban villages than in cities. Suburban office, industrial, and retail rents are far lower than downtown rents are. Although construction costs are approximately the same in the two areas, land outside cities costs less—$10 to $50 per square foot in suburban office locations versus $50 to $1,000 per square foot downtown. The cost of building parking spaces is also lower in the suburbs, where most parking is located in above-ground structures or surface lots, both of which are much cheaper to construct than a downtown office tower's subterranean garage. The difference in land and parking-garage costs shows up in office rents—$15 to $24 per square foot per year for prestige high-rise suburban space versus $18 to $42 for comparable downtown space in most American metropolitan areas.

These factors might only have encouraged more suburban sprawl if another factor had not come into play: most Americans like cities and the concentration of services that they provide. A critical mass of employment and housing is necessary to support desirable everyday services such as a good selection of shops, restaurants, and hotels. But this critical mass is achieved at well below the size of the prewar downtowns. For instance, it takes about 250,000 people within a three-to-five-mile radius to support a modest regional mall and about 20,000 middle-class people within an equal area to keep a good restaurant in business. And it requires roughly 2.5 million square feet of offices (about fifty typical three-story suburban office buildings) to support a 250-room hotel.

Since people do not want to drive very far for these services, particularly office workers looking for lunch or business travelers needing a hotel, a degree of concentration much greater than that of a low-density suburb is necessary. Only so many potential locations have the necessary highway access and visibility to permit this concentration, a fact that has led to the focusing of postwar suburban sprawl into urban-village cores. When prime urban-village cores become congested, developers and employers hopscotch to outlying locations where land is cheaper and commuting is easier.

Thus our metropolitan areas have expanded tremendously. In the past twenty-five years, for example, metropolitan New York's sphere of direct influence has tripled in size. The metropolitan area extends as far north as New Haven, Connecticut, and into once-rural Dutchess, Ulster, and Sullivan counties, in New York State. On Long Island the metropolitan area now encompasses Yaphank and Brookhaven, in the middle of Suffolk County, and fades away just miles from the Hamptons. Metropolitan New York has swept southward past Princeton and into Ocean County, New Jersey, and Bucks County, Pennsylvania. In several years New York's suburban growth will collide with that of slower-growing metropolitan Philadelphia.

Los Angeles has always had an enormous metropolitan area; it was knit together before the Second World War by the world's largest trolley system. By 1960 metropolitan Los Angeles's sphere of influence stretched to Santa Monica in the west, Pasadena in the east, Long Beach in the south, and the San Fernando Valley in the north. Now metropolitan Los Angeles extends westward into Ventura County, east to San Bernardino and Riverside, and south across most of Orange County to span an area many times the metropolitan area's size twenty-five years ago. Other cities—among them greater San Francisco, Phoenix, Dallas, Houston, Boston, Chicago, Atlanta, and Washington, D.C.—are also expanding quickly.

As outlying urban-village cores become more urban, with their high-rise office buildings and hotels, increasingly sophisticated shopping, and high-density housing, the center cities are becoming more suburban. This emerging trend extends beyond the fast-food restaurants on downtown business streets and festive retailing complexes like South Street Seaport, in Lower Manhattan, and Harborplace, in Baltimore. Suburban-style shopping centers—where you buy children's clothes and household appliances, not just clever T-shirts and scented soaps—have recently opened in several American cities, and are doing quite well. Large, privately financed apartment and condominum projects have been built in cities without a downtown residential tradition, including San Diego, Detroit, Atlanta, and Los Angeles.

HOW URBAN VILLAGES AFFECT CITIES

Few cities have been transformed by the urban-village phenomenon as rapidly as Atlanta, whose metropolitan region has a population of 2.2 million, having gained 90,000 residents last year and 78,000 the previous year. In 1980 downtown Atlanta was the metropolitan region's unchallenged center for all kinds of office employment. Although urban-village cores were emerging around shopping malls at the intersections of major highways, office space was limited. And neither in the appearance of the building nor in the quality of the tenants were the fledgling urban-village cores any match for the downtown high-rises. Most of the buidings were two-story wood-frame structures, and many of their tenants were banks and insurance companies renting inexpensive space for back-office operations.

By 1985—just five years later—this pecking order had changed completely. Downtown Atlanta had gained 4.3 million square feet of new office space, but Perimeter Center, at the I-285/Georgia 400 intersection, due north of downtown, had gained 7.6 million, and Cumberland/Galleria, at the I-285/I-75 intersection, northeast of downtown, 10.6 million. Many of the new buildings were gleaming, architecturally distinguished high-rises of the kind that once had been bulit only downtown. If present trends continue, the amounts of office space in both the Perimeter Center and the Cumberland/Galleria urban-village cores will easily surpass the amount in downtown Atlanta by 1990.

Downtown Atlanta, like downtown Los Angeles, is losing its metropolitan hegemony and becoming just another one of the region's urban-village cores. The urban villages around it, like those around downtown Los Angeles, are gaining distinct identities. Government, professional services, finance, wholesale trade, and the convention business are still in downtown Atlanta. However, as prestigious high-rises have been completed at Perimeter Center and Cumberland/Galleria, those areas have lost many of their price-conscious back-office banking and insurance tenants and have become instead favorite Atlanta locations for corporate headquarters and business services. Nearly every major accounting firm and many downtown law firms have offices in Perimeter Center and Cumberland/Galleria, the better to serve their clients. The regional corporate headquarters of Northern Telecom and HBO are in Perimeter Center.

Despite their importance as business locations, Perimeter Center and Cumberland/Galleria do not boast that traditional downtown symbol, a soaring skyline of office buildings. The high-rises in these two urban-village cores are not clustered. Instead they are widely separated by parking lots and heavily landscaped open space. In keeping with the non-urban mood, the developers have even neglected to add sidewalks connecting the buildings, and most people use their cars instead of walking. From the top floor of a seventeen-story high-rise in Perimeter Center the view is mostly trees and lawns, not other office buildings. Like many (though not all) urban-village cores, these relatively low-density north Atlanta developments visually blend into the suburban landscape. Yet the explosive rise in the number of office workers and their cars at Perimeter Center and Cumberland/Galleria has affected once-peaceful residential neighborhoods, including the Spring Mill area, near Perimeter Center.

One surprising result of perimeter center's boom has also been seen in suburban Dallas, Chicago, and Washington, D.C., and may be a trend. Residents of middle-class subdivisions near high-rise office clusters are banding together in corporations, getting their neighborhoods re-zoned for offices, and selling large "assembled" land parcels to real-estate developers, who either raze the houses or move them to new sites and build high-rises on the land. The suburban homeowners see the chance to leave an increasingly congested neighborhood where house prices are not keeping pace with those in comparable but quieter subdivisions nearby. Often they can sell their lots for twice what the homes on them are worth.

One such group formed three years ago in Spring Mill, a subdivision abutted on three sides by the high-rise office complexes of Perimeter Center. Randy Campbell, a former Spring Mill resident, recalls seeing a notice in a neighborhood newsletter in July of 1983 asking residents if they wanted to sell their property to office developers. "If you were interested," Campbell says, "you tore off a strip of paper on the bottom of one page, filled in your name, and sent it to the head of the neighborhood association. All but three or four of the 117 homeowners mailed in that piece of paper, so we got together in the community room of the Perimeter Mall. The average Spring Mill house sold for $70,000 to $75,000 then, but we learned that the half-acre lots were worth three times that figure to office-building developers. We didn't know how to go about the re-zoning or sale. So all the neighbors appointed a committee of five men and four women—small-business owners, several attorneys, a clergyman, an accountant, and a stockbroker. That was me."

For the next few months the committee met almost every night, learning about real-estate development and discussing strategy. By the end of the summer Alston & Bird, Georgia's largest law firm, had agreed to represent the homeowners on a contingency basis, and the firm drafted an option agreement to create Kingsborough-Spring Mill, Inc., a nonprofit corporation. By signing this option agreement a homeowner gave Kingsborough-Spring Mill's three-person board of directors power of attorney to accept a minimum of $225,000 in cash for his home, net after any expenses. It was understood that no deal would be final until the county commissioners changed Spring Mill's zoning from single-family homes to high-rise offices and multi-family units.

The night that Alston & Bird presented the agreement, in October of 1983, eighty homeowners signed it, and by January of 1984 the figure had risen to 114 of the 117. "Once the neighborhood was 'tied up'," Campbell says, "we looked for a buyer. They weren't knocking the doors down, because we wanted top dollar and all cash. As we explored various deals, all the neighbors wanted to know what was happening. So we prepared a newsletter almost daily, ran if off on our office copying machines, and stuffed it into everyone's mailboxes during our lunch hours."

In October of 1984 the board of directors located a buyer. "Albritton Development wanted to redevelop the twenty-seven-parcel Lake Hearn Place subdivision adjacent to Spring Mill," Campbell says. "But the Dekalb County commissioners rejected the re-zoning, reportedly because the project was too small, too piecemeal. So we made a deal with Albritton, which combined Lake

Hearn's twenty-seven lots, the hundred and fourteen lots that we represented, and the three other Spring Mill owners that didn't join the neighborhood corporation. That made eighty-two acres—hardly piecemeal."

In June of 1985, however, the Dekalb County commissioners denied the re-zoning request for the site. Six months later Kingsborough—Spring Mill, Inc., the Lake Hearn Place homeowners, and Albritton Development sued the county commissioners. By last June the case had ended up in the Georgia Supreme Court, which ruled six to one in favor of the homeowners.

Baltimore is another kind of city being reshaped by the urban-village trend. Despite all the good news about Harborplace and the renovation of nineteenth-century row houses, the Baltimore metropolitan area is essentially stagnant. The population of the metropolitan area, as defined by the City of Baltimore and Baltimore County, declined from 1.53 million in 1970 to 1.44 million in 1980, and it is projected to drop to 1.40 million in 1990. Employment is growing only modestly—from 771,300 jobs in 1980 to 782,400 in 1985 and to 813,400 estimated for 1990. The Baltimore metropolitan area hardly seems like the place for a surge in construction. Yet there has been a boom in the outlying districts of Towson, Owings Mills, White Marsh, and perhaps most notably Hunt Valley.

Hunt Valley, twenty miles north of downtown Baltimore, was farmland until recently. Now it has a prestigious high-rise office and hotel core, which is surrounded by surface parking and low-rise industrial and office space. The low-rise industrial and office complexes came first, when well-to-do Baltimore businessmen decided to move their plants and offices closer to their homes in the north and northwest sections of the metropolitan area. After Hunt Valley was established as an employment center that offered low rents, the major landowner—McCormich Properties, owned by the family that owns the McCormick spice company—upgraded Hunt Valley by building high-quality mid- and high-rise office space and a Marriott Hotel. The PHH Group, Westinghouse, Burroughs, and Allstate Insurance promptly moved in. Now Hunt Valley is Baltimore County's top corporate address.

Even New York City has fallen under the sway of decentralizing urban-village development—which is surprising, considering that New York is the classic example of the traditional pre-Second World War city and that Manhattan has never been prosperous nor its skyline more crowded with construction cranes. But although Manhattan gained 23.5 million square feet of office space from 1982 to 1985, during this short time its share of the metropolitan area's more than half a billion square feet of office space fell from 67 percent to 60 percent. This year, too, 7.6 million square feet is schedule to be completed in Manhattan, bringing its total to 317.6 million—and 16.1 million square feet of office space is scheduled to be added to suburbs within a sixty-mile radius of Times Square, bringing the suburban total to 216.1 million.

In the past the typical new office building in New York's suburbs was part of the campus-like headquarters of a major corporation in Westchester

County or southwestern Connecticut. Today, though, the exodus of corporate headquarter from Manhattan—to the suburbs or other parts of the country—has slowed. Now the typical new building is built on speculation, in an urban-village, core such as Stamford or Danbury in Connecticut, the Route 110 corridor on Long Island, or Morristown or Princeton in New Jersey. One kind of tenant is a corporation's back-office operations or an entire divison that can be separated from the Manhattan headquarters at considerable savings in rent, taxes, and business services but linked to the main office by computers and telecommunications.

As more and more jobs move to the suburbs (most of the metropolitan region's employment growth in the next decade is expected to take place not in New York City but in northern New Jersey), housing construction is booming in outlying towns. At Brookhaven, in Long Island's Suffolk County, 4,000 homes were built last year and another 8,000 are planned. At Dover, New York, a Dutchess County town of 7,2000 residents northeast of Poughkeepsie, developers have filed porposals with the planning board for 1,200 new condominium units. The transformation is likely to be even more rapid at Chester, New York, fifty-five miles northwest of New York City. The population is 7,000, and the town board has received proposals for 3,000 new condominiums and homes.

In attractive communities near urban-valley cores the prices of existing homes are rising fast, owing to the sudden demand for employees of relocated companies. That is one reason that metropolitan New York City experienced a 24.6 percent increase in the price of housing between the first quarter of 1984 and the same period in 1985. In Stamford, for example, a typical three-bedroom house now sells for at least $250,000.

The Princeton, New Jersey, area illustrates the impact of the urban-village trend on metropolitan New York. Seemingly overnight a metropolis-sized urban-village core has emerged several miles east of this tree-shaded college town on U. S. Route 1. It is estimated that by the mid-1990s the Princeton corridor will have 12 to 15 million square feet of office space—surpassing the amount currently in downtown Milwaukee, downtown Newark, or southwest Connecticut's I-95 corridor between Greenwich and Stamford. Two luxury hotels have opened in greater Princeton; a Marriott and a Radisson are planned. The number of housing units is expected to double, with over 25,000 scheduled for completion by 1990. If current growth projections prove to be accurate, greater Princeton will become New Jersey's largest city in ten years.

HOUSING FOR ALL URBAN-VILLAGE WORKERS

None of the far-reaching social, economic, and political issues raised by urban villages has a broader and deeper impact than housing. Urban-village cores generally arise at the outskirts of cities and grow in a limited number of directions. Nearly all of Atlanta's urban villages, for example, are in the northern suburbs, leaving the southern part of the city largely unaffected. Dallas is growing mostly toward the north and northwest. St. Louis, not exactly a booming metropolitan area, is heading west and northwest. The vast majority of America's urban villages, in fact, have one thing in common: they are growing in white, upper-middle-class areas. Executives and business owners usually make the decisions about office locations and industrial sites. Most of them are white and upper-middle-class, and they usually decide to bring their offices or industrial plants nearer their homes.

This arrangement is time- and energy-efficient for executives and business owners but not necessarily for clerical, light-assembly, and service employees, nor for the employees of the stores, restaurants, and gas stations near offices and plants. These workers cannot afford the executive-priced single-family houses, townhouses, and condominiums near most urban-village cores. They face a long—and often expensive—car or bus commute to the suburban or city home they can afford to live in.

The scenes in the parking lots of north Atlanta's Perimeter Center at 5:00 show the results of this geographical mismatch. Executives and professionals get into their Cadillacs and BMWs for the relatively easy drive home or a visit to one of the nearby "formula" restaurants for a drink. At the same time, many black employees are walking through the parking lots—Perimeter Center has few sidewalks—on their way to the bus stops, which are little more than a pole with a bus sign on top, planted on a flat, grassy spot that usually turns into mud when it rains. Atlanta's working-class black sections are south of downtown, fifteen to twenty miles from Perimeter Center, but the roads and bus lines to them don't follow a straight line. With one or maybe two transfers, many bus passengers endure a one-to-two-hour ride twice a day.

Suburban fast-food restuarants, hotels, department stores, car washes, and cleaning services are already experiencing a shortage of low-paid service employees. In some fast-growing suburbs businesses are offering bounties for new employees, raising the starting pay from the $3.35 minimum wage to $4.00 or more, and offering raises after several weeks rather than a year. One solution to the shortage is to bring affordable housing closer to the employment in the suburban urban-village cores.

This idea is one of the most polarizing issues in local politics' today. Most suburbs have raised the drawbridge against housing for the poor and the working class, as epitomized by new guarded and gated "secure" communities. Affluent families do not want low-income housing in their one-acre-lot neighborhoods. Nor do they need to allow it. Our urban villages now encompass areas the size of many pre-Second World War cities. They have more than enough land for all kinds of housing—and socio-economic groups—if zoning boards permit affordable housing.

The key to truly affordable, non-subsidized suburban housing is allowing higher density, primarily in the form of rental apartments, so that fixed land costs are spread over more units. Building suburban apartments in smaller than average sizes, to reduce rents, should be permitted. Empire West, a Tucson developer, has built smaller than average apartments in the Southwest. Situated in 250-to-500-unit projects, the apartments range from 320-square-foot studios (equivalent to a 14-by-23 foot space), to 615-square-foot two-bedrooms (a 20-by-30-foot space). All apartments have complete kitchens, bathrooms, and closets, and are outfitted with scaled-down furniture.

These complexes usually offer the lowest rents in their local markets—typically from $225 to $400, at least $25 to $50 a month below rents for the cheapest conventional apartments. These rents fit the budgets of clerical employees, security guards, and maintenance workers who make less than $15,000 a year. Yet the complexes still offer "luxury" features like swimming pools, tennis courts, volleyball courts, and social events.

Of course, upper-middle-class suburbanites fight almost any kind of high-density housing. But attached housing complexes don't have to be built in affluent neighborhoods. They have been successful near commercial districts, where they create a buffer between a single-family-home neighborhood and an office-and-retail district. Another logical place for apartments is next to an urban village's commercial core. It seems likely that the vast surface parking lots of regional shopping malls will eventually be redeveloped as high-density housing and commercial buildings, with the cars going into above-ground or even underground structures. When these apartment and condominium developments are built near the shopping centers that form the focal point of so many urban villages, the members of more age groups and economic classes will be able to live near where they work and shop.

NEW TRANSPORTATION PROBLEMS

In most metropolitan areas the suburban rush hour now rivals downtown traffic, and in some surpasses it. The 1980 census reported that 27 million Americans commuted from one suburb to another, whereas only half that number traveled from suburbs to downtown cities. The imbalance has increased sharply since 1980, owing to the boom in suburban employment. Today's suburban highways are so overcrowded that once-easy five- and six-mile commutes take forty-five minutes of stop-and-go driving in locations as widely dispersed as north Dallas; Contra Costa County, east of San Francisco; and northern Virginia, near Washington, D.C. And the emergence of urban villages at the edges of the metropolitan areas—like Chesterfield, twenty-two miles west of downtown St. Louis—have brought bumper-to-bumper traffic to two-lane country roads that still run past farms and cow pastures.

Living up to its image, Los Angeles has the most heavily traveled freeways in the nation. The three busiest points on them are now near urban villages—miles from downtown Los Angeles and its four-level interchange, which once held the city's traffic record but has now fallen to fifth place. To reach their record-setting traffic levels, these Los Angeles freeways—like the slow-moving highways outside other cities—do not have a morning rush hour in one direction and an evening rush hour in the other. Commuter traffic now comes to a standstill at morning and early evening in both directions, and slowdowns occur frequently throughout the day.

Unfortunately, some city planners examining metropolitan-area congestion do not acknowledge the traffic near urban villages and repeat time-worn clichés about automobiles destroying cities. Others recommend the expansion of highways and mass-transit systems, but continue to think only of access to downtown, just as transportation needs soar in the suburbs. Still others recognize the rise of urban-village cores in the suburbs but propose transportation plans that are years behind the demand.

Planners and governments are working under a serious handicap in alleviating suburban traffic problems. Most urban-village cores are being created by private developers in a series of unrelated, uncoordinated decisions. To make matters worse, the process of formulating a transportation plan, gaining its approval, finding the funds, and completing the project is so time-consuming that traffic problems are bound to become painfully obvious before any action is taken.

How can government alleviate suburban traffic congestion near urban-village cores? The answer is not the construction of subways or elevated trains. Fixed-rail projects (unlike light-rail construction—for example, trolleys) justify their enormous capital expenditure and high operating costs "only when you have both trip originations and destinations highly clustered," according to Peter Muller, a professor of geography at the University of Miami and one of the earliest observers of the urban-village phenomenon. "We are building 1920s-style mass-transit systems for our 1990s metropolitan areas," he says.

One example of the gap between urban image and reality is Walnut Creek, California, near San Francisco, which has experienced considerable office and light-industrial development in the past decade. Although most of the new buildings are near the BART mass-transit station connecting the town to San Francisco, Walnut Creek now has such severe traffic congestion that last year it enacted an office-construction moratorium. Muller suggests that mass transit failed to help the town avoid traffic problems because the vast majority of the local workers commute by car regardless of a mass-transit option. "A great many of the people who ride the new fixed-rail mass-transit systems in cities like San Francisco, Atlanta, or Miami used to ride buses," he says.

Nonetheless, fixed-rail transit remains a gleam in many a downtown businessman's or politician's eye. Some proponents may imagine that it takes a subway for a city to be a city, and some may have in mind that the federal government has paid up to 90 percent of the bill in the past (though it will not in the future, if the policies of the present Administration remain in place). Even Los Angeles—the capital of California's "car kingdom"—wants to build an eighteen-to-twenty-mile "metrorail" subway system, at a cost of $3 billion. According to an official of the Southern California Rapid Transit District, the metrorail system "may come close to breaking even on an operating basis, but there is no way that it will pay for the construction of the system." Of course, this statement is based on predictions of operating cost and ridership. The reality of Atlanta's new MARTA fixed-rail system is that the fare box brings in only 35 to 40 percent of the annual operating budget, which would more than double if a reasonable amortization of the capital cost were included. Moreover, the proposed Los Angeles subway route serves downtown and only two of the sixteen other urban-village cores in the metropolitan region.

Even buses do not fully meet the needs of the emerging urban-village cores, because serving all the possible permutations between where people live and work would require too many routes to be economically feasible. In the new urban-village cores office and industrial buildings are not closely packed together, as they are in midtown Manhattan or Chicago's Loop; they are widely separated by parking lots and landscaped areas.

This lack of density means that only a few people will find it convenient to walk to any given bus or subway station, and that office workers must have a car during the day if they want to visit a client or go out to lunch. And where internal transit programs using buses and mini-vans have been tried, such as Tysons Corner, Virginia, and south of the Los Angeles International Airport, they have quickly failed for lack of patronage. In fact, even an official of the federal Urban Mass Transportation Administration asserts that conventional fixed-route transit systems will never carry more than a small percentage of non-downtown commuters. The official thinks that the only kind of mass transit with any chance of significantly decreasing commuter traffic is ride-sharing, and particularly car-pooling.

That leaves us with our automobiles. The new office complexes provide more room for cars than for the workers who drive them. The typical building allots approximately 200 square feet of space for each employee. The usual planning assumption is that one out of five workers uses a car pool or public transportation, and so a building will require four parking spaces for every 1,000 square feet of office space. Each parking space requires approximately 325 square feet for the space itself, ramps (if a garage, rather than a lot, is planned), and aisles. That means the developer must set aside 1,300 square feet of parking for every 1,000 square feet of office space.

Two years ago some of the nation's leading urban planners gathered in Scottsdale, Arizona, and searched for solutions to suburban traffic congestion. The planners expressed considerable disagreement, presented a myriad of solutions but little data to support the ideas, and were greatly concerned about finding the money to pay for any transportation improvements. The planners did for the most part agree on an approach to the problem: government working together with developers. This is already happening. In north Atlanta's fast-growing Buckhead urban-village core, for instance, developers paid $1 million toward a new $5.2 million bridge over a highway interchange. And in Montgomery County, Maryland, developers will pay half of the $109 million needed for road improvements around Germantown, an emerging urban-village core. Many developers favor this new spirit of public-private cooperation, because they realize that the alternative could be government-ordered construction slowdowns or moratoriums.

PLANNING FOR URBAN VILLAGES

Local governments can resist developers, but they can't resist them indefinitely or protect their communities from the bad effects of rapid growth. They are decentralized nineteenth-century entities trying to deal with far-reaching twenty-first-century issues. Greater Los Angeles, for instance, consists of well over a hundred cities and five counties. Metropolitan Atlanta has forty-six cities and seven counties. How can so many governmental jurisdictions possibly coordinate their actions to handle the amorphous growth of urban-village cores, which, like multinational corporations, know no loyalty to any single governmental body?

Real-estate developers most often deal directly with local governments. In this high-stakes game the developers hold many of the cards, because they can play one jurisdiction off another. Many local governments, in their desire to expand their tax bases, generate employment, an welcome "progress," try to entice developers with liberal zoning, temporary tax abatements, and improvements to such things as roads and sewers at little or no cost.

Contrary to what their residents often think, well-to-do communities cannot totally control growth on their own, unless they are geographically isolated, like Santa Barbara, California. When a town is part of a metropolitan area, its fate is inextricably tied up with the region's, no matter how wealthy its residents or how stringent its growth regulations.

Even Beverly Hills finds itself relatively powerless to resist the relentless development pressures from Los Angeles, which almost completely surrounds the 5.7-square-mile city. Several years ago Four Seasons Hotels wanted to build a high-rise hotel on Wilshire Boulevard near Rodeo Drive, in the heart of the city's commercial Golden Triangle. The issues of building density and hotel-generated traffic congestion became so controversial that the city council refused to vote on the Four Seasons project and instead placed a hotel-development ordinance on the ballot. Last year Beverly Hills voters overwhelmingly defeated the ordinance, thereby eliminating the possibility that another major hotel will open in the city. That didn't stop Four Seasons. Earlier this year the company started construction on a 287-room, 15-story hotel on Doheny Drive and Burton Way, literally across the street from the Beverly Hills city line. Beverly Hills lost the new hotel's sales and occupancy tax revenues but will still be affected by the traffic.

What can local government do about growth pressures? Probably the best solution is the creation of strong, effective, multi-city and multi-county agencies, and perhaps even entire government structures, that correspond to the actual economic and psychological boundaries of a metropolitan area and its urban villages. Such broad-based government has already been pioneered by Indianapolis (Marion County) and Miami (Dade County.) This governmental reorganization is not easy, because it can involve jealously competing towns, townships, cities, and counties—not to mention states, as are involved in greater New York, Philadelphia, Washington, D.C., St. Louis, and Kansas City. But much as the number of school districts nationwide has shrunk from 100,000 in 1940 to 15,000 now, the number of planning jurisdictions can be brought down.

One possible first step toward the creation of more-effective local governments is metropolitan-wide zoning and planning boards that have real clout in granting zoning changes and allocating improvements to the region's physical facilities. Some metropolitan areas are already moving in this direction. In 1979 Phoenix—the nation's ninth largest city, according to the most recent census—took the revolutionary course of adopting the urban-village concept in its planning process. In 1985 the city finished its General Plan to guide the manicipality's growth, and this established nine urban villages. "Each [urban village] would become relatively self-sufficient in providing living, working and recreational opportunities for residents," the General Plan said. The city also encouraged the "concentration of shopping, employment and services located in the village core."

Phoenix is using zoning approvals and planned allocations of money for physical improvements to try to match employment with housing. The city has clearly specified goals. In 1980 only two of the now defined urban villages had more than half as many jobs as people, indicating that they had moved beyond being bedroom communities. Phoenix hopes that by the year 2000 five urban villages will have more than half as many jobs as people, with the remaining four urban villages having between one fourth and half as many, clearly showing that jobs and people have moved closer together.

Elsewhere, developers, employers, and local governments have created transportation-management associations (TMAs) to find solutions to transportation problems. What are perceived locally as transportation crises have brought TMAs into existence in Baltimore's airport area; Warner Center, in Los Angeles's San Fernando Valley; north Dallas; and Tysons Corner, Virginia. According to Kenneth Orski, the president of Urban Mobility Corporation, which is a TMA proponent based in Washington, D.C., TMAs "fill a vacuum in suburban areas."

Besides organizing ride-sharing and van pools, promoting staggered work hours, and lobbying for government-funded capital improvements, some TMAs are expanding their role into child care, private police, and other services for their geographic areas. It is imaginable that TMAs, born of the traffic-congestion crisis, could mature into an echelon of government well suited to the realities of our emerging urban villages.

Considering what has happened so far, is the urban-village trend a good or a bad thing for our cities and our citizens? To a certain extent the question is irrelevant, because the trend is already so advanced that it is irreversible. But it is hard to imagine the ideal urban village as being anything but very good. The opportunity for all kinds of Americans to live, work, shop, and play in the same geographic area—while retaining easy access to other urban-village cores with specialized features that their own district lacks—seems almost too good to be true.

As the nation's cities reshape themselves along the urban-villages model, we must ask ourselves. What are the design features, housing policies, transportation solutions, and governmental structures that will make our cities and the lives of their residents more productive and satisfying? Unfortunately, the questions and opinions vastly out-number the solutions and directions. A great deal of study, experimentation, and planning needs to be done—and done quickly—to help cities that are being rebuilt almost from scratch.

36 *Changing Industry and Regional Shifts: The Case of Silicon Valley*

ANNALEE SAXENIAN

> If small is beautiful, that mandate is fulfilled not chiefly in windmills and solar cells, but in California's vale of cubistic new factories across the bay from San Francisco—Santa Clara's Silicon Valley, where worlds indeed unfold in grains of sand.

George Gilder's words vividly convey the current mystique surrounding Silicon Valley (Gilder, 1981). The image of a pristine, futuristic seedbed of innovative industry which holds the promise of renewed vigor for America's lagging economy has captured the public imagination. Yet this gilded image bears little relationship to the realities of the California region to which it has been attributed.

The growth of Santa Clara County during the years since World War II has been explosive. It has also been highly contradictory. Prior to World War II, Santa Clara County was a sparsely populated agricultural valley. In the thirty years between 1940 and 1970, the county's population increased by almost one million people, from 174,949 to 1,066,421. Countywide employment doubled itself during each successive decade, resulting in the creation of about 350,000 new jobs. And personal and family incomes rapidly surpassed both state and

Source: Annalee Saxenian, "The Urban Contradictions of Silicon Valley: Regional Growth and the Restructuring of the Semiconductor Industry," *Sunbelt/Snowbelt: Urban Development and Regional Restructuring,* Larry Sawers and William K. Tabb (eds.), (New York: Oxford University Press, 1984), excerpts from pp. 163–197.

national averages, with median family income in the county a full 30 percent above that of the United States as a whole by 1969. During the seventies, strains began to appear. While job growth accelerated (240,000 new jobs were created during that decade alone), population growth failed to keep up. By 1975 Silicon Valley was distinguished by unbearably congested freeways, dangerous levels of air and water pollution and a no-growth movement calling for a halt to further industrial expansion. Today the county also boasts among the highest average housing prices in the nation.

The story of Silicon Valley's growth is also the story of the birth and development of a new industry, the semiconductor industry. The electronics-related companies which established operations in the area during the 1940s and 1950s to gain access to war-related markets paved the way for the birth of the first semiconductor companies. The subsequent agglomeration and expansion of this new industry in the valley provided the impetus for its accelerated growth. Almost all of the country's leading semiconductor companies can trace their roots to Santa Clara County. Today, the leaders in this rapidly maturing industry remain headquartered there, surrounded by a dense concentration of smaller electronics-related suppliers and producers. Yet the country's largest semiconductor and electronics companies are no longer expanding production in the region. Plagued by a variety of urban problems, they have started to disperse their manufacturing operations to distant locations.

Regional growth and industrial development have thus been inextricably linked in Silicon Valley. This essay examines the relationship and interactions between these two processes—arguing that neither the region's growth nor the development and spatial patterning of the semiconductor industry can be understood in isolation. For at the same time that the expansion of semiconductor and electronics production generated rapid economic growth, it also transformed the local class structure and influenced the patterns of residential settlement and urban development in Santa Clara County. I argue here that the urban problems which emerged in the region during the seventies--breakdowns of the housing market and transportation systems, along with environmental degradation and the no-growth movement—all are direct outcomes of the evolution of this industrially determined spatial structure. As these urban problems began to raise the costs of production in the region and to threaten already existing activities, they in turn triggered the decentralization of manufacturing operations out of the valley. In short, the young semiconductor firms which brought economic growth to Santa Clara County also shaped the region's urban geography. As the industry developed, the contradictions of this urban form became increasingly pervasive, eventually feeding back into the locational behavior of the very same companies. The current decentralization of production in turn will affect both the nature of growth in Silicon Valley and in the new host regions as well as the future development of the semiconductor industry itself.

After a brief discussion of the theoretical antecedents for this analysis—

found primarily in the literature which examines the processes of regional change, but also related to the theory of product cycles and industrial location—this essay is divided into five sections. The first describes the basic characteristics of the young semiconductor industry as it emerged during the 1950s and 1960s in Santa Clara County. This information is then used in the following section to describe the way in which the nature of production and the class structure generated by the industry shaped the county's urban landscape, particularly its patterns of industrial and residential development. In the third section, the urban problems which emerged in Silicon Valley during the 1970s are examined and shown to be a direct outcome of the evolution of this particular spatial structure. The response of the industry to these urban contradictions, and the consequences of this industrial restructuring for future patterns of regional growth, are then discussed in the fourth section. Some concluding comments are proffered in the final section.

THEORETICAL BACKGROUND

As the pace of growth and decline of regions around the world has accelerated, with dramatic changes in the fortunes of particular regions attracting both popular and academic attention, traditional approaches to the study of regional change have been seriously challenged (see, for example, critiques by Holland, 1976; Massey, 1975, 1978a; Walker and Storper, 1980). Neoclassical theories of interregional equilibration appear increasingly weak. Empirical evidence casts doubt, for example, on the view that we are not witnessing a convergence of growth rates between the Northeast or Midwest and the Southwest regions of the United States. The related regional growth theories, while each providing insights into certain aspects of the process of regional change, lack comprehensive analytical power. Demand-based approaches, such as export-base theory, isolate external demand for local output as the sole determinant of a region's growth and unrealistically assume that the existing composition of its industrial base remains fixed. On the other hand, supply-based approaches such as growth pole or agglomeration theories explain the fate of a region as a function of the local supply of land, labor, and capital, thus restricting causality to a region's predetermined characteristics. In addition, these theories all neglect the crucial role of macroeconomic forces in regional change, and little if any attention is given to the impact of changing locational strategies of firms on the growth and decline of regions. Location theory, which purports to fill this latter gap, provides insights into a role of comparative costs and revenues influencing a firm's location decison—focusing primarily on the supply of factors of production and the location of markets. Constrained however by its focus on marginal decisions within a static and analytically narrow framework, location theory lacks the capacity to comprehend the forces underlying shifts in the comparative factor costs in a region or changes in behavioral motivations specific to particular firms and industries. Furthermore, social and political factors which significantly influence industrial location decisions are only artificially reflected, if at all, in the cost mechanism.

In recent years a more comprehensive and analytically powerful approach has been developed which identifies industrial restructuring as the key force underlying regional change. This new brand of regional analysis, pioneered by Massey and Meegan (1978), views regions as the outcome of processes of accumulation and industrial development. Industrial behavior both past and present is the object of the analysis. Its method comprises examination of both the evolving characteristics and needs of production in particular industrial sectors and the way in which economic activity responds to geographic variations in the conditions for accumulation. In examining the decline of the old industrial inner cities in Britain, for example, Massey and Meegan demonstrate that technological innovation and cost-cutting reorganizations of the production process in the electrical engineering and electronics industries underlie significant shifts in the regional distribution of employment. They trace this rationalization and industrial restructuring process to the pressure on firms to boost profitability in the face of declining international competitiveness. Similarly, in an analysis of the economic transformation of the New England region in the post-World War II period, Bluestone and Harrison (1980, 1981) use industrial case studies to detail the technological changes and reorganizations of the production processes which allowed the shifting of production tasks out of the region. While stressing the role of changing ownership patterns (which have increased centralization and concentration of control) in facilitating this relocation, their analysis also documents the historical importance of labor militancy in accounting for the timing of these instances of rationalization and industrial dispersion.

In applying this method of regional analysis to Santa Clara County, a new element must be added to our understanding of the process of regional change. Previous researchers have forcefully illustrated the importance of viewing a region's fate within the context of the interregional and international division of labor (see also Frobel, 1980); and they have documented the role of shifting industrial strategies, technologies, and ownership structures in regional change. Yet they have failed to examine the impact of production on patterns of urban development *within* a region, and the ways in which this industrially determined spatial structure in turn influences industrial behavior. In this essay, I use the case of Santa Clara County to argue that processes of regional change can only be fully understood by examining the impact of production on the local class structure and urban geography of a region, and the subsequent feedback effects which this particular spatial form has on future industrial behavior.

Santa Clara County provides a rare opportunity to examine the causal links between industry, urban spatial structure, and regional development. The county, which lies at the southern tip of the San Francisco Bay, encompasses a valley flanked by low coastal mountains. Prior to 1940 the valley was devoted entirely to agricultural uses. Today, the electronics industry completely dominates the region, with over one-third of the county's workers employed in electronics-related companies. Many of the remainder are employed in occupations that service or support the electronics complex. Most of the region's recent growth can be attributed to the evolution of this sector alone. Similarly, as there was very little prior urban development, the current organization of space in the county can be traced directly to the characteristics of production in the industry. While these relationships between production, local geography, and regional growth occur everywhere, it is unusual to find a case in which spatial and regional outcomes are so clearly attributable to a single industry.

On a very general level, the spatial behavior of the semiconductor industry in the United States has conformed to the predictions of the product cycle theory of industrial investment associated with R. Vernon (1960, 1966). The theory postulates initial agglomeration in the development of a new product—allowing flexibility and the exploitation of the benefits of communication and external economies. This is followed by a large-scale dispersal of production facilities to lower-cost locations as the product is standardized and demand for it grows—enabling the achievement of scale economies through mass output and long production runs. During this process of industrial maturation, new technologies develop which allow the "decomposition" of production processes and their successive transfer outward.

As the material presented here illustrates, Santa Clara County clearly functioned as a seedbed region for the young semiconductor industry, the place where virtually all fledgling firms agglomerated to exploit the benefits of proximity to each other in a new and rapidly changing industry. Furthermore, the subsequent dispersal of production activities out of the valley has indeed coincided with the maturing of the industry. Vernon's observation that movement along the development cycle—involving expanded scale, product standardization, and a declining need for flexibility—renders an industry increasingly footloose is, in this case, a valuable insight. Yet beyond predicting these very general trends, the product cycle approach is of limited value in explaining the locational behavior of the semiconductor industry.

Product cycle analysis provides no explanation of the specific *causes* of the industrial dispersal process, and thus lacks the power to account for the precise timing or destination of these movements as well. Basing its spatial analysis on the simple concept of a decentralization to lower-cost regions, it fails to comprehend the forces underlying both shifting costs within a single region and cost differentials between regions. In the absence of such explanatory power, individual processes of industrial relocation which differ dramatically from case to case are simply homogenized into an overly generalized stage theory. As this paper illustrates, the forces underlying the decentralization of semiconductor manufacturing out of Silicon Valley were not those of the traditionally cited U.S. textile and apparel industries moving south in search of low wage labor. Nor were they the same as those macro-economic shifts which spurred the locational shifts of the electrical equipment and electronics industries in the United Kingdom. Likewise, while the timing of the technological rationalizations and the dispersal of shoemaking and textiles out of the New England region can be attributed in many cases to upsurges of labor militancy, in the case of Santa Clara County, the timing is related to the growing severity of its urban problems. Only through in-depth analysis of specific industries and regions can the complexities of the industrial location and regional change processes be fully understood.

THE YOUNG SEMICONDUCTOR INDUSTRY

World War II stimulated the emergence of the semiconductor industry and shaped the character of its early development. The industry's birth, generally associated with the invention of the solid state transistor at AT&T's Bell Laboratories in 1948, was spurred by military demand for more versatile components for use in weaponry. Commercial production of transistors began in 1951. During the next two decades, fueled initially by massive defense-related support from the federal government and later through expansion into computer and industrial markets, this new industry literally exploded under the steam of ongoing technological innovation. Its growth was phenomenal: the value of shipments by U.S. firms jumped from $5.1 million in 1954 to over $500 million in 1960, and by 1970 it had reached $1.5 billion.

The young semiconductor industry was characterized by unusually low barriers to entry—resulting from the accelerated pace of technological change, the speedy diffusion of innovations, the absence of economies of scale, and the ready availability of finance capital—which in turn allowed for a proliferation of small, intensely competitive firms. These characteristics also made the spatial clustering of firms highly advantageous. While many large, already established electrical companies and vacuum tube producers began commercial production of transistors during the 1950s, it was the small new firms formed specifically to produce semiconductors which had the flexibility to exploit new innovations in production. By the mid-1960s, they had come to dominate the industry. And it is these new enterprises, formed almost exclusively by scientists and engineers with no prior business experience and extremely limited capital resources, that clustered in Santa Clara County.

World War II also laid the foundation for industrialization of the bucolic agricultural fields of the Santa Clara valley. Hostilities in the Pacific theater greatly stimulated California's economy and generated a massive influx of population into the region for war-related industries such as tank, aircraft, and missile manufacture. The Moffet Field naval air station in the northern part of the county drew in thousands of military personnel for training. Most critical for Santa Clara County, however, was the genesis of a technological watershed at Stanford University, located in the northwesternmost corner of the valley, in the city of Palo Alto. The initial flow of federal funds to Stanford's laboratories for the development of military components and equipment during the war years ballooned into a substantial and ongoing source of support for basic electronics research and development during the years of the Korean War and throughout the ensuing cold war period. In addition, Pentagon contracts for ballistic missiles and other new weapon systems significantly boosted production levels of the west coast aerospace industry, which in turn required large quantities of electronic inputs. During the 1940s and 1950s, a number of large national electrical and electronics companies thus established both manufacturing and research operations in Santa Clara County in order to gain proximity to these sizeable new markets.

While the war stimulated the region's economy and the dramatic growth of government military and aerospace contracts to California-based firms bolstered local electronics-related production and research, Stanford was quick to step up its electrical engineering programs. The number of advanced degrees in engineering awarded annually in the region increased rapidly during these years, as other local educational institutions in turn established special electronics and electrical engineering programs. By the middle of the 1950s, the region was thus distinguished by a rich and supportive technological milieu consisting of high-caliber universities, research institutions, and older technology-based firms, along with an unusually large supply of skilled technical workers. Santa Clara County had become an ideal environment for innovative, science-based industry.

The first semiconductor firm was established in the county in 1955. William Shockley, one of the three inventors of the transistor, left Bell Laboratories and moved to Palo Alto to form the Shockley Transistor Company. He thereby set a precedent for the "spinoff" process, soon commonplace in the valley, whereby engineers and scientists simply left their employers in order to start their own semiconductor companies. In 1957 eight of Shockley's best scientists in turn splintered off and gained financial backing from the local Fairchild Camera and Instrument Company to start their own firm. Fairchild Semiconductor soon became known as the grandfather of the industry, spawning ten new firms by 1965, and almost fifty new semiconductor companies over the next decade and a half. All of these new firms remained in Santa Clara County. While the industry's low barriers to entry and minimal initial capital requirements made this spinoff process possible (many firms were started with only $1 million), the pace of technological innovation and the constant need to remain abreast of new developments ensured this spatial clustering in the valley. By the mid-1960s, this original concentration of production had begun to act as a powerful centripetal force for the continued agglomeration of new semiconductor firms in the region. External economies, including an enlarged supply of both skilled and unskilled workers, specialized inputs and services, and a social, cultural, and educational environment, made the region particularly appropriate for semiconductor production. The proliferation and explosive growth of these new firms in turn fueled the region's economic boom. By 1970 the county was known as Silicon Valley, the worldwide capital of the semiconductor industry and the densest concentration of high technology and electronics enterprises in the nation.

As it grew, this new industry shaped the patterns of urban development in Santa Clara County. The key to the region's current spatial structure lies in the nature of semiconductor production and the unique class structure which it generates. The semiconductor industry is distinguished by the fact that it is a science-based industry, an industry which involves ongoing scientific investigation and the application of scientific knowledge to the process of production (Noble, 1977). Not only does scientific research contribute to the development of new technologies and novel applications, but science is also integral to the production process itself. This becomes evident through examination of the process of producing semiconductors, a process which has three phases: (1) product development, (2) wafer fabrication, and (3) assembly and testing.

Product development involves the original conception of a semiconductor device, the extremely complex design and engineering of circuits to meet given product specifications, the transfer and reduction of the final artwork representation of the circuit design onto miniature glass photomasks, and protoype production runs for adjustment of both the design and the manufacturing process. This initial phase of production requires a firm's most highly skilled scientists and design engineers along with some technical assistance. Wafer fabrication (or advanced manufacturing) has been described as the most complex manufacturing technology adaptable to mass production. Basically it is a batch process with a large number of precise mechanical, chemical, and electrical operations performed in sequence. First the photomask circuit design is transferred onto a thin silicon wafer (ranging from two to five inches in diameter) through the use of high-intensity light or electron beams. Selective impurities are then introduced into the silicon to impart the desired electrical conducting properties. This begins with ion implantation and other chemical processes inside a high-temperature furnace and is followed by a series of metalization, passivation, oxidation, washing, etching, and diffusion processes. Each step produces irreversible changes in the silicon wafer, so that any error in the overall sequence means the entire batch must be discarded. Errors are not uncommon, as uniformity of temperature, extreme cleanliness, and precise dimensional control (tolerances of up to seventy-five millionths of an inch are normal) are all essential to the fabrication process.

Once prepared, the wafers are tested and divided into thousands of identical chips. A yield of under 5 percent usable chips is not unexpected, especially in the early phases of a product's life. The complex and sensitive nature of this phase of production means that semiconductors cannot be produced routinely by production workers under the supervision of a few managers, as can most manufactured products. Skilled engineers and technicians who are versatile with the scientific principles involved must measure and control the variables at each stage, making sure that the overall sequence is in order. While closely controlled by engineers, this process still requires unskilled and semiskilled production workers to perform routine loading, monitoring, processing, and cleaning tasks. Thus, wafer fabrication has typically required a work force consisting of both skilled technical workers and minimally skilled production workers in a ratio of 1 : 2 or 1 : 3.

The final assembly and testing phase of semiconductor production involves the highly routine process of hand-bonding very fine wire leads or connectors to the chips (under a microscope) and the final sealing of the chips in ceramic, metal, or plastic protective packages. This assembly process is identical for all product lines, varying only in the number of wires to be welded. The final testing and sorting of the chips is a highly capital-intensive, often computerized process. No prior training or education is required to perform these assembly and testing tasks, and the work force thus consists solely of unskilled workers. Furthermore, these tasks form a completely discrete step in the production process which can be physically separated without technical difficulties. As a result, virtually all semiconductor assembly is now being performed outside of the United States, in plants in Asia and Mexico where wage rates are only a fraction of those at home. Most semiconductor facilities in the United States today are thus devoted primarily to either/or both of the first two phases of production, product development and advanced manufacturing operation.

An unusually top-heavy occupational structure reflects the scientific nature of semiconductor production. In fact, the industry has among the highest percentages of nonproduction workers of all industries, with 40 percent of the total U.S. semiconductor work force in 1972 classified as nonproduction workers, compared with 16 percent in the production of motor vehicles and only 13 percent in the apparel industry (Mutlu, 1979). A 1971 occupational survey classified 27 percent of the semiconductor work force in professional and technical positions alone (including engineers and other scientists, skilled technicians, and draftsmen), with another 13 percent in executive, administrative, marketing, and supervisory positions (U.S. Department of Commerce, 1979).

Not only is the employment structure top-heavy, but it is also severely dichotomized. While 40 percent of the industry's 1971 work force was in these managerial and professional positions, 48 percent was in production and maintenance occupations—four-fifths of which are listed as semiskilled or unskilled positions—with the remaining 12 percent in secretarial and clerical occupations. In other words, the industry employs an unusually large proportion of highly educated professional and managerial employees alongside an equally large, but minimally skilled production work force, with very few medium-skilled workers in between.

In Santa Clara County, a region dominated by semiconductor and other electronics-related production, this bifurcated employment structure generated an equally bifurcated class structure, and patterns of residential settlement and urbanization in turn have served to replicate this class structure on the county's urban landscape.

URBAN GEOGRAPHY OF SILICON VALLEY

It was logical that Santa Clara County's first semiconductor enterprises should locate near Stanford University. For these technologically sophisticated but experimental ventures—many of which were started by Ph.D. scientists and engineers leaving academia for the first time—the university environment provided a familiar and supportive intellectual climate. This initial pattern was consolidated with the establishment of the Stanford Industrial Park on 770 acres of land adjoining the university campus in the early 1950s. Leases in the park were granted only to high-technology firms. Seeking spatial proximity in a rapidly changing and highly competitive industry, those firms which failed to find space in the industrial park also located in or around Palo Alto. The adjacent towns of Mountain View and Sunnyvale, and later Cupertino and Santa Clara, soon recognized the tax revenue benefits of a strong industrial base. Following the Stanford model, they too established industrial parks and provided land and financial and infrastructural benefits to attract electronics companies. In this manner new industrial development clustered in the northwest corner of the valley, gradually drifting southward from Palo Alto, city by city. The result is a striking imbalance in the county's land use, with the electronics industry overwhelmingly concentrated in the five northern cities—the North County. By 1970, the ratio of electronics employment to city population was 1 : 4 in Palo Alto, 1 : 5 in Mountain View, 1 : 9 in Sunnyvale, 1 : 7 in Santa Clara, and only 1 : 50 in San Jose—the county capital, located about twenty miles southeast of Palo Alto (Keller, 1979).

As industry and jobs continued to flow into the North County, San Jose followed a different route. Explicitly committed to "making San Jose the Los Angeles of the North," the city's administration—supported by a coalition of landowners, realtors, contractors, road builders, speculators and bankers—promoted rapid urban expansion and rampant residential and commercial development. The city government rezoned large tracts of land; provided easy access to credit; aggressively annexed nearby territory; extended sewers, storm drains; and roads to peripheral areas; and helped to gain access to the freely flowing federal funds for freeways and to FHA mortgage financing which made tract development of inexpensive, single-family homes highly profitable. San Jose thus grew from an agricultural processing and distribution center of only 17 square miles in 1950 to a sprawling 147-square-mile metropolis in 1975. Today the city of San Jose alone encompasses half of the county's total incorporated area and is the home for almost half of its population.

Immigration swelled the population of Santa Clara County dramatically during the postwar years, generating growth rates that far surpassed those of both the United States as a whole and even the booming state of California (see Table 36-1). Of the over one million new residents recorded in the county between 1940 and 1975, a large majority were in-migrants. (Natural increases, the excess of births over deaths, accounted for only one-quarter of the annual

Table 36-1 Population Trends: Santa Clara County and Other Geographical Regions

Region	*Percent Increase* 1940–1950	1950–1960	1960–1970
United States	14.5%	18.5%	13.3%
Western States*	40.4	38.9	24.1
California	53.0	48.5	27.0
Santa Clara County	66.0	121.1	66.0

*Includes Washington, Oregon, California, Idaho, Nevada, Utah, Arizona, Montana, Wyoming, Colorado, New Mexico, Alaska, and Hawaii.

SOURCE: U.S. Department of Commerce, Bureau of the Census.

population growth during these years.) Coming in a dual stream, these immigrants reflected the electronics industry's unique labor requirements. On the one hand, an influx of professionals and highly skilled workers responded to the industry's unusually large demand for scientists and engineers. Close to 50 percent of the adult migrants to the region during the 1950s and 1960s had some college training, so that between 1940 and 1970, the proportion of college-educated adults in the county doubled from 20 percent to 40 percent (Keller, 1979). At the same time, the industry's demand for production workers stimulated an equally large in-migration of unskilled, predominantly minority workers. This included displaced agricultural workers from California and the Southwest (primarily Chicanos and some Filipino-Americans), foreign-born Mexicans and Filipinos, and smaller numbers of U.S. blacks and Native Americans. Chicanos and Mexicans alone accounted for 25 percent of the population increase during the 1960s, and by 1970 minority groups represented approximately 25 percent of the county's total population.

The settlement of this dual stream of immigrants into socially and economically segregated residential communities in turn shaped the urban geography of Santa Clara County. For analytic purposes, the county's fifteen cities can be divided into four homogeneous spatial clusters, each representing a specific social and/or economic function in the county.

Santa Clara County's most affluent professionals and executives reside in the Western Foothills. The five cities in this cluster are the newest in the county and are solely residential. They were originally settled during the 1950s by electronics industry entrepreneurs and scientists who desired both the spacious natural beauty of the foothills and the proximity to their new workplaces. The residents of these foothill cities earn incomes which dwarf those in the rest of the valley, they have the highest average levels of educational attainment in the county, the majority are in professional or managerial occupations, and virtually all are white.

The five cities of the North County form the heart of the Silicon Valley industrial complex, where virtually all of the county's electronics companies are clustered. Half of the county's jobs are located in these five cities alone. While sharing this common industrial base—which is reflected in municipal property values—the residents of the North County represent an intermediate socio-economic mix. In terms of incomes, education levels, occupational status, and racial mix, Palo Alto clearly belongs with the affluent professional and white foothill cities, as does the far western city, Cupertino. Moving southward from Palo Alto city-by-city, the socioeconomic level of the local population declines steadily, the occupational mix becomes increasingly dominated by craftsmen and operatives, and the minority population rises dramatically.

This trend culminates in the three older cities of the Central and East Valley, where over half of the county's population resides. Dominated by San Jose (which alone accounts for 47 percent of the population), these rapidly expanding cities served as a landing spot for the influx of unskilled, minority immigrants who were attracted by the electronics boom. With incomes that are below the county average, the residents of these Central and East Valley "bedroom communities" are considerably less well-educated and far more likely to be of minority descent than their neighbors to the north and west.

The cities of the South Valley and the unincorporated areas are the home of the poorest tenth of the county's population and a disproportionately high percentage of its minority residents. Primarily agricultural and rural, these cities are only marginally connected with the electronics complex.

Silicon Valley's urban landscape has thus come to mimic the semiconductor industry's bifurcated class structure. The large and affluent professional-managerial strata of the industry's work force are insulated in the north and west—in the foothills, in Palo Alto, and to a lesser extent in the other North County industrial cities—with easy access to the electronics complex. Meanwhile, the industry's large, low-income production work force is concentrated further to the south and east—primarily in San Jose and the adjacent communities. Comparison of a few statistics for Palo Alto and San Jose underscores this social-spatial dichotomy. In 1970, 50 percent of Palo Alto's employed residents were in professional, technical, and managerial occupations, and 42 percent of the city's adults had four or more years of college education, while only 29 percent of San Jose's employed residents were in professional, technical, and managerial occupations, and a mere 15 percent were college-educated. On the other hand, 25 percent of San Jose's employed residents were employed as craftsmen or operatives, and 22 percent of its population was of Spanish-American descent, while only 10 percent of Palo Alto's residents were craftsmen or operatives, and under 6 percent were Spanish-Americans.

Severe fiscal imbalances between the cities in the north and south of the county both reflect and in turn reproduce these differences. The vastly different municipal tax bases—the 1974–1975 assessed net valuation per capita in Palo Alto was $7,200, two and one-half times San Jose's—have generated wide disparities in the quality of social services and the nature of both public and private institutions such as schools, parks, libraries, recreation centers, and other social and cultural amenities across the county (Bernstein et al., 1979).

CONTRADICTIONS OF THE SPATIAL STRUCTURE

This urban form evolved during the 1950s and 1960s to enable the differential reproduction of the two classes of labor required for electronics production within a single metropolitan area. It did not take long, however, for the contradictions of this spatial structure to manifest themselves. By the 1970s Silicon Valley was plagued with skyrocketing housing prices, overly congested freeways, unhealthy levels of air pollution, and an active no-growth movement—all of which were rooted in the region's distorted pattern of urban development.

Housing

Housing price inflation is a nationwide phenomenon, but the imbalanced social-spatial structure in Silicon Valley has generated housing that is among the costliest in the country. During the 1970s the region's housing supply failed to grow in pace with its accelerated job growth, resulting in price rises which outpaced both national and statewide increases. By 1980 there were over 670,000 jobs but only 480,000 housing units in Santa Clara County, and the average home price had surpassed $100,000—almost double the U.S. average.

This shortage of housing is a direct outcome of the restrictive land use policies and exclusionary planning practices imposed by local governments desiring to preserve the "quality of life"—the differential nature of reproduction—in their jurisdictions. In pursuit of an expanded tax base, each of the already industrializing North County cities chose to rezone extensive tracts of land from residential to industrial usage. Between 1965 and 1975 alone, such rezoning reduced the countywide housing capacity by 43 percent, a loss of 417,000 potential units (Bernstein et al., 1979). Similarly, low-density zoning and residential growth-management policies imposed by the affluent northern and Western Foothill cities in order to preserve the spacious and rural nature of their communities have further limited the county's potential residential land. In 1972, for example, Palo Alto zoned all of its foothill areas exclusively for open space (Greenberg, 1974). Further, virtually all residential construction in the Western Foothills is restricted to single rather than multiple unit housing, and it is generously spread out. There is, on average, only one dwelling unit per acre of residential land in these communities. In the very exclusive city of Los Altos Hills, there are two acres for every home (SCC Manufacturing Group, 1979). As a result, if all local land use plans were now built out according to their designated uses, there would be 325,000 employed people in the county without the possibility of housing at any cost (SCC Industry Housing Management Task Force, 1979).

In this context the electronics industry's top-heavy employment structure directly fuels housing inflation. Since an unusually large proportion of the new jobs created through an expansion of production are for engineers and scientists or managers, any industrial growth automatically draws more highly paid employees into the county. With incomes that are often ten times those of the industry's production workers, these newcomers have simply asserted their superior buying power in competition for the valley's scarce housing, bidding prices up to levels that only their incomes can bear and often displacing lower-income residents in the process.

This process has a spatial component as well. The concentration of industrial location and expansion in the North County cities has continued, yet virtually all of the remaining land in the county zoned for residential uses is located in east and south San Jose, a long distance from these new jobs.

Although inflationary pressures originated in the socially desirable cities of the Western Foothills and the North County, the limit of this housing supply was rapidly reached. Demand by the higher-income work force thus began to filter into north San Jose and other communities with easy access to the electronics complex. As a result, the least costly homes and apartments in the county—and thus the only residences affordable by the average electronics production worker—are increasingly located in east and south San Jose and the South Valley cities, often as much as fifty miles from the industrial belt. With incomes that have failed to rise in pace with the inflation of housing prices, the county's low-income population is being gradually forced further south, frequently out of the county altogether.

Transportation

The spatial imbalance of jobs and housing location and the sprawling nature of postwar residential development have always meant long commutes for many of Santa Clara County's residents. More than thirty years of sustained population growth have exacerbated the congestion of the county's roadways. But it was not until the 1970s, when the dynamics of the housing market began to force the industry's large pool of production workers to locate (or relocate) yet further from the "jobs belt" in the north, that the transportation problem in Silicon Valley reached crisis proportions.

With no viable mass transit alternatives—municipal buses typically take twice the driving time—local residents are dependent upon private automobiles for transportation. By 1975 over four million automobile trips were being taken daily in the county. Of these, a large proportion were home-based work trips, and the imbalanced location of jobs and housing meant that most of these commutes were long, slow, and costly. An average ten-mile commute at rush hour now takes thirty minutes, and workers living in the eastern and southern parts of the county face round-trip commutes of over three hours daily (SCC Planning Department, 1979).

According to a transportation study conducted by the Santa Clara County Planning Department (1979), within a decade there will be more than 130,000 cars reaching destinations in the northwest part of the county during morning rush hours—that is, more than ten times the number of vehicles that can be accommodated during a single hour on a six-lane freeway. The study concludes that "as now planned, the land use patterns and growth rates of Santa Clara County will overwhelm the capacity of the major roadways in the urban area by 1990."

The negative effects of congestion and overburdened traffic networks fall most heavily on those who commute from the southern and easternmost reaches of the county—almost exclusively lower-income workers who can least afford the rising fuel costs. However, all commuters now suffer delays and frustrations, and the overflow of traffic onto residential streets has brought noise, accidents, and disruptions to even the more affluent neighborhoods in the North County.

Environment

All of this has taken a toll on the environment of Santa Clara County. Auto emissions have generated the brownish-orange cloud of smog which now hangs over the valley, where one or more federal air quality standards are violated at least 10 percent of the time (SCC Industry Housing Management Task Force, 1979). In the Central and East Valley cities which face severe shortages of tax revenues, improvements in sewers, roads, parks, and other basic public works have failed to keep up with the demands of local population growth and development. Repeated breakdowns of San Jose's sewage treatment plants in recent years have seriously damaged the aquatic life and water quality of the San Francisco Bay. The fruit orchards and agricultural fields have disappeared. With the exception of the most affluent foothill communities, the entire valley is covered over with strip development, shopping centers, industrial parks, parking lots, and freeways. In 1971 the city of San Jose maintained only eight acres of open space per thousand residents, compared to thirty-five acres in the nearby San Francisco area and twenty-five acres in New York City (Stanford Environmental Law Society, 1971).

No-Growth Movement

As late as 1974, an observer described the environmental and social attractions of Santa Clara County for the electronics industry's professionals as follows:

> It's a particularly pleasant place to live and work—a beautiful landscape of hills and plains, a bounteous garden of nature where fruit trees and wild flowers bloom even in February. . . . Few places on earth so agreeably mix hedonistic delights with the excitement of urbanity. Outdoor sports and recreation are year round attractions. . . . The area boasts 4,000 PhDs. There are also at least 12,000 horses, some kept by those PhDs right on their home acreages, which are often within minutes of work. And within an hour's drive are the shops, restaurants and cultural offerings of San Francisco. (Bylinsky, 1974.)

Yet the qualities of life which once made the valley so desirable an environment for the industry's executives and scientists are rapidly being undermined. Even living in the isolated cities of the Western Foothills no longer provides protection from the region's urban disamenities. The no-growth movement in Santa Clara County, which was mobilized by these well-educated, middle-class professionals, is another manifestation of the urban contradictions generated by the particular form electronics production has taken. It is their response to the environmental degradation and urban problems wrought by the region's distorted spatial development, as these urban problems now threaten the very lifestyles which this spatial form arose to preserve.

This no-growth movement includes local environmentalists, planners, housing activists, women's groups, elected officials, and individual residents. It accumulated strength gradually during the 1970s, as community concern developed over a number of seemingly isolated issues. The work of several studies and countywide task forces examining the region's environmental, transportation, and housing problems culminated in a recognition of their interconnectedness. "Living Within Our Limits: A Framework of Action for the 1980s," a report by the Santa Clara County Industry and Housing Management Task Force (1979), identified uncontrolled job growth as the primary cause of the county's urban and environmental problems. In the interest of preserving the county as "a desirable place to live and work," the task force advocated the imposition of strict governmental controls over the rate, amount, type, and location of further industrial development.

Santa Clara County thus became the first area in the country to legislate public controls over industrial growth. The task force's most sweeping recommendation—for countywide planning of job growth, revenue sharing to redress the fiscal disparities between cities, and taxation of local industry to improve housing and transportation systems in the region—all were defeated at the county level. However, ameliorative measures were enacted by some of the industrial North County cities. In 1980 the Sunnyvale city council rezoned substantial amounts of land from industrial to residential usage, imposed fees on all new industrial development in order to improve the transportation system, and legislated strict regulations on the job density in local plants. Similarly, Palo Alto now requires fees for all industrial development to contribute to a fund for low- and moderate-income housing in the city. And most of the North County cities have imposed strict regulations concerning the environmental impacts and physical appearance of new development.

INDUSTRIAL RESTRUCTURING AND REGIONAL REALIGNMENT

By the mid-1970s, Silicon Valley's high-technology companies began to feel the strains of these urban contradictions. According to local semiconductor producers, their operations in the region were seriously threatened by two problems. First and foremost, the inflated price of housing had undermined their ability to attract the professionals who are crucial to their operations. With housing costs more than 50 percent above those in other Western states and close to double those in the rest of the United States, even the most prestigious semiconductor firms could not attract experienced engineers and top caliber scientists from outside the region. Second, all suffered from a shortage of production workers and high turnover rates among their production work forces which they attributed to the lack of affordable housing and the increasingly long and expensive commutes to work. As the personnel manager of one firm noted, "We're in trouble with our commuting patterns. Eventually the local labor force isn't going to be able to get here because of the crowded highways." By 1980 there were over ten thousand unfilled jobs at all skill levels in Santa Clara County (SRI International, 1980).

The initial response of local firms was to raise wages, salaries, and benefits. A survey of sixty-eight Silicon Valley companies showed the salaries of engineers rising 15 to 20 percent a year during the late 1970s while benefits were expanded to include everything from stock options, dental insurance, and group legal services to paid sabbaticals and scuba diving lessons in the company pool (Garcia, 1979). Most companies also began to offer "bounties" to their professional employees, paying up to $1,000 per new recruit. In competition for the local labor force, these firms were also forced to raise hourly wage rates and benefits. Some devised gimmicky reward schemes to lure new workers (and keep the old ones). A few companies even began to provide bus tickets or van pool services for their workers.

Relocation, however, soon emerged as the preferred long run solution for these companies. In contrast with past policies of concentrating all manufacturing growth within the valley, by 1980 Silicon Valley's largest semiconductor and electronics companies had started to expand their advanced manufacturing operations in distant locations. As the spokesman for American Microsystems, Inc.—which recently opened a plant in Pocatello, Idaho—commented to the author, "Housing costs here are outlandish compared to elsewhere. You can get an awfully good home for $85,000 in Pocatello. . . . The cost of living is 30 to 40 percent cheaper." National Semiconductor has turned its fifty-five-acre plot in Santa Clara into a park for its employees while dispersing the growth of its wafer fabrication activities to new plants in Tucson, Arizona, and Salt Lake City, Utah. Intel is building a new manufacturing facility in Albuquerque, New Mexico, having recently completed construction of fabrication plants in Aloha, Oregon, and Chandler, Arizona. Signetics is now in Orem, Utah, and Advanced Micro Devices is in Austin, Texas. Hewlett-Packard's most recent expansions have been directed to sites in Idaho, Oregon, Colorado, California, and Washington. None of these companies intend to expand manufacturing operations in Santa Clara County any further.

This sudden dispersion of semiconductor manufacturing activities out of the Santa Clara Valley was thus triggered by the contradictions of the spatial structure which became critical in the late 1970s and which were experienced by local firms as massive shortages of labor and escalating production costs. Yet it is only the simultaneous evolution of the industry itself which allowed firms to undertake this decentralization strategy. While the precise timing of the dispersion can only be explained as an outcome of the urban processes in the region, the consolidation and restructuring of the semiconductor industry—which began during the 1970s—provided the capability for the shift.

Dramatic increases in the technological sophistication of semiconductor devices and the growth of huge new markets for these products have fundamentally altered the nature of competition in the industry. Miniaturization—the process of squeezing more and more electronic components onto a single chip—has raised the effective barriers to entry significantly in recent years. The newest breed of circuits is essentially immune to copying by rival firms, the R&D and design costs required to keep abreast of new technological developments have risen precipitously, and the increased chip complexity demands fabrication equipment which is far more sophisticated and costly than in the past. Driven thus by rapidly escalating capital needs along with the pressures of growing international competition, the small independent semiconductor companies of the past are rapidly being acquired or forced into mergers with larger electronic systems producers of conglomerates. Only the largest, most financially stable, and technologically advanced producers have escaped this amalgamation process. Meanwhile, the emergence of the mass-consumer, automotive and telecommunications markets for semiconductors has enabled these still independent firms to exploit economies of scale in both production (through product standardization, longer production runs, and automation of the fabrication process) and in marketing—thereby further boosting their competitive advantage and reinforcing their positions of dominance. Capital formation, labor productivity, cost minimization, and marketing have thus replaced the former preoccupation with technological advance as the main concern of these leaders in an increasingly concentrated industry.

By the late 1970s, Silicon Valley's major semiconductor firms had thus changed from small, intensely competitive, technology-dominated ventures to large, mature, marketing-oriented corporations. The five leading companies in the valley, all of which rank among the top eight worldwide, now dwarf the remainder of the county's producers in terms of both sales and employment. Their increased size has given them the financial ability to disperse their advanced manufacturing operations, and in conjunction with the changing nature of competition in the industry, has freed them from the need to agglomerate in Silicon Valley. (Meanwhile, numerous smaller firms which lack the scale and financial ability to decentralize their manufacturing activities, and which still remain dependent upon the external economies of the region, are placed at a further disadvantage by the escalating costs of production in the area.)

Yet the industry is not leaving Silicon Valley altogether. Hewlett-Packard recently completed a new international corporate headquarters building in Palo Alto, and Advanced Micro Devices is starting operations at its new Technology Development Center in Sunnyvale. As manufacturing growth is directed elsewhere, the region is gradually being transformed into a high-level control center—the site of corporate headquarters and sophisticated research, design, and development activities. Eventually only the highest-paid professional and top managerial segments of the industry's work force will remain in an increasingly expensive and exclusive white-collar enclave. The region's population growth has already slowed down dramatically. As condominium conversions speed the displacement of the county's low-income population, the urban landscape is being upgraded as well, through newly imposed environmental regulations and other programs to preserve the region's urban and rural amenities.

Ironically, it appears that the no-growth movement in Silicon Valley will thus end up serving the long-run interests of growth. It is now advantageous for the large local firms to follow the exhortations of these no-growth activists—to stop the local growth of manufacturing and to devote ample resources to upgrading the urban environment—in the interest of promoting future growth of a leaner, more elite sort in the region. Meanwhile, expansion of production is proceeding rapidly elsewhere in the country.

New hubs of regional growth are now booming throughout the South and the West as a result of this dispersion of semiconductor and electronics manufacturing. These new production sites are all medium-sized urban areas, with populations in the five hundred thousand to one million range. In contrast to the final highly labor-intensive assembly phase of production—which is now located in low-wage areas of the Third World—this advanced manufacturing phase of semiconductor production still requires a sizeable proportion of engineers and electronic technicians along with an unskilled production work force (in a ratio of at least 1 : 3). New locations must thus provide both a sizeable and tractable pool of unskilled workers, and more importantly—since the first is commonly available throughout the Sunbelt in the form of low-wage, often immigrant, female labor—a social and cultural environment that will ensure an adequate supply of engineers and other professionals. Such locational factors as proximity to a university, cultural and social activities, and availability of recreational facilities as well as affordable housing are thus key elements in the locational calculus for a new semiconductor fabrication facility.

One observer has even predicted the emergence of a "silicon desert" in Arizona, a "silicon prairie" in Texas and a "silicon mountain" in Colorado (Marshall, 1980). While it is unlikely that any one region will face the sort of agglomeration that exists today in Santa Clara County, this possibility is not so farfetched. New centers of semiconductor production in Roseville, California, in Phoenix and Tucson, Arizona, in Salt Lake City, Utah, in Austin, Texas, and in Colorado Springs, Colorado, all are experiencing economic booms. All also face dramatic inflation of land and housing values along with increasingly severe transportation congestion and environmental damage. As the industry constitutes its labor force in these new regions, by recruiting professionals and management personnel from afar and by hiring unskilled production workers locally, it is replicating the top-heavy and bifurcated class structure of Santa Clara County. And with it, we can expect familiar urban problems.

CONCLUDING COMMENTS

An entire cycle has thus been completed. We have seen how agglomeration and expansion of the semiconductor industry transformed Santa Clara County from an agricultural community into one of the fastest-growing and most affluent regions in the nation within a few decades. We have also seen the way in which semiconductor production shaped the class structure and the organization of space in the region, and how the contradictions of this spatial form in turn have translated into significant costs which have caused the industry to restructure. As advanced manufacturing is decentralized, we are witnessing the creation of a new interregional division of labor within the semiconductor industry. While Silicon Valley is upgraded as an elite control and research center, new cities throughout the West and South are growing as manufacturing regions for the industry.

The case of Silicon Valley demonstrates that a region's spatial structure, itself a product of the past and present industrial base, can significantly alter the profitability of a particular region for production. It is important to stress here that this dispersion of semiconductor manufacturing out of Santa Clara County is not a response to some abstract or general congestion costs in a city that is too big or has too many people. Nor are its urban problems simply those of misplanning or overly rapid growth. Rather, they are due to the specific class and urban structures produced by the industry and the social and functional contradictions inherent in them. Semiconductor production generated a bifurcated class structure in the county, one which was distinguished by a large proportion of highly skilled engineers and managerial personnel alongside an even larger number of minimally skilled manufacturing and assembly workers. A highly segregated residential pattern evolved to accommodate the vastly different nature of social reproduction required for these two dominant classes of labor power. As the industry expanded, it became increasingly difficult to accommodate and reproduce both segments of this dichotomized work force within the same metropolitan area. Inflation of housing prices, transportation congestion, labor shortages, and the no-growth movement all are manifestations of the limitations of the local spatial structure for accommodating the industry's bifurcated work force. These urban contradictions eventually caused the industry itself to restructure, thus preserving Silicon Valley as a site for headquarters, high-level research, and prototype production activities.

Many residents recall the days, not long ago, when Santa Clara County had clean air, an abundance of affordable housing, and uncongested roads. The whole cycle—the transition from a rural agricultural region to a highly urbanized manufacturing center, and the ensuing emergence of urban contradictions which spurred the industry to restructure and once again transform the nature of the region's growth—took only three decades. The case of Silicon Valley exemplifies the drastic acceleration in the pace of regional change today. As new regions boom through the dispersion of the industry's manufacturing operations, production will once again shape the local class and spatial structures. While it is rarely possible to isolate the effects quite so neatly as in Silicon Valley, similar analyses of the impact of industry on the organization of space—with particular reference to the location of residential communities and jobs—and the eventual feedback effects which this spatial structure has on industrial behavior, will enrich our understanding of the processes of regional change in the future.

REFERENCES

AXELRAD, MARCIE. "Profile of the Electronics Industry Workforce in the Santa Clara Valley." Mountain View, Calif.: Project on Health and Safety in Electronics, 1979.

BERNSTEIN, ALAN; DEGRASSE, BOB; GROSSMAN, RACHEL; PAIN, CHRIS; and SIEGEL, LENNY. "Silicon Valley: Paradise or Parodox?" Mountain View, Calif.: Pacific Studies Center, 1979.

BLUESTONE, BARRY, and HARRISON, BENNETT. Papers from research project, "Private Investment, Public Policy, and the Transformation of Older Regions: An Analysis of the New England Economy." Cambridge, Mass.: Harvard-MIT Joint Urban Studies Center, 1979.

———. *Capital and Communities: The Causes and Consequences of Private Disinvestment.* Washington: Progressive Alliance, 1980.

Business Week. "More Elbowroom for the Electronics Industry." March 10, 1981:94–100.

———. "Rolling with the Recession in Semiconductors." July 21, 1980.

BYLINSKY, GENE. "California's Great Breeding Ground for Industry." *Fortune* 89 (June 1974):128–35.

CASTELLS, MANUEL. *Sociologie de l'espace industriel.* Paris: Editions Anthropos, 1975.

———. *The Urban Question.* Cambridge, Mass.: MIT Press, 1977.

FROBEL, FOLKER. *The New International Divison of Labor.* Cambridge, Eng.: Cambridge University Press, 1980.

GARCIA, ART. "Silicon Valley Seen Shifting Its Emphasis." *Journal of Commerce,* January 17, 1979.

GILDER, GEORGE. *Wealth and Poverty.* New York: Basic Books, 1981.

GREENBERG, DOUG. "The Dunbarton Bridge: A Study in Regional Conflict." Unpublished paper, University of California, Berkeley, 1974.

HARRISON, BENNETT. "Rationalization, Restructuring, and the Industrial Reorganization in Older Regions: The Economic Transformation of the New England Region Since World War II." Cambridge, Mass.: Harvard-MIT Joint Center for Urban Studies, 1981.

HOLLAND, STUART. *Capital Versus the Regions.* London: Macmillan, 1976.

KELLER, JOSEPH. "Industrialization, Immigration and Community Formation in San Jose, California: Social Processes in the Electronics Industry." Paper delivered at the Rackham Graduate School, University of Michigan, Ann Arbor, November 12, 1979.

LEFAVER, STUART. "Will Success Spoil Silicon Valley?" Unpublished paper, Santa Clara County Planning Department, 1980.

MARKUSEN, ANN, and WALKER, DICK. "Causal Forces in Regional Development of the Western United States: A Study of Industrial Structure and Location." Research proposal, Institute of Urban and Regional Development, University of California, Berkeley, 1980.

MARSHALL, MARTIN. "Silicon Valley Is Filling Up." *Electronics* 53 (February 28, 1980):98–100.

MASSEY, DOREEN. "Toward a Critique of Industrial Location Theory." In *Radical Geography,* ed. R. Peet. Chicago: Maaroufa, 1975.

———. "Regionalism: Some Current Issues." *Capital and Class* 6 (1978a):196–225.

———. "Capital and Locational Change: The U.K.

Electrical Engineering and Electronics Industry." *Review of Radical Political Economics* 10 (fall 1978b):39–54.

———. "Industrial Restructuring as Class Restructuring: Some Examples of the Implications of Industrial Change for Class Structure." Working paper. London: Centre for Environmental Studies, 1980.

———, and MEEGAN, RICHARD. "Industrial Restructuring Versus the Cities." *Urban Studies* 15 (1978):273–388.

MUTLU, SERVET. "International and Interregional Mobility of Industrial Capital: The Case of the American Automobile and Electronics Companies." Ph.D. diss., University of California, Berkeley, 1979.

NOBLE, DAVID. *America by Design.* Oxford: Oxford University Press, 1977.

ROBINSON, ARTHUR L. "Giant Corporations from Tiny Chips Grow." *Science* 208 (May 2, 1980): 480–84.

———. "Perilous Times for U.S. Microcircuit Makers." *Science* 208 (May 9, 1980):582–86.

———. "New Ways to Make Microcircuits Smaller." *Science* 208 (May 30, 1980):1014–22.

———. "Problems with Ultraminiaturized Transistors." *Science* 208 (June 13, 1980):1246–49.

———. "Are VLSI Microcircuits Too Hard to Design?" *Science* 208 (July 11, 1980):258–62.

Santa Clara County Economic Development-Job Needs Project. *Training and Jobs: Ways to Reduce Unemployment in Santa Clara County.* San Jose: Santa Clara County Planning Department, 1978.

Santa Clara County Housing Task Force. *Housing: A Call for Action.* San Jose: Santa Clara County Planning Department, 1977.

Santa Clara County Industry Housing Management Task Force. *Living Within Our Limits: A Framework for Action in the 1980s.* San Jose: Santa Clara County Planning Department, 1979.

Santa Clara County Manufacturing Group. *Report on Estimates of Job Growth and Building Expansion of Sixty Santa Clara County Companies, 1979–1985.* N.p.:1979.

———. *Vacant Land in Santa Clara County. Implications for Job Growth and Housing in the 1980s.* N.p.:1980.

Santa Clara County Planning Department. "Assessed Valuation for Selected Fiscal Years, 1959–1971, and Estimated Assessed Valuation per Capita, 1959–1970, by City." Information Fact Sheet 389, November, 1970.

———. "Socio-Economic Characteristics of Cities, April 1, 1970." IFS 469, November 1972.

———. "Assessed Valuation by City, Selected Years, 1959–1976." IFS 565, January 1976.

———. "Total Population by Age and Sex: April 1950, 1960, 1970 and 1975." IFS 591, June 1978.

———. Components of Yearly Population Increase, 1950–1979." IFS 660, February 1980.

———. "Transportation/Land Use Planning Within the Present General Plans Structure." 1979.

SAXENIAN, ANNALEE. "Silicon Chips and Spatial Structure: The Industrial Basis of Urbanization in Santa Clara County, California." Working Paper 345. Berkeley, Calif.: Institute of Urban and Regional Development, University of California, 1981.

Stanford Environmental Law Society. *San Jose: Sprawling City.* Palo Alto, Calif.: Stanford University, 1971.

Stanford Research Institute, International. "The Mid-Peninsula in the 1980s." Palo Alto, Calif.: SRI, 1980.

U.S. Bureau of Labor Statistics. "Industry Wage Survey: Semiconductors." Bulletin 2021. Washington: U.S. Government Printing Office, 1977.

U.S. Department of Commerce, Industry and Trade Administration. "A Report on the U.S. Semiconductor Industry." Washington: U.S. Government Printing Office, 1979.

VERNON, RAYMOND. *Metropolis 1985.* Cambridge, Mass.: Harvard University Press, 1960.

———. "International Investment and International Trade in the Product Cycle." *Quarterly Journal of Economics* (1966):190–207.

WALKER, RICHARD, and STORPER, MICHAEL. "Capital and Industrial Location." *Progress in Human Geography* 3 (1981).

PART VI

Policy Alternatives to Meet the Challenges of the Structural Transformation of the Economy

Referring specifically to technological change and the future but applicable to the current economic transformation in general, Barry Jones has said:

> The Western world is passing through a period of technological change, more far-reaching—and much faster—than at any other time. This rapidity and intensity presents massive challenges. . . . We face an extraordinarily ambiguous future. Technology can be used to promote greater economic equity, more freedom of choice, and participatory democracy. Conversely, it can be used to intensify the worst aspects of a competitive society, to widen the gap between rich and poor, to make democratic goals irrelevant, and institute a technocracy. We must evolve policies in response to the current era of rapid technological change (Jones, 1982:254).

These themes—the uniqueness of this historical turning point, the ambiguity of future direction, and the clear need for planning to meet these challenges—guide our inquiry.

The selections in this book have chronicled the effects, mostly negative, of the economic transformation. We have seen, for example, the occupational restructuring that has occurred as several million workers have moved out of the smokestack industries since 1975 or so. Work has been fragmented and, for the most part, deskilled. Many workers have been forced to leave jobs as their employers closed plants permanently or relocated them. Many hundreds of thousands of these unemployed have become permanently displaced—about 60 percent of older workers aged fifty-five to sixty-four; and women, Blacks, and Hispanics are twice as likely to remain unemployed as white males (Boehlert and Dubrin 1986). Despite this nagging problem with displaced workers whose jobs have permanently disappeared, there are more jobs overall. The problem is that while the number of jobs has risen, the number of "good jobs" with good pay and security has fallen. The result of this trend and related ones is that there is greater economic inequality now than at any time since 1969. In that year the poorest 20 percent of the population had 5.6 percent of the total income, a proportion that has since slipped to 4.7 percent. The richest 20 percent, on the other hand, had 40.4 percent of the total income in 1967 and 42.9 percent in 1986 (*New York Times*, 1986). The gap is especially wide for women compared to men and racial minorities compared to whites.

There are many other problems resulting from the structural determinants of this strategic turning point. Among these are the loss of competitiveness worldwide, the rapid deterioration of basic industries domestically, communities gutted by runaway corporations, and the like. Are these negative consequences resulting from the economic transformation short-term ones? Is this just a period of transition where there are dislocations that will right themselves over time such as was the case when society shifted from an agricultural economy to one based on manufacturing? We assume that this situation is different from that at any time in history. The conditions now are unprecedented in magnitude and velocity. We assume, further, that we cannot expect society's economic problems to be solved magically by the mechanism of the marketplace (Adam Smith's "invisible hand"). In short, there must be national economic planning to meet the challenges of the new economy.

The notion of economic planning often meets resistance, however. This is ironic because planning is accepted by corporations, governments, the military, and other organizations. As Faux has put it: "National economic policy remains the only significant organized human activity in America where planning ahead is considered irrational" (Faux, 1982:12). Also the resistance to economic planning is misleading, since national economic planning occurs regardless of the political party in power and whether liberals or conservatives are in control. The New Deal policies of Franklin D. Roosevelt clearly intervened in the marketplace as did the supply-side economic policies instituted by President Reagan and a willing Congress in the 1980s (Harrington 1984; Zeitlin 1983).

We argue that planning to solve economic crises is imperative. The question, then, is what kind of planning? How do we rationally confront the current profound, and in many ways, unprecedented crises? Most critically, we ask: Can the government intervene in the public interest, or as has been the case historically, Will the intervention be in the interest of the powerful? Can we assume that state legislatures, Congress, and the chief executive will respond fairly to the simultaneous pressures of the disemployed, the poorly paid, and the corporations who cannot exist without profit?

Some representative questions that planners must address are: (1) Should basic industries be allowed to fail if they cannot compete internationally? (2) Should multinational corporations be allowed freely to drain the United States of productive investment? (3) Should corporations be allowed to shut down a plant without sufficient warning (and compensation) to affected workers and communities? (4) Must an industry be kept alive artificially through subsidies so that workers and communities have sufficient time to adjust? (5) Who is responsible for the welfare and/or retraining of displaced workers? (6) Should taxes be levied on robots? (7) Should we have a government/corporate goal of full employment? (8) What is fair compensation for labor? We cannot address each of these and other related questions here. We will concentrate on policy formation in five important areas: how to democratize the allocation of capital, how to invigorate "sunset" industries, how to halt plant closings, how to be competitive in the new global economy, and how to provide full employment with "good" jobs for all—including minority groups.

REFERENCES

BOEHLERT, SHERWOOD and RICHARD J. DURBIN. 1986. "A Remedy for the Displaced-Worker Problem," *The New York Times* (January 25):19.

FAUX, JEFF. 1982. "Who Plans?" *Working Papers for a New Society* 9 (November/December):12–16.

HARRINGTON, MICHAEL. 1984. "A Case for Democratic Planning," *Alternatives: Proposals for America from the Democratic Left*, Irving Howe (ed.). New York: Pantheon Books.

JONES, BARRY. 1982. *Sleepers, Wake! Technology and the Future of Work.* Melbourne: Oxford University Press.

New York Times. 1986. "Warnings Over Widening Income Gaps," (July 13):F11.

ZEITLIN, MAURICE. 1983. "Democratic Investment," *The Center Magazine* 16 (July/August):16–36.

37 *The Democratization of Investment Decisions*

MAURICE ZEITLIN

America's basic industries are crumbling, hit by a wave of plant closings. In steel mills and auto factories, canneries and copper mines, chemical plants and sawmills, workers have suddenly found themselves out of a job, out of a company, and out of a future. Why?

It's more profitable for General Motors to import Opels and Isuzus from its affiliates abroad than it is to produce Chevettes at home. For Anaconda it makes sense to ship concentrated copper ores to Japan for smelting while shutting down its smelters in Montana. U.S. Steel makes more profit by spending $6.2 billion to take over Marathon Oil than by investing in modern plant and equipment to create jobs.

A handful of huge conglomerates straddling a host of unrelated industries dominates the economic life of the biggest cities and the smallest towns in America. The two hundred largest nonfinancial corporations control more than sixty percent of America's net capital assets. Since 1950, there has been a steep rise in economic concentration. The top twenty companies alone now get more than forty cents of very dollar of manufacturing profits. Of the country's fourteen thousand banks, the one hundred largest hold half of all deposits; the top fourteen alone control about one-fifth.

Source: Maurice Zeitlin, "Democratic Investment," *The Center Magazine* Vol. 16 (2) (July/August 1983), pp. 16–20. Reprinted by permission of the Robert Maynard Hutchins Center for the Study of Democratic Institutions.

The misinvestment and disinvestment of these giant corporations are at the root of America's economic jeopardy and industrial deterioration.

Insulated from domestic competition and ruling the economic roost abroad until recently, our complacent and comfortable corporations could "administer" their prices, as economists politely put it, to assure themselves of an effortless flow of profits. Their managers went right on paying high dividends to stockholders and bonuses to themselves while letting America's competitive technological edge dull. The conglomerates used their high profits in one line to drive out competitors in another. Draining subsidiaries of the capital to modernize, they then sloughed off and shut down these newly "unprofitable" enterprises (and took lucrative tax write-offs).

Less and less of the money invested by American companies remains in this country: in 1960, twenty-one cents of every new investment dollar went overseas, but by 1980 the figure had more than doubled. For the biggest multinationals, with production facilities in dozens of countries, the effect of their investment decisions on this country and its institutions is not considered.

It's time for a change: the flow of capital can no longer be held by private hands alone. Left to themselves, these huge corporations will let our basic industries go down, one after the other.

The private right to earn profit will have to be balanced with the public right to decide democratically how to reinvest some profit in the public interest. The place to begin is here. California is already experiencing a spate of plant closings from Sacramento to South Gate. Since 1979, eight hundred plants in the state have shut down. Nearly half a million workers have lost their jobs.

What California needs is a democratically constituted Public Investment Reserve System. PIRS (pronounced "purse") should have a pool of capital large enough to allow it, by selective allocating of financing, to steer the investment process in the public interest: toward expansion of employment, encouragement of new industries and modernization of "mature" ones, construction of affordable housing and health-maintenance centers, and fulfillment of other social priorities.

The system could be funded from four main sources:

- *The state's "temporarily idle public funds," amounting to about eighteen billion dollars, for leveraged capital.* These short-term taxpayer funds, invested or deposited by public agencies, are roughly analogous to personal checking accounts and liquid accounts of less than a year's maturity.
- *California's public-employee pension funds, roughly thirty billion dollars more.*
- *Voluntary pooling of union pension funds, which add up to another thirty billion dollars.* These deferred workers' earnings constitute an immense pool of capital that could be productively reinvested in California in their interest, under government guarantees of principal and a rate of return.
- *Private capital that might otherwise be invested in wasteful mergers and speculative ventures or sent overseas.* To operate in California, every large corporation would have to make a tax-deferred "security deposit" in the reserve system of a specified percentage of its net investment and annual added value. If a big

> corporation misinvested or disinvested, its shareholders, not the public, would bear the costs. the system's funds would be selectively allocated and redistributed within the business community itself, as loans and equity capital, to applicant companies whose investments were productive, job-creating, and socially responsible.

The system would be administered by a board consisting of appointed and elected directors who would represent the citizenry at large and public and private employees, whose pension funds would be a major source of capital. Dividing the system into districts would enable local residents to participate fully in investment decisions.

By establishing such a system, we would of course risk the possibility that some major firms would decide not to make new investments in California or that others might be tempted to leave. Few businesses, however, would want to antagonize a population—and market—as large as ours and face consumer boycotts and other public-relations problems in California for the mere price of a security deposit.

In fact, most firms, especially small and medium-sized companies, would stand to gain from a pool of public capital far bigger than their individual deposits. Smaller firms are particularly starved for capital by the biggest financial institutions. Yet they are typically much more cost-efficient and innovative than giant corporations, and they create more jobs per dollar invested. The fact that smaller manufacturers could compete with the giants more effectively by having access to reasonable public financing would make the giants wary of moving their operations out of state and relinquishing the chance to get similar funding.

If California showed the way, other states would soon follow in stimulating economic growth through new democratic forms of public investment for the public benefit.

For those of us for whom capitalism is not an intrinsically just society—let alone the best of all possible worlds, the question is, what should be done? I address that question, not just to the Left, but to everyone who is trying rationally to confront the present profound, and in many ways unprecedented, economic crisis.

What is it that has changed in our society that provides the objective rationale for attempting to think through a way to achieve the democratic allocation of public capital for public purposes?

Basically, what there was of a competitive capitalist economy in the United States has long since been relegated to the dust bin of history. This is now a highly concentrated capitalist economy in which fewer and fewer and increasingly larger and more centralized bureaucratic units of capital make the most basic economic decisions. These enormous corporations confront citizen and government alike, worker and investor alike, as relatively autonomous forces, controlling the flow of capital in our society. This is true in the marketplace, in the stock market, and throughout the system of production and distribution.

So this structural situation raises concretely the question of who should make the decisions concerning the flow of capital in a society in which the "marketplace," as such, surely does not make them. It was not possible to ask that question concretely when a relatively diffuse and abstruse marketplace made economic decision-making invisible.

Simultaneously, the rise of the multinational corporation and its increasingly dominant place in the global economy is one of the most central and politically relevant facts of our time. The transnationalization of capital and the near-control of the global flows of capital by giant units of production have now made transparent for Americans, and not only for the underdeveloped world, the inherent contradiction between what is good for the multinational corporation and what is good for the nation.

That contradiction now has direct and immediate relevance for the citizens of the United States. What is characteristic, crucial—and ironic—about the moment we live in is the tendency toward the increasingly rapid deterioration of America's basic industries as the result of the decreasing rate of investment of American multinational corporations at home and their accelerating rate of investment abroad. Less and less of every dollar of net investment made by the largest U.S. corporations is being put to work in this country.

The transnational flows of capital are controlled by, at most, a few hundred giant corporations based in the United States, plus a handful of others in Japan, West Germany, France, and Sweden. These corporations are able to move capital around at an extraordinarily fast rate into any areas of the globe almost at will. What we now have is an internationalization of capital flows that results in disinvestment in the United States. It is almost as if the red-lining of the inner cities has now been apotheosized at the level of the whole society, albeit unconsciously, and without premeditation or coordination.

So the decision as to who invests where, in whose interests, is now on the immediate historical agenda, not only because of the concentration of capital, but also because of the consequences of that concentration for our country.

Should American multinational corporations be allowed freely to drain the United States of productive investment? Or is there a democratic public rationale for political intervention, not only in the "marketplace," but in the most basic decisions affecting the allocation of private capital? The political economic question of the democratic allocation of capital is *the* question in the present historic moment. If the allocation of capital is not to be democratically decided, then there will still be a new form of explicit state intervention in the investment process. Of that I have not doubt.

• • • •

What I am talking about, what the Campaign for Economic Democracy has been talking about for a long time, and what the Ohio Public Interest Campaign and the Massachusetts Fair Share programs have all been talking about is the democratic rather than authoritarian allocation of capital. The question is not whether there will be state or other forms of public intervention in the allocation of capital, but whether that intervention will be democratic, whether it will serve the public interest or merely the common class interest of capital.

It is in this concrete political context that I propose my alternative of an autonomous public investment reserve system, PIRS. It is one possible method of democratic participation by citizens in allocating a portion of the profit stream in their interest rather than in the sole interest of private investors.

38 *The Revitalization of "Sunset" Industries: The Case of Steel*

DAVID BENSMAN AND ROBERTA LYNCH

A sense of the inevitable has clouded all discussions of the industrial crisis confronting our nation. Policy analysts of divergent opinions have found common ground in the view that the steel industry cannot and should not be saved. Most see its demise as intimately linked with the decline of our entire manufacturing sector: the death of smoke-stack industries. Some believe that no matter what America does, it can no longer compete as an industrial power against the low-wage countries of the Third World. Others hold that unless the steel industry can stand on its own, it does not "deserve" to survive. In this Darwinian economic model, the fittest would emerge, trim and tough, from the rigors of economic competition.

Underlying these views are two unexamined premises. The first is that the economic environment of the early 1980s—the strong dollar, high interest rates, huge budget deficits, and surging imports—provided a fair test of the viability of the steel industry. The second is that a sound manufacturing sector is no longer essential to the future of the United States.

Both Premises Are False The temporary economic problems of the early 1980s made it impossible for U. S. steelmakers to compete, thus seriously exaggerating the extent or effects of the industry's backwardness. And the

Source: Excerpted from David Bensman and Roberta Lynch, *Rusted Dreams: Hard Time in a Steel Community*, McGraw-Hill Book Company, 1987, pp. 202–235.

future prosperity of the nation will be endangered if we allow our industrial strength to wither. It is possible and necessary to revitalize America's steel industry. By doing so, we would also strengthen the entire durable goods manufacturing sector, for which cheap, high-quality steel is essential.

A program to revive steel would require large public expenditures and substantial change in economic policies. Government leaders have made a number of half-hearted attempts to save the steel industry. But all the recent federal steel programs are doomed to failure because they are piecemeal responses to crises of the moment. Since political leaders hold fast to the idea that the steel industry must compete in a free marketplace, they propose only limited intervention. Their programs attack one aspect of the industry's problems without addressing the others. As a result, they cost the government a lot of money, but don't enable U.S. steelmakers to become world class.

If the steel industry is to survive as more than a shell of its former self, there is only one solution: a comprehensive strategy that coordinates a series of remedial programs.

The Demand Side

A comprehensive program for the steel industry should begin with the demand side. Demand for American steel fell from 100 million tons shipped in 1979 to only 74 million tons in 1984, a 26 percent decline. Without a rebound, domestic steelmakers will have no incentive to modernize.

The decline in steel sales was caused partly by a rise in imports, to more than 25 percent of the U.S. market in 1984. But that was only one part of the story. Equally important was the sharp rise in imports of cars, tractors, subway cars, buses, machine tools, etc. Steel-containing imports reduced the demand for American steel by 13.7 million tons per year in 1985, according to a study by Locker/Abrecht Associates.

A third part of the story was declining exports. The U.S. lost almost six million tons of steel exports in the years 1970–83 because of the overvalued dollar and slumping Third World economic growth. When overseas sales of Caterpillar tractors or Harvester trucks decline, so does demand for American steel.

A comprehensive program should bring down the value of the dollar, restrict imports, rebuild America's aging physical facilities, support research in product innovation, and transfer capital spending from military to domestic uses.

The sharp decline in demand for American steel during the years 1979–84 was partly the result of the strong dollar. From 1980 to 1983 the dollar's value increased 82 percent against the French franc, 78 percent against the Italian lira, and 41 percent against the West German mark. The dollar's new strength had a clear effect on U.S. competitiveness, according to a Data Resource Inc. study: "German labor compensation costs have moved from about

25 percent above U.S. costs in 1980 to about 20 percent below U.S. costs in 1982." This means that an American mill producing steel 25 percent cheaper than a German mill in 1980 might very well have lost orders to that rival in 1984, even if the domestic facility were more efficient. If U.S. manufacturing industries are to survive, the federal government will have to give priority to keeping the dollar stable and realistically aligned with currencies of our chief trading partners.

Even if the dollar stabilizes at realistic levels, the government will probably have to impose strong trade restraints to ensure that the demand for American steel is sufficient to induce domestic producers to modernize.

Protectionism is a dirty word in American discourse. Yet the reality is that "free trade" hardly exists anywhere in the world today. Even the United States, bastion of free trade ideology, manages trade in many economic sectors. The prosperity of the American auto industry in 1983–4 was based on Japan's voluntary agreement to limit exports to this country.

But America's sins against free trade are minor compared to those of our trade partners. Wolfgang Hager, economic consultant to the Common Market, estimates that 75 percent of the free world's trade is subject to quotas, export subsidies, barter arrangements, and other government restrictions. The growth of Third-World industry has made the need for protection more urgent. During the past decade, multinational corporations have built modern manufacturing plants in the poor regions of the world. At the same time, governments around the globe have been promoting basic manufacturing industries like steel, auto, and electronics. The result is enormous overcapacity, and competitive pressure to cut taxes and to lower wages.

Some Americans who accept the necessity of protecting our industries from competition by subsidized Japanese or European producers believe that nothing can or should be done about competition from countries like Korea, Mexico, or Brazil. Now that these rivals have combined modern technology with low labor costs, the argument goes, there's no way U.S. manufacturers can compete with them.

Such thinking rests on questionable assumptions. South Korean and Brazilian mills do have modern technology, but their productivity still lags far behind that of U.S. mills. The difference can be seen in these statistics: American mills use 3.6 man hours to make a ton of hot rolled band steel, while Korean mills need 5.5 hours and Brazilian mills 5.9. The reason is primarily that American steel-workers are more experienced and knowledgeable in the art of steelmaking.

Third-World steelmakers do have one important advantage over U.S. manufacturers: low wage rates. In most of the Third World, severe political repression enables export-oriented producers to keep workers' wages so low that they can never hope to buy the things they produce. Machine guns and police spies provide the foundation of the "labor cost advantage" that is eliminating the jobs of steelworkers in Southeast Chicago.

But isn't protectionism unfair to the poor workers of the Third World? This question is more complicated than it seems. There is evidence that export-platforms like Korea's steel mills and Mexico's auto parts plants, financed by the International Monetary Fund and the World Bank, are not unmixed blessings for their local economies. They draw people off the farms and make for slums in the cities, yet don't seem to generate the kind of self-sustaining economic growth that creates steadily improving living standards. Encouraging the export of Third-World manufactured goods to the U.S. may be a way of helping to ease the international debt crisis, but it's disastrous for U.S. steelworkers, and of dubious benefit to the poor of the Third World.

To protect U.S. steelworkers without making the workers of the Third World carry the burden, the U.S. should enact an import tax based on the differential between what American workers need to live decently and what Third World workers are paid. The taxes collected could be used for two purposes—to modernize American steel mills and to fund Third World development projects. This would provide an incentive for countries like South Korea to develop home markets. They would have a choice of paying the tax or raising their workers' wages, thereby generating domestic demand. This would help U. S. steelworkers one way or the other, and might give workers in the developing world a shot at jobs that are not completely dependent on the vicissitudes of America's economy.

Most Americans are not likely to welcome permanent import restraints. We instinctively believe they will foster inefficiency and will cost consumers billions. The example of Great Britain's nationalized and protected steel industry—whose losses were subsidized for thirty years, with the mills becoming more and more backward—is what we think of first. But it doesn't have to be that way. Japan's protected steel and auto industries have outstripped our own; there's no reason why a protected American steel industry couldn't be modern and efficient too.

But protection will have to extend beyond the steel industry. It would not make much sense to exclude foreign steel if we allowed foreign manufacturers to flood U.S. markets with products containing a great proportion of steel. To do so would leave American steel mills without markets. A program to protect steel only makes sense as part of a larger policy to protect American manufacturing.

At the same time that our government limits competition from abroad, it must also increase domestic demand. One of the most dramatic opportunities to expand demand for steel would be a program to rebuild our infrastructure.

Since 1965, America's investment in public infrastructure has dropped 50 percent. Decaying roads, bridges, and railroad tracks increase transportation costs and energy consumption and decrease productivity. In October 1983, Clifford Garvin of Exxon and Lane Kirkland of the AFL-CIO released a report estimating that in the years 1983–1995 the cost of rebuilding America's highways and bridges, urban water supplies and wastewater treatment facilities would exceed planned spending by $8.5 billion to $10.8 billion *annually*. If to this we add America's deteriorated local roads, railroad beds, and urban mass transit systems, we end up with even larger shortfalls. A thorough going rebuilding program would generate demand for more than 10 million tons of steel per year, according to the Midwest Center for Labor Research.

If America initiated an infrastructure rebuilding program without enforcing trade restraints, the orders generated by the program would be filled by foreign steelmakers. And if we instituted trade restraints when the dollar was overvalued, the price disparities between domestic and foreign steel would eventually topple the import barriers. Only coordinated action to ensure high levels of steel consumption would minimize the program's costs and maximize its effectiveness.

Investment in Modernizing Steel

Increased demand for American steel must be linked to a program to make sure that the steel companies modernize. Stimulating investment in steel will require governmental action because the American economy discourages long-term investment in manufacturing industry.

In recent years, America's financiers and corporate executives have developed means of making profits largely unrelated to social purposes. In the new system of what Seymour Melman has termed "profits without production," financial expertise seems more important than production.

Government tax policies and a revolution in banking practices have encouraged a corporate flight from productive domestic investment. Robert Kuttner notes that, "at the same time that our government pressures Japanese automakers to restrict exports to the U.S. 'voluntarily,' the U.S. tax codes gives Ford tax incentives to shift production from Detroit to the Mexican border." Tax code preferences have other bizarre consequences: they provided U.S. Steel with incentives to shut its mills so that it could take over Marathan Oil.

The current tax overhaul leaves in place many of these incentives. A genuine reform would eliminate the tax incentives for unfriendly mergers, conglomerate takeovers or unrelated businesses, and plant shutdowns.

But tax reform with reform of our credit markets is only half a solution. In the 1960s, banks became aggressive buyers and sellers of money; their search for depositors drove up interest rates—and the steel companies had to pay more to borrow the money needed for modernization.

The transfer of billions of dollars from Western consumers to OPEC accelerated the banking revolution by creating a huge pool of highly mobile capital available for investment in high-yield, short-term instruments. Frenzied buying and selling of new forms of paper drove up interest rates, further increasing the cost of industrial modernization.

The banking revolution and the rise of paper entrepreneurialism should have alerted Federal officials to the need for increased supervision of the banking system, but the opposite occurred. "Deregulation" was the name of the game. In the meantime, new forms of speculation appeared every month; and the gains to society were hard to fathom.

Congress and the Federal Reserve have the power to discourage bank financing of hostile takeovers, excessive foreign investment, and commodity speculation, but they fail to exercise it. (The Federal Reserve Board's move against "junk bond" financing of hostile takeovers is a lonely exception.)

To strengthen the incentive for private capital to modernize industry, the federal government should use public investment as a lever. Such a strategy could shift billions of dollars to private funds into modernization projects.

Pension funds are one available source of public investment. $700 billion of America's capital lies in pension funds. The private capital markets use this huge pool of public capital to finance corporate takeovers and other forms of "profits without production." But pension fund assets could be used to create jobs and modernize industry; indeed, the Building Trades Department of the AFL-CIO already has developed a program to funnel pension funds into the mortgage market to stimulate construction activity.

The military budget is a second potential source of funds for leverage on private capital. Currently, federal defense spending harms American industry rather than helping it. Cost-plus Pentagon contracts have diverted manufacturers from efficient practices, thereby rendering them unable to compete with foreign producers. Tens of thousands of engineers are taken out of civilian employment to work on military production. National spending on research and development is skewed to defense.

If America cut back its military budget by just 10 percent, the savings would reach nearly $30 billion per year. That sum could be used to rebuild basic industry and infrastructure and at the same time leverage greater sums of private capital.

If all these parts of a comprehensive steel program were implemented, the steel industry might be able to make it on its own. But if this program were not implemented in toto, the steel companies would need direct federal subsidies. According to a study by the American Iron and Steel Institute, $3–4 billion annually would be required in the absence of a comprehensive program.

Most Americans don't want their tax dollars spent subsidizing firms like U. S. X. and Armco. When they got tax breaks in 1981, they used their windfalls to diversify, not modernize: why give them more money now?

Such suspicions are entirely well-founded. American corporations have a long and sorry history of finding ways to enjoy the benefits of public support while escaping public accountability.

Moreover, this entire program cannot be viewed apart from political realities. Such a massive public investment program is unlikely to be prompted by the decline of our basic industry. American politics is not governed by some benign intelligence that analyzes critical problems and then seeks to address them. It is power, after all, that makes the wheels go round.

The steel companies themselves, blinded by their antipathy to government intervention, are not likely to advocate such a comprehensive program.

Pressure must come from other places. The potential coalition for such change remains inchoate. But several important elements have been taking shape in recent years, the most vital being the response from steel communities and steelworkers. It began half a decade ago, with the attempt to organize a worker/community buyout of the shuttered LTV steel mill in Youngstown, Ohio. Then the locus of struggle shifted to the Mononghahela Valley where there have been pitched political battles over issues of worker ownership, eminent domain, and corporate responsibility.

A second part of this unfolding coalition has been the United Steelworkers union (USWA). Once highly conventional in its tactics and skittish about "radical elements" and ideas, the USWA has gradually realized that it must fight with every weapon at hand. As a result, the union has begun to show a greater emphasis on technological improvements in its internal communications and its external public relations, as well as a greater willingness to ally itself with worker/community groups arguing for public or worker ownership.

Church groups have also help to bring strength to this burgeoning coalition.

Fourth, local elected officials have been forced to respond to these forces and have become, however shakily or involuntarily, allies in the search for solutions. Chicago Mayor Harold Washington established a Steel Task Force that has developed a major report on methods to revitalize the region's steel industry. Pittsburgh's Mayor Richard Caliguiri has been required to address such previously unimaginable issues as the public takeover of a steel mill. And even a Republican mayor, George Voinovich of Cleveland, has called for a public jobs program to provide work for the unemployed.

Whether such a loose-knit and geographically spotty confederation can articulate and press for major national policy shifts is difficult to predict. Two things are certain, however. First, if such a movement is to succeed, it cannot focus only on steel, but must have a base in and a program for the manufacturing sector as a whole.

Second, it must have as its centerpiece not just a program of centralized planning of public investment in steel, but an insistence on corporate accountability. Public investment in steel should be only granted in exchange for corporate commitments to modernize old facilities or build new ones.

If such an industrial policy is to be truly democratic, then the terms of the bargain—how much public investment in exchange for how much modernization—should be determined by a process with widespread public participation. Finally, if public agencies invest in the steel industry, they should be given equity stakes, so as to share in the steelmakers' profits.

Saving steel is vital to our national interest. Saving jobs is vital to our economic well-being. But it is only worth undertaking such an uphill battle if it is done with the recognition that this crisis was created by an arrogant industry bloated with power. A revitalized steel industry must be characterized not only by greater efficiency, but by a higher regard for the public interest.

39 *How to Fight Plant Closings*

ROBERT GILOTH AND SUSAN ROSENBLUM

On December 4, 1984, the city of Chicago sued Hasbro Industries, Inc., for closing its Playskool toy factory on Chicago's westside and laying off 750 mostly Hispanic and women workers. Playskool symbolized the predicament of many cities. Chicago had lost 110,000 manufacturing jobs between 1979 and 1984 from firm deaths and outmigrations. Whether explained as a result of corporate disinvestment, economic structural change, regional and international competition, or technological adjustment, these losses disproportionately affected blue-collar, minority workers, their neighborhoods and cities. Yet most cities do not publicly question plant closings.

During 1983–85, Chicago's Department of Economic Development (DED) funded the Midwest Center for Labor Research (MCLR) and the Center for Urban Economic Development (CUED) at the University of Illinois at Chicago to design and implement a plant-closing project on Chicago's westside. The purpose of this project was to go beyond crisis-initiated responses to plant closings and establish an early-warning system founded on public, labor, and community cooperation that could interpret early warning signals of impending plant closings (e.g., employment changes and buyouts) and formulate appropriate responses. To facilitate information gathering and responses, MCLR and CUED established the Westside Jobs Network (WJN)—a community/

Source: Robert Giloth and Susan Rosenblum, "How to Fight Plant Closings," *Social Policy* 17 (Winter 1987), pp. 20–26. Published by Social Policy Corporation, New York, New York 10036. Copyright 1987 by Social Policy Corporation.

labor coalition that became the major actor in Chicago's early-warning demonstration.

Chicago's experiment with early warning was in part a response to the closing of Wisconsin Steel and the partial closing of U.S. Steel South Works in the early 1980s with the loss of nearly 10,000 jobs. Both closings engendered widespread public concern. The Save Our Jobs Committee of Wisconsin Steel workers sued to protect their pensions and final paychecks and fought the federal Economic Development Administration (the current owner of Wisconsin Steel) to return the mill to steel production rather than demolish it.

Illinois Attorney General Neil Hartigan and steelworkers filed suit against U.S. Steel for a "breach of contract" in obtaining city, state, and union concessions and incentives for a rail mill they had promised to build at the South Works site. U.S. Steel reneged on their commitment. The public threatened eminent domain and called for an employee buyout but ultimately settled for an industrial park and U.S. Steel's commitment to reinvest in its remaining beam mill. Slowly, parts of South Works and Wisconsin Steel have been sold off to steel fabricators or finishers (many of whom are non-union) and to nonsteel users such as bulk storage.

Much has been learned from the cases and lessons of the Chicago early-warning demonstration. Although unions hesitated to participate in early warning, workers provided key information about plant investment. Active community, labor, and government involvement expanded the accepted range of plant-closing responses. Although early-warning information often proved elusive and routine responses were seldom adequate, the demonstration increased local knowledge about plant closings and strengthened community-labor networks.

The frequency of economic dislocation is not likely to slow in years to come, even with economic "recovery." Advance notice of plant closings like Playskool makes negotiation, development, and planning dislocated workers' programs possible. There is time to act. Federal plant-closing legislation is the preferred formal method of advance notice because it creates a level playing field of business incentives; municipal and state laws are weak or may engender interjurisdictional competition, few union contracts contain notification provisions, and voluntary business notification is minimal, despite several notable examples.

But formal notification of plant closings is not sufficient. European countries that have plant-closing notification have learned that public, private, and labor cooperation is needed to make notification a meaningful part of industrial policy and planning. In the United States, neither federal plant-closing legislation or labor-management cooperation is expected in the near future.

Without a federal policy, communities must rely upon informal notice or early warning as in the Chicago demonstration. The absence of a legislative framework, regulated information, and public-private bargains means that early warning occurs in an environment dominated by business ideology, bad information, and shifting political concerns. As a result, effective plant-closing responses in this context are found to be discrete and incremental, realistic but innovative about available political and economic resources, and dependent upon the capacity of locally and often temporally unique coalitions of communities, business, labor, and local government. Even the best of these responses, however, whether private or employee buyouts, negotiated reopenings, increased employee severance, or innovative legal actions (e.g., eminent domain) are overshadowed by many heroic but inconclusive efforts.

Chicago's early-warning demonstration concluded in June, 1985. Early warning in Chicago now includes union education about plant closings and economic development, expanded contracting with neighborhood industrial councils to visit firms, a program to identify and resolve waste and energy problems of electroplaters and food processors, and regular review of early-warning cases by staff from DED and the Mayor's Office of Employment and Training.

Planned initiatives include a targeted steel-visitation program and business-community intervention in selected early-warning cases. One important legacy of the Chicago demonstration is that three firms, employing over 5,000 people, are now being monitored by public, community, and labor groups on an ongoing basis.

EARLY WARNING CASES

The Chicago demonstration used worker and union contacts, community research, and existing business information to monitor 75 firms on Chicago's westside, including ten closings or near closings that accounted for 4,000 jobs. These firms were large, unionized, and absentee-owned, and cited multiplant excess capacity, markets, or local profitability as the reasons for shutdown. Three cases are presented below. Actual shutdowns were not prevented; rather, jobs were extended, benefits increased, and retraining and new development options explored.

Ludwig Drum

Ludwig Drum showed all the early-warning signs of a plant closing. By early 1984, employment had declined to 125 workers from more than 400 a few years before. After a maintenance worker had tipped city officials that a closing was imminent in late 1985, the West Side Jobs Network (WJN) monitored Ludwig for more than a year before the parent company, Selmer, officially notified workers that the plant would close on August 9, 1985.

At one time a family-owned manufacturer of high quality musical drums, Ludwig had been acquired in 1981 by the Selmer Company of Elkhart, Indiana, a subsidiary of the Dutch conglomerate North American Philips.

Selmer, known for its cheaper line of musical instruments, began to shift Ludwig's production to a new plant in North Carolina and imported parts from Taiwan for assembly at the Chicago plant.

Ludwig workers kept the WJN informed about equipment being disconnected and transported, management memos, and rumors. Workers not only described the dismantling of production, but also had information about the problems Ludwig was having with its new unskilled workforce and imported parts. The Carpenters Union represented the Ludwig's workers, but organizational problems and changing leadership limited their ability to respond to early-warning signals.

Ludwig's impending closing inspired multiple explanations; all were plausible but none suggested a strategy to challenge the closing. Ludwig's management assured DED that they were not abandoning Chicago. All indicators, however, pointed to a shutdown. The question was when it would occur. And could it be prevented?

One explanation emphasized corporation disinvestment. Research discovered an international network of angry unions and a European research group that monitored the disinvestment activities of North American Philips' plants. This conglomerate seemed to fit the model of the rapacious multinational that set unreasonable "hurdle" profit rates for its subsidiaries, moving them like pawns if they did not meet corporate expectations.

Other explanations of the Ludwig closing were: 1) Ludwig lost its quality-market niche due to American and Japanese competition and new electronic product lines, 2) the end of the baby boom resulted in decreasing demand for musical instruments, 3) federal and local cutbacks undermined school music programs, and 4) international competition made domestic production in high-wage areas unprofitable.

Hope for meaningful intervention in the Ludwig closing renewed when the WJN discovered an English drum manufacturer, Premiere, that had been formerly owned by Selmer. A management buyout had saved this firm from disinvestment. When Mayor Harold Washington visited London in winter, 1985, his Commissioner of Economic Development contacted Premiere management, who expressed interest in the U.S. market and a Chicago location. That spring DED visited Selmer to explore a possible buyout. Selmer's president shrugged off the possibility, claiming that Premiere was near bankruptcy. Enthusiasm to pursue a buyout diminished.

By fall, 1985, Ludwig was nearly closed. The Mayor's Office of Employment and Training conducted a prelayoff assistance workshop in cooperation with the union and counseled workers, while community organizations began to think about reusing the Ludwig facility for other job-generating uses. The WJN, however, was not able to put together a coalition of community, labor, and business groups to fight the closing. Organizations were overextended, and no organizing strategy seemed viable.

In October, 1985, a developer bought the Ludwig buildings to redevelop as multiple-tenant manufacturing and upscale service space, in part serving photography firms displaced from loft districts directly north of Chicago's Loop. This redevelopment plan prompted a community coalition to meet with developers about employment for displaced Ludwig workers.

Trinity Bronze

This westside manufacturer had originally only requested the city to move a light pole in August, 1984. Yet the firm was in serious trouble; its sales had drastically declined since the late 1970s, and it was losing money. The WJN learned that Trinity Bronze was on the verge of collapse at a September, 1984, meeting with the International Molders Union that represented the 50 Hispanic and Black workers at the plant. Management had failed to meet union health-fund obligations, and a severe cash-flow crisis had forced a temporary layoff.

It was later learned that the IRS was about to foreclose on Trinity Bronze, and that OSHA wanted to reinspect the plant. After learning about the WJN and methods to prevent plant closings, the workers, despite ethnic and union conflict, requested Trinity Bronze management to participate in a joint labor-management assessment of the firm's problems and potential for recovery. Workers returned to work in good faith. Management had little choice but to agree to the study; they contributed \$3,500 on behalf of owners and workers and DED contributed \$1,500. A taskforce made up of management, workers, WJN, and DED oversaw the study. The taskforce obtained valuable help from volunteer market, technology, management, and industrial health experts. Research examined past, current, and potential markets, finances, management, worker productivity, competitors, technology, and reorganization options such as worker ownership.

Poor management was at the heart of the company's problems. Management had not initiated new products nor had they sought new markets in response to concentration, economies of scale, and product diversification in the grave-marker market. They were blind to the fact that business as usual was not enough. In particular, management cut back drastically on their sales force and failed to meet the delivery and servicing needs of customers. Nevertheless, management stubbornly asserted that high wages and lack of worker productivity had brought this formerly successful company to ruin.

But the research also identified strategies to turn the company around—building upon its quality designs and skilled workforce. There were untapped parts of the cemetery business, and specialty plaques offered new product and market potential such as plaques for public monuments and historic buildings.

Unfortunately, although the owners agreed with the evaluation, they refused to pursue the report's recommendation for restructuring management. What started as labor-management problem-solving ended with the owners asserting their private perogative, no matter how incompetent. Workers sadly accepted what they knew too well and could not change.

The WJN continued to monitor the situation but decided that a fightback strategy was inappropriate. While not stopping this closing, the WJN retained 50 jobs for a year and cooperatively explored how to salvage a once-healthy company. Trinity Bronze limped on for six months, only to be sold in June, 1985, to a competitor, Williamsburg Bronze.

At first there was a show of corporate good will and talk of reinvestment. DED and the WJN met with Williamsburg management and offered their resources and services to help salvage Trinity Bronze. Good will, however, was shortlived. Williamsburg had purchased Trinity Bronze for its valuable grave-marker designs, not for its skilled workforce, equipment, or building. A final shutdown occurred in winter, 1986.

Playskool

The Playskool closing ignited a broad-based community campaign that ultimately pushed the City of Chicago into court. Hasbro Industries' announcement on September 19, 1984, that they were closing Playskool, laying off 750 mostly Hispanic workers and relocating to Massachusetts, stunned not only workers, city officials and community groups, but also the local business community. It was particularly controversial because Playskool's former owner, Milton Bradley, had received a $1-million Industrial Revenue Bond (IRB) from the city in 1980, pledging to retain 1,100 and create 400 additional new jobs.

In early 1984, an anonymous letter to Mayor Washington claimed that Playskool, a toy manufacturer and mainstay on Chicago's westside, would be closing. The Greater North Pulaski Development Corporation ultimately received this letter because Playskool's general manager sat on their board. When confronted with the closing rumor, he replied that the company would definitely be staying in Chicago, and that Playskool's sales had soared from $60 million in 1982 to at least $80 million in 1983 and an expected $125 million in 1984.

Hasbro offered several public rationales for the Playskool closing: corporate consolidation of Hasbro and Milton Bradley facilities, the Playskool plant's inefficiency and poor condition, and the need for cheaper labor and fewer work rules. In a more sensitive vein, the president of Hasbro Industries, Inc., Stephen Hassenfeld, stated that he had bought a shutdown decision that he did not agree with and had delayed the closing for six months. Hasbro knew that they might incur the wrath of the city and community, but as the Playskool general manager said, "We'll take our licks."

Playskool's closing created a dynamic of its own. During the fall of 1984, the WJN led a vigorous organizing and education campaign against Hasbro Industries and gained support from close to 100 community, business, religious, and labor groups. This effort built upon several years of community organizing that had questioned the real jobs payoffs from IRBs and other public investments.

A national toy boycott as the holiday season neared brought widespread local and national media attention, and Chicago newspapers editorialized against Hasbro's "breach of trust." There was not, however, the time or know-how to build substantial community involvement or rank-and-file participation of Playskool workers, many of whom felt estranged from the union.

The city had no established responses for Playskool, and was cautious while it examined it's options. At first there did not appear to be any reasonable recourse, until a city attorney, encouraged by public interest lawyers, came up with a legal angle that minimally would get the city and Hasbro to the negotiating table.

On December 4, 1984, Mayor Harold Washington, pressured by the community uproar, announced that the city had filed a lawsuit against Hasbro Industries for violation of its 1980 IRB agreement. The lawsuit was the first time that a municipality had taken a company to court for violating its IRB promises. Although the Playskool suit never came to trial, the publicity, its legal-ethical issues, and growing national attention forced Hasbro to negotiate an out-of-court settlement with the city.

The settlement reached on January 31, 1985, and described as a "comfortable compromise," contained limited victories for the city and Playskool workers: a commitment to keep 100 production jobs at the plant for another year, the company's "best efforts" to help the city identify a new user for the plant, the establishment of a $50,000 emergency fund for dislocated workers; and a Playskool Jobs Center to assist laid-off workers. Hasbro denied that the settlement was a "deal," but there was little doubt that it had resulted from public action.

The court suit and final negotiations with Hasbro Industries raised difficult questions about what social costs from this plant closing deserved most attention. Should better severance or job training and placement take center stage? Should the redevelopment of the Playskool site have precedence? These approaches were not mutually exclusive, although they were argued to be so during final bargaining.

By no means did the court settlement end the Playskool closing. Monitoring the performance of the Playskool Jobs Center by community organizations showed that few blue-collar workers obtained job placements, particularly female, Hispanic workers. Only 26 percent of these workers had obtained jobs by June 1, 1985, according to a community study. Hasbro gave out little performance data on the Jobs Center, and showed no leadership in coordinating their employment training and placement efforts with those of the city.

On the development side, no users came forward to purchase the entire Playskool facility. DED funded the Greater North Pulaski Development Corporation's marketing plan that recommended that Playskool be subdivided to house small manufacturers as in industrial parks or incubators. Following this recommendation, DED identified a major industrial developer who is now building a phased industrial park in conjunction with Greater North Pulaski. Hasbro Industries is donating the property and contributing a small portion of the development costs.

An unanswered question is whether early warning might have averted the Playskool shutdown or provided more time to leverage a better settlement. Prior knowledge, however, does not always produce action as we have seen. But there were signs.

Playskool's employment had dropped consistently since they received the industrial revenue bond. Workers reported that Milton Bradley executives were urged to rent rather than buy homes as early as 1981. Milton Bradley had overbuilt with the surge of the electronic game business in the 1970s, accumulating excess capacity in the 1980s, when the market dropped off. Equipment purchased with Chicago's industrial revenue bond actually made toy production less labor intensive. Finally, in February, 1984, the union was only able to negotiate a one-year contract, not three years as was expected.

The aftermath of the Playskool closing continues. In early 1986, Hasbro flew an ex-Playskool worker to Springfield, Massachusetts, to teach workers there how to operate the puzzle machine, formerly located at the Chicago plant and purchased with Chicago's industrial revenue bond.

THE THEORY AND PRACTICE OF EARLY WARNING

Chicago's early-warning demonstration was ambitious. It worked on multiple fronts that had been neglected or unexplored in the past. This was both its strength and weakness; building networks, diagnosing complex cases, working with city government, and mounting public campaigns stressed its fragile foundation. By the end of the demonstration, key early-warning actors were hardly on speaking terms. Yet much had been accomplished.

The early-warning demonstration supplemented a city industrial representative program in existence since the mid-1970s. Visiting over 6,000 Chicago manufacturing firms each year, this program involves face-to-face contact with management, and has produced referrals for financing and a host of infrastructure, safety, and labor force problems encountered by firms. This approach to industrial retention had three drawbacks: management reluctance to reveal investment-disinvestment plans to the city, no analysis of firm changes (i.e., employment, ownership, or technology), and the absence of a company tracking system. In response, DED has increased representative training, contracted with neighborhood councils to visit firms, and initiated an early-warning demonstration.

Compared to other early-warning approaches, Chicago's demonstration contained two novel components—labor and community information and creative interventions by community, labor, and government actors. The Chicago demonstration assumed that worker knowledge about plant investment-disinvestment decisionmaking was extensive, untapped, available, and more revealing in comparison to that of management sources. The role of a community-labor network (the West Side Jobs Network) was also critical to the Chicago demonstration; it would gather, interpret, amplify, and transmit early warning information to DED as well as promote plant-closing interventions.

Research

The West Side Jobs Network produced an industrial profile of the area—a labor-market analysis of where residents worked, a manual on "how to do" early-warning research, and a compendium of early-warning indicators. In the second year of the demonstration, 50 firms were selected for intensive early-warning contact on the basis of size, sectoral growth or decline, or prior early-warning indicators.

Early-warning researchers surveyed community organizations and unemployment offices, contacted unions and workers at particular firms, and visited stores and restaurants near plants to tap local information on plant activities. Extensive corporate and market research was conducted on four westside firms, including the three case study companies. Other research included studies of prominent plant closings and a comparative analysis of firms that had moved with those that had stayed.

Early warning must employ multiple research methods and tap varied sources of local intelligence to make up for the deficiencies of any one information source: early-warning information is fragmented, difficult to interpret, and rarely generalizable. In Chicago, workers proved knowledgeable about plant activities, but community groups often lacked knowledge about neighborhood firms. Nonunionized firms were particularly difficult to gather information about, and discussions with their workers at times raised the spectre of union organizing. Corporate research was provocative but often inconclusive.

West Side Jobs Network

The West Side Jobs Network assumed that locally-based groups—community, labor, and business—would be able to identify residents who worked at area plants, hear about layoffs and potential closings, and play a decisive role in responding to plant closings. Network information gathering and organizing were more difficult than expected.

Community residents did not typically work in neighborhood factories, thus minimizing local intelligence and a level of self interest necessary to sustain active involvement in the network. Coalition participation went up dramatically when neighborhood residents were employees, as in the Ludwig, Trinity Bronze, and Playskool cases. And from the outset, tensions between labor and business, workers and unions, and neighborhood groups and unions surfaced. In spite of these drawbacks, the West Side Jobs Network was able to mobilize an impressive number of organizations around the Playskool and Trinity Bronze closings.

City Involvement

Mayor Harold Washington has promoted an economic-development strategy based upon equity and openness and, in contrast to previous Chicago administrations, has made a commitment to retaining Chicago's existing manufacturing firms. His Department of Economic Development (DED) has been the primary vehicle to achieve these ends.

Early warning turned up cases that raised new expectations about the city's response system. DED's normal response to plant closings included referrals from industrial representatives, critical cases handled by industrial troubleshooters, and then an uncharted dimension of commissioner and city administration special responses.

DED typically could not help firms that suffered from poor management; they were told to come back when they got their acts together. DED also had no strategies to retain absentee-owned firms or to cope with corporate disinvestment. On the other hand, despite uneasiness about funding a community-labor coalition, DED sued to keep Playskool in Chicago and participated in several worker-buyout feasibility studies.

Early-Warning Information

Successful plant closing responses require that government, community, and labor coordinate their action. How open and sustained this cycle is depends upon local politics, the strength of community-labor coalitions, and the opportunities for responses.

Local economic planning requires both community-specific information about business activities and facilities and information about multinational business strategies, locations, and markets. Early-warning intelligence, therefore, requires several building blocks.

Business visitation by city, neighborhood, or business representatives provides information from the perspective of management. Some experts believe that business visitation can engage management in shared problem identification and solutions if proper training and targeting is done in advance. Utilities, media, banks, business organizations, competitors, industry analysts, realtors, accountants, and lawyers also turn up information on company investment and disinvestment. These "peer" networks should be nurtured.

Worker-union knowledge is a new building block that has not been tapped in many cities: workers hear rumors, see plant disinvestment, understand technology and production processes, have a sense of work quantity and quality, and negotiate contracts. All of these primary sources of new information should be interpreted with the use of published data on firms and economic sectors.

Plant Closing Diagnosis

Early-warning diagnosis is ongoing as new information provides a better understanding of the closing problem and the opportunities for response. Are firms closing, relocating, disinvesting, starting new product lines, expanding, or shrinking? Early-warning problems or opportunities, however, are defined differently by labor, the community, local government, and business.

Three frameworks clarify firm decisionmaking. Models of business transition, the first method, identify basic patterns of business change that result in job loss, relocation, or business failure: aging-owner liquidation, conglomerate disinvestment, technologically mature or emerging industries, sectors vulnerable to international competition, and industries experiencing various combinations of market growth or decline, investment, and work-process restructuring.

Each model suggests questions and responses. For example, does a family-owned firm have its business on the market? What is happening to the other plants of a multiplant corporation? How vulnerable is a firm's product line to international competition? What is the corporate strategy of the parent company? How does it differ from the strategies of its competitors?

The second method is early-warning indicators derived from case studies of plant shutdowns. Several general categories of indicators are: labor force changes (number, skills, layoffs, and overtime), ownership changes (merger, sale, takeover, receivership, and bankruptcy), and disinvestment (no maintenance, movement of machinery or product lines). Indicators include, for instance, the length of recently negotiated union contracts, whether relocated executives are renting or buying, or whether parts have arrived to reline a blast furnace.

Indicators themselves, however, do not tell the story; they must be interpreted in the context of the models of business transition. Job loss and a conglomerate merger suggest capital-shift disinvestment. Job loss, shrinking markets, and international competition may indicate a more complex and indeterminate scenario.

The third method narrows the focus. Given limited resources, government agencies may target their monitoring to the largest firms, absentee-owned firms, or to firms in growing or lagging sectors. Unionized firms or those receiving direct public incentives (e.g., IRBs, UDAGs, tax abatements, or defense contracts) may also be targeted because of their contractual and reporting obligations and because of worker knowledge about firm decisionmaking.

A final targeting approach focuses visitation or responses to specific economic sectors, for example, to small electroplaters that are affected by waste-removal regulations, or to printers that are experiencing rapid technological change or displacement from inner-city loft districts.

Images of possible plant-closing response raise questions that should also guide diagnoses and information gathering. Does a company use public incentives? Did it commit to job creation-retention goals? Does the company have other plants? Is the company shifting production from a minority labor area? Is there a pattern of closings? Who is on their board of directors? Are directors tied to other local companies or banks? Who owns stock in the corporation? Answering these questions not only clarifies the dynamics of a specific closing, but also suggests ways to affect closing outcomes.

Plant Closing Response

The response of local or state governments once they obtain an early warning of a plant closing has not received enough attention. In general, jurisdictions only make their normal marketing and financial incentives available, often in overabundant packages; other public agencies have interdepartmental "swat" teams, negotiate with firms, apply corporate peer pressure, or in a few cases invoke regulations, court action, or eminent-domain powers. In other cases, however, local and state governments may help firms extract labor or community concessions.

To assume that jurisdictions can effectively react to early-warning signals may be wrong, particularly when a "crisis" has not yet been precipitated. As with other types of economic targeting, early warning raises expectations about government intervention that often exceed municipal capacities and underestimate the bureaucratic frictions that may inhibit aggressive action when a plant closing occurs. But community-labor coalitions can expand the accepted range of plant closing responses if they can mobilize public sentiment and resources, thus lowering the risks to government of innovative action and increasing the risks of non-action.

Both routine and non-routine responses have been used by local governments, community-labor coalitions, and businesses once a plant closing has been diagnosed. Routine responses are those in which the plant closing is not challenged; they include presentation of financial incentives, employment training and referrals, and real estate listings for empty buildings.

Non-routine responses challenge the closing and attempt to stop it or obtain ownership. There are three non-routine, innovative responses to plant closures: fightbacks, buyouts, and new development. Fightbacks include community-labor campaigns, boycotts, legal suits, or the hard negotiating of city and state governments, in an attempt to stop a closing by making the costs of leaving unacceptably high. Buyouts include management buyouts, new entrepreneurs, employee stock ownership plans (ESOPs), worker ownership, or non-profit or public ownership. Finally, new development replaces the lost economic activity with industrial parks, incubators, or commercial/institutional uses, and, in the best of cases, provides jobs or business opportunities for dislocated workers.

Another dimension of non-routine plant closing responses goes beyond discrete cases and attempts to build local institutions that can facilitate strategic responses. In Pittsburgh, municipalities have joined with labor and the community as the Mon Valley Steel Authority and hope to raise sufficient financing so that they can acquire idle factories using eminent domain. In Jamestown, New York, a labor/management committee was established in the late 1970s to coordinate reinvestment in its local manufacturing base, aiding worker and management buyouts. Under Mayor Washington, Chicago established a Task Force on Steel and Southeast Chicago that has developed a long-term strategy for revitalizing steel while responding to several steel closings.

THE LESSONS LEARNED

Chicago's early-warning demonstration explored strategic information gathering and plant-closing responses. Generating early-warning information proved easier than doing something about closings. Routine responses were often inadequate. Fightbacks, buyouts, and new development were long shots. But two years of experimentation has better positioned the city, labor, and community organizations to pursue more timely and viable strategies in the future.

A number of lessons emerge from the Chicago demonstration. First, timing is critical; a response mode is needed before a crisis has hit to develop a plan, monitor change, and build union and community constituencies. Early warning can make a difference. Second, early warning requires diverse information sources such as workers and unions. Early warning must be ongoing and guided by models of business transition. Third, each plant closing is distinctive and requires responses appropriate to the firm and the type of closing. Fourth, plant-closing responses are limited because business owners have established rights and resources regarding their firms' location while the rights of the public are ambiguous.

Playskool's shuttered buildings remain. Given the current U.S. political environment, strategic responses can only partially mitigate the devastating social costs of plant closings, occasionally facilitating a buyout. Surely this is better than passive acceptance of supposed market outcomes. But more responsible public-private partnerships are needed if a dynamic economy is also to be an equitable one.

40 *The New Opening for a Full Employment Movement*

SHEILA COLLINS AND FRANK RIESSMAN

In the last decade the demand for full employment has not been fashionable. Job-creating programs like CETA were discredited by policymakers, despite many positive achievements. Yet a new opening to build a full employment movement may be on the political horizon. There are at least five reasons for our optimism, stemming from structural changes in the international economy not fully recognized or experienced until recently.

First, in the last five years, millions of blue-collar workers and family farmers have lost their jobs due to the decline in manufacturing, plant closings, disinvestment, and federal farm policy, creating both regional and sectoral devastation. Added to the unemployment rolls are not only those directly affected by plant closings and farm foreclosures, but those holding jobs in the ancillary businesses that support major industries like steel, auto, and agriculture. Most of the laid off who do find jobs are working for lower wages and benefits in sectors that provide no long-term security. Those who remain employed are far less secure in their jobs. Added to this is the growing fear of technological unemployment in an automated age.

Second is the increased participation of women in the labor force and their concentration in the secondary labor market in low-paying service jobs.

Source: Sheila Collins and Frank Riessman, "The New Opening for a Full Employment Movement," *Social Policy* 17 (Winter 1987), pp. 2–4. Published by Social Policy Corporation, New York, New York 10036.

Women are expected to account for about .63 percent of the new entrants into the labor market between 1985 and 2000.

Third is the extremely high unemployment rate that persists even in a period of economic recovery. Although the wild swings in unemployment that characterized earlier periods are gone, there appears to have been a steady rise in official unemployment rates since 1970, so that periodic drops in the unemployment rate reach nowhere near the levels they did in the 1950s and 1960s, suggesting endemic unemployment. Moreover, since 1982, there has been a doubling of the ranks of workers so discouraged about job prospects that they have dropped out of the labor market altogether.

Fourth is the growth of involuntary, part-time, and temporary employment and the contracting-out of an increasing proportion of business services.

Fifth is the Roman Catholic Bishops' Pastoral Letter on the U.S. economy, which is both a moral/pastoral recognition of the human suffering caused by these structural changes and a call to do something to ameliorate it. The Bishops' letter calls current levels of U.S. poverty a "social and moral scandal," declares that people have "economic rights," as well as civil and political rights, and call for "more far-reaching reforms that go beyond the present system."

These reasons constitute an objective base for building a new movement for full employment. But for a full employment program to have resonance today, it must be responsive to the vastly different economic and social environment than existed in 1946 when the first full employment bill was adopted. *Equally important,* is the need to address a perennially ignored problem: namely, the minority unemployment and wage gap that has persisted irrespective of the national employment level.

New full employment initiatives must take account of the following:

1. The decline of large-scale manufacturing, the inability of labor unions to prepare for, prevent, or adjust to plant closings and disinvestment, and declining rates of unionization resulting in a greatly weakened labor movement.
2. The increase of two (or more) wage-earner families in order to maintain a standard of living formerly maintained by one wage-earner, along with the reluctance of the private sector to incur day-care costs.
3. The growth of relatively low-paying, low-skill, non-unionized jobs in the service and high-tech industries.
4. Large budget and trade deficits together with an enormous consumer debt.
5. A shrinking middle class—both in terms of real income and as a percentage of the population.
6. The low rate of productivity increase and the economic stagnation of the past decade.
7. The internationalization of the workforce.
8. The possibility of arms reductions and the fear of its economic consequences.
9. The widespread use of immigrant labor at sub-minimum wages and the resulting tensions between native and immigrant workers.
10. The demand for jobs among the newly politicized disabled.

For a new approach to become effective, however, it has to overcome some of the basic difficulties experienced by previous full employment initiatives, particularly those of the 1970s, which were focused around the Humphrey-Hawkins Balanced Growth and Full Employment Bill. Two problems from this period stand out.

First, was the overreliance on public service employment as the *direct* employer of last resort. In the 1970s, CETA became the model, with an enlarged CETA as the vision of full employment. The CETA program has been criticized as a channel for political patronage, as a "make-work" program, and as providing no long-term security or upward mobility for its employees.

Second, the unemployed themselves were not organized by the full employment movement, even though they were its natural constituency. Moreover, broadbased grass-roots activist support was neither sought nor generated. Rather, the support came from the leadership of large organizations, particularly labor, with a wide range of other groups providing only paper support.

NEW INITIATIVES

In dealing with these two difficulties, Europe may offer some useful models. In France, for example, voluntary associations and private firms hire unemployed youth via government subsidies. It is estimated that 45 percent of the unemployed in France are employed through such *indirect* government-aided mechanisms. Another approach would be to provide *job vouchers* to the unemployed. In the United Kingdom unemployed workers pool their prospective unemployment benefits as investment capital to start new enterprises, thus, in effect, increasing the demand for labor. There are a host of similar examples, all of which avoid direct government involvement in job creation, thus sidestepping the current anti-bureaucratic, anti-big government atmosphere and providing more flexibility and decisionmaking for the workers and unemployed themselves.

With regard to organizing the unemployed and underemployed, there are several approaches that are currently underway or have the potential of succeeding if given the proper support and encouragement.

In many ares of the country, local councils of the unemployed have been formed to provide mutual aid and support, information on job opportunities, development of new employment strategies, and advocacy for new employment policies at the local and national levels. Many of these groups are brought together through the National Unemployed Network.

A regional approach to the unemployment crisis is illustrated by the Tri-State Conference on Steel, which operates in Pennsylvania, Ohio, and West Virginia. Tri-State brings together entire communities that have been affected by plant closings—unemployed workers and their families, the religious community, small businesses, local and state elected officials—to build a movement that gives priority to social impact and the public interest in plant-closing decisions, over private profitability. Another of Tri-State's goals is to democratize the entire area of economic development and, in the process, to create new popular institutions that connect the public interest to economic issues.

One of Tri-State's innovations is the Steel Valley Authority, a public entity invested with the power of eminent domain. Composed of representatives of the councils of mill towns, who previously had no say over the economic decisions that affected their lives, the authority can take over a mill that is in the process of leaving the area and, with sufficient capital, turn it into a worker-community-owned enterprise, thus saving jobs and revitalizing the community. Tri-State is now in the process of creating a capital fund for the purchase of plants about to be abandoned.

Different approaches, of course, need to be elaborated for different groups of the unemployed and underemployed. We need to distinguish between those who need retraining and those who need training in the first place, between the long-term unemployed and the underemployed youth entering the workforce.

Approaches that deal effectively with the vast number of unemployed and underemployed women who have been in and out of the workforce require an intensive day-care program and related family support services. Women know best what kinds of services and support they need and ought to be directly involved in the planing of such new initiatives.

There is a need to involve groups not directly affected by unemployment in discussions and education on the effects of unemployment on every other aspect of our social, economic, and political life, so that those with jobs will not be pitted against those without jobs. A full employment strategy has economic benefits for *all* taxpayers. The addition of jobs in the economy, for instance, increases the taxes collected and reduces unemployment insurance costs.

A moral-cultural ethos similar to that created around the homeless and foreclosed farmers needs to be developed for the unemployed. A beginning has already been made with the bishops' Pastoral Letter, but much more needs to be done to utilize the creativity and commitment of cultural workers and the media. Perhaps a call for a national Economic Bill of Rights would stir the imaginations of people if linked to the upcoming bicentennial celebration of the U.S. Constitution and its core assumption of the "inalienable right to life, liberty, and the pursuit of happiness."

The Roman Catholic bishops of Maryland have inaugurated a comprehensive statewide program to apply the findings of the Pastoral Letter to every sector of the population in the state. Starting with individuals in parishes, they call on them to reexamine their priorities in terms of life-style, work, investment, and consumption habits. They also address the state legislature, calling upon it to revise its regressive taxation policies, to invest in new job creation, to step up its investment in housing, and to work for increased federal benefits for the poor. The bishops also direct specific proposals to the business community, in particular, calling upon the private sector to join in a new effort to improve housing for the poor. Such a model, based on concrete conditions in a particular state, could become an ecumenical project.

Finally, an authentic full employment approach suited to the 80s and 90s needs to make very clear that the work provided produces *quality products and services,* at *decent wages,* with built-in guarantees for *income security.*

41 *Minorities and the Labor Market: Twenty Years of Misguided Policy*

RICHARD MCGAHEY AND JOHN JEFFRIES

Industrial policy advocates have said very little about the particular economic problems of Black Americans. This failure to discuss Black problems in more depth is surprising, as Blacks suffer disproportionately from America's sustained economic problems and are concentrated in many of the industries that have declined during recent years. Although some of this silence can be attributed to a general rhetoric about the widespread nature of current economic problems, it also reflects the unconscious influence of the mismatch idea.

Advocates of industrial policy stress retraining and education policy in the context of macroeconomic growth, in terms rather consistent with U.S. manpower policies of the past 30 years. We suspect that many industrial policy advocates would address Black unemployment problems with education and training policies based on a mismatch explanation for Black unemployment, an approach that we question.*

INDUSTRIAL POLICY, EDUCATION, AND TRAINING

Industrial policy programs for Black employment are conditioned by the overall view of American economic and social institutions. Reich (1983) attempts to square a traditionally American circle by claiming that social spending can foster

* The mismatch hypothesis asserts that individuals are out of work because they lack skills or education.

Source: Richard McGahey and John Jeffries, "Minorities and the Labor Market: Twenty Years of Misguided Policy," *Social Policy* 17 (Winter 1987), pp. 5–11.

economic growth, rather than see social programs as a deduction from national wealth, as something we can afford only if economic growth is first ensured. He argues that the new world economy will be based on more flexible production and more mobile and better educated labor.

Government programs to create this flexible "human capital" will, in fact, be creating the conditions of new economic growth. Workers in changing firms can be given more training and adapted to new jobs; welfare should be replaced by grants to businesses that hire the chronically unemployed; traditional training programs and public-service employment should be replaced by vouchers that the unemployed can use to get on-the-job training at private firms. The economists to the left call for more use of affirmative action programs and community control of resources (Bluestone and Harrison 1982, endorse producers' cooperatives at the local and state level), but even their attention to minority workers is subsumed under the general problem of industrial and regional adjustment.

There are three areas in which we believe industrial policy falls short of addressing Black employment problems: the nature of firms where Blacks work, the role of schools and training in the labor market, and shifts in the composition and number of available jobs. In all three areas, we see a grimmer picture than many other policy analysts, and this picture makes us question much of industrial policy's current relevance to Black employment problems.

Firm Size, Job Ladders, and Black Employment

Most industrial policy writing depicts a world of potential cooperation between business, labor, and government leaders, were large, politically powerful firms with unionized work forces, capital intensive production, and extensive layers of managerial personnel are battered by international competition. But a large part of the labor market for Blacks, especially in inner cities, is made up of rather different firms and job experiences. Some economists have described a "dual" or "segmented" labor market, where distinct types of firms offer distinct types of jobs:

> The primary market offers jobs which possess several of the following traits: high wages, good working conditions, employment stability and job security, equity and due process in the administration of work rules, and chances for advancement. The other, or secondary sector, has jobs which, relative to those in the primary sector, are decidedly less attractive. They tend to involve low wages, poor working conditions, considerable variability in employment, harsh and arbitrary discipline, and little opportunity to advance. The poor are confined to the secondary labor market. (Piore 1977, p. 93–94.)

This picture of a dual market captures an important facet of the American labor market, one to which most industrial policy writers pay little attention. Indeed, much job growth in the past decade has been precisely in firms and industries that are low-skilled and low-paying, with erratic employment opportunities (Jeffries and McGahey, forthcoming).

When Reich calls for firms to administer new programs, or Rohatyn endorses linking schools to jobs and private firms, or Bluestone and Harrison want to restrain capital mobility, they all seem to have in mind the primary sector. That is, industrial policy advocates are focusing on large firms with internal mobility through job ladders, often with unionized work forces.

They do not seem to be thinking about the small, secondary firms that provide a good deal of inner-city employment. Training and skills are not very important for workers in these firms, as they are often labor-intensive service or manufacturing firms with few skilled jobs. Benefits are marginal or nonexistent, and it is hard to imagine these secondary firms helping their displaced workers or hiring new ones from the ranks of the chronically unemployed. Perhaps industrial policy advocates hope economic growth will eventually aid these firms and workers, but this would sound disturbingly like the supply-side economist' "trickle-down" theory.*

The School-to-Work Transition

Rohatyn states, "We have all heard that high technology creates higher requirements for skills and few requirements for labor." He endorses adapting training programs and "even our education system to what we need," which might include "inner-city schools that will be, more closely connected with job opportunities" (1983, pp. 46–49).

The concern about the school-to-work link is nothing new in American policy. School-to-work transition problems are periodically recycled as a major explanation for youth employment problems. This usually leads to calls for more direct ties between jobs and school in the form of more involvement by private firms in curriculum design, ties between apprenticeship programs and schooling, or cooperative ventures in which part of the school day is spent at a private-sector job.

Osterman, however, argued that "much of the policy focus on the 'school-to-work transition' is misguided. . . . The ineffectiveness of schools is not a recent phenomenon . . . and the essential characteristics of the youth labor market make this inevitable" (1980, p. 154). Youth are not ready to stabilize in the labor market, and firms that provide long-term job opportunities and stability of employment do not hire them.

A recent variation of the school-to-work issue has been a refocus on improving basic skills of high-school graduates. Dramatic figures about functional illiteracy are cited along with employer complaints about the low level of

*Bowles et al. do call for raising minimum wages to force modernization in secondary firms, driving some of the least productive out of business (1983a, pp. 280–81).

skills that new graduates possess. The Center for Public Resources, in calling for a program to improve basic skill acquisition in school, cited evidence that 44 percent of Black youth leaving high school at age 18 cannot read above the fourth grade level, that scores on the Scholastic Aptitude Tests have been declining since the 1970s, and that the problem is compounded by the high number of school dropouts (1982, p. 2). Much Black youth unemployment has been blamed on inadequate schooling. A U. S. General Accounting Office Study (1982) summed up this view in its title, "Labor Market Problems of Teenagers Result Largely from Doing Poorly in School."

The level of attendance at public schools in general, and minority schools in particular, is also a serious problem. The New York City public school system, for example, admits to a 45 percent dropout rate between ninth grade and graduation; Calitri (1983) claims that the overall rate is even higher (68 percent), with a Black dropout rate of 72 percent.

There is a natural potential tie between the stress on schools and industrial policy's emphasis on productivity and training. Industrial policy advocates concentrate on restoring productivity which they feel can be accomplished more easily with a better educated work force. The emphasis on declining education and skills is often coupled with an assertion that job requirements have risen dramatically; it is this "mismatch" between low skills and education and high demands of jobs that is seen to foster much Black unemployment.

Although high levels of functional illiteracy and substantial school dropout are cause for serious concern, we question whether the mismatch analysis is an adequate explanation of Black employment problems. The jobs that have grown the fastest in the U.S. economy are not highly skilled jobs requiring much training. According to a recent Bureau of Labor Statistics forecast, there will be 19 million or so new jobs by 1990, 3.5 million of which will be professional and technical, a category that includes some menial low-skilled work but generally higher-paying jobs. Low-wage service and clerical work will account for almost 7 million new jobs.

Many analysts (for example, Kuttner 1983; McCahey 1983) have pointed out that most recent and projected job growth in the United States is concentrated in lower-skilled, lower-paying jobs, rather than technically demanding jobs with higher pay and higher skills. Rumberger found some evidence that overall skill requirements did increase between 1960 and 1976 "but at a slower rate than earlier periods" (1981, p. 578). In New York City, a leader in the transition from manufacturing to white-collar service work, the real skill demands of jobs have not changed in the last decade (Sullivan 1982).

The problem remains, as many analysts point out, that there are simply not enough employment opportunities in urban areas. Human capital policy does not seem an adequate solution for this problem; although education and training are consistently associated with higher earnings, they are not consistent predictors of whether someone will find a job in the first place. And human capital programs by themselves do not create jobs.

The Changing Labor Market

The growth in the number of low-skilled jobs is a central issue for many industrial policy analysts. One solution envisioned is the growth in the high-tech sector. "High-tech" refers to both the production of highly sophisticated commodities and the use of highly sophisticated production techniques such as robots, computer-aided design (CAD), computer-aided manufacture (CAM), and the integration of all of these elements into flexible manufacturing systems (FMS). But high-tech enterprises do not necessarily require highly skilled labor. And since these processes are highly capital intensive, they tend to reduce the overall number of workers needed, while retraining smaller numbers of highly skilled workers.

The new labor demand generated by high-tech enterprises does not seem likely to take up the slack of workers displaced, much less provide substantial numbers of new jobs for the underemployed. Forecasts by the Bureau of Labor Statistics and by Data Resources, Inc. (DRI) show that in the next 10 years, high-tech jobs will replace less than half of the 2 millions jobs lost in manufacturing in the last 3 years. And most jobs in high-tech industries will not be highly skilled. DRI estimates that fewer than one-third of the jobs will be for engineers and technicians; the remainder will be in the traditional occupations (*Business Week* March 28, 1983).

Economist Ann Markusen has concluded that "dramatic employment growth" in high-tech "may be shortlived" and that the jobs created will not be stable blue-collar production work (1984, p. 8). Even traditional jobs are under the threat of displacement from high-tech. To cite one example, since 1973 DuPont has doubled its output capacity for synthetic fibers, but owing to mechanization and robotics the number of production workers required has risen only 4 or 5 percent (*New York Times* May 16, 1983).

This poses a dilemma for some advocates of industrial policy. Innovations such as those at DuPont dramatically raise productivity by integrating more sophisticated production technology, but they cut the number of jobs required in production in equally dramatic fashion. Unless the economy can boom in some unforeseen way, it is hard to see how enough labor demand will be generated to reemploy displaced industrial workers, much less hire those who are chronically unemployed.

The U.S. Department of Labor (November 30, 1984) has found that of 5.1 million workers who lost their jobs because of economic dislocation between 1979 and 1984, only 60 percent were reemployed when surveyed in 1984; over 40 percent of those with jobs in 1984 were earning less than they had in 1979. (Reich's stress on faster technical innovation and shifts to more specialized production, which requires fewer workers per unit of output, seems to ignore the problem of aggregate job growth.) The focus on generating aggregate economic growth through restoring productivity is thus problematic; the numbers and types of jobs that are likely to be generated will not meet the employment needs of many distressed Americans, especially minorities.

FUTURE POLICY OPTIONS

Several employment and training policy options emerge from the industrial policy debate. In this section, we comment on their implications for minorities and on the role of more traditional policies, such as affirmative action.

Retraining Policy and Displaced Workers

One of the central themes of industrial policy involves the retraining of workers displaced by economic changes. This policy is also advocated in different form through the Reagan Administration's Job Training Partnership Act, which has a separate title for displaced workers.

Retraining implies that someone had a previous job; it does nothing for the long-term unemployed. Further, most retraining models concentrate on large primary firms and often involve union cooperation. Minority workers are at risk in these firms, as they often have less seniority than white workers—and retraining along will not create new jobs. Industrial layoffs have hit minorities in certain firms and industries especially hard. When Atari, Inc., laid off 1,700 production workers while moving production to Asian plants, union organizers claimed that Latinos and Asians made up almost 90 percent of those laid off. According to the United Auto Workers, layoffs in the auto industry alone accounted for a 1 percent drop in the national average income for Blacks (*Los Angeles Times* June 2, 1983).

Minorities in the secondary labor market face even bigger obstacles in obtaining meaningful training and job placement. Small firms are even less likely than large ones to administer training programs or find new jobs for redundant workers. The lack of unionization for many workers in small firms means that there are no institutions to coordinate retraining and transitional services.

A final concern is that retraining funds may be taken up by more politically powerful primary workers displaced from major unionized industries. Indeed, some analysts endorse this as an explicit goal. Harvard economist D. Quinn Mills, in a 1981 conference on industrial policy, stated:

> It is time to loosen the connection between training and job placement activities and the disadvantaged, and to redirect public employment and training efforts to a broader segment of the population. . . . Funds are scarce for large-scale public programs to assist the unemployed. Should there be a reallocation of

> effort from the disadvantaged to the industrially displaced? The answer is yes. Training and relocation programs have done only limited good for the disadvantaged. They promise a far higher payoff with the industrially displaced (1983, pp. 276–77).

Mills's endorsement of shifting resources from the disadvantaged is not explicitly adopted by any advocate of industrial policy. But the stress on increasing interfirm and geographic mobility of workers through retraining programs and on administering these programs through firms and unions runs the danger of leaving many Blacks out of retraining programs. Reich stresses that the costs of adjustment should not fall too heavily on Blacks, but the policies that he has advocated thus far do not directly address this problem.

Our focus here is on a subset of unionized workers—production workers in blue-collar primary industries. Although Blacks are unionized in many sectors, including public employment, health care, and some secondary labor-market sectors, the political power of nonindustrial unions in fighting for retraining funds is relatively weak when compared to unionized workers who have primary jobs in sectors like steel, automobiles, and construction.

Retraining, in any case, may be of little value if the real problem is too few jobs. Retraining policy extends the mismatch analysis to workers who have been steadily employed,; as they cannot be said to have bad work habits or attitudes, the analysis concentrates on their obsolete skills. But displaced workers are often removed from their jobs by the very factor called for in industrial policy: increased productivity through technical innovations in production and capital substitution. Mills notes that such "reindustrialization" may only slow the loss of manufacturing work without adding any new jobs (1983, p. 264).

A more jaundiced view of retraining is that workers will simply be removed from the labor market for some period of time, while participating in programs. In this view, retraining programs will perform one of the latent functions of all schooling—keeping people out of the job market. Retraining may also be coupled with receipt of transfer payments.

Some analyses suggest that the real barrier to displaced workers' finding new jobs is the workers' reluctance to accept work that pays substantially less than their previous jobs (*Business Week* September 19, 1983). Many retraining programs feature group meetings for displaced workers, parts of which consist of "reality therapy"—getting workers to accept that their futures have changed—and then teaching them how to obtain jobs, fill out resumes, make calls on employers, and so on. Ironically, these methods, now extended to prime-age industrial workers, are the same as those directed at chronically unemployed minorities in the 1970s.

David Swinton has criticized the retraining thrust of current policy, claiming that the real problem is that "there are not enough jobs to fully employ those people who want to work. By training people in the absence of jobs, we will train people who remain unemployed, or the trained will displace the untrained" (*Los Angeles Times* June 2, 1983).

In the face of the weak projections for new jobs in technically sophisticated work and the possibilities of technological displacement in manufacturing and services, advocates of retraining must address the question of where new jobs will come from. Will the microeconomiic effects of productivity and management changes produce new jobs? In what industries, occupations, and regions of the country? It may be entirely possible to get rises in productivity without substantial overall gains in output and without creating new employment. In this situation, what will displaced workers do? And what is the role of retraining policy for those in the secondary labor market, especially those who are chronically unemployed? Mills's frank endorsement of shifting resources away from the disadvantaged to displaced industrial workers raises a grim vision of competition among different groups of the unemployed for increasingly limited government funds.

Infrastructure Rebuilding and Public-Service Employment

Some industrial policy programs, notably those proposed by Rohatyn (1983), involve substantial commitments to rebuild the infrastructure in cities, especially in declining older urban areas. This may hold some promise for Black employment, but its possibilities would be enhanced by expanding government hiring. The public sector accounted for much improvement in Black employment during the 1970s. Rumberger estimates that, in 1980, 35 percent of Black males and 45 percent of Black females owed their jobs either directly or indirectly to government spending; 54 percent of Black male college graduates and fully 72 percent of Black female graduates had jobs attributable to federal spending (1983, Table 5).

And Black college-educated professionals were not the only ones to benefit from an expanding public sector; short-term public-service employment also aided less-educated Blacks. Ellwood and Wise found that the federal Summer Youth Employment Program (SYEP) provided almost 45 percent of all summer jobs for minority youth aged 14–19 in July 1978; three percent of white jobs were due to the program. Fifty percent of white youths had private-sector non-SYEP jobs, while only 18 percent of minority youth did (1983, pp. 56–62).

Infrastructure programs may not provide extraordinary assistance to Black men, since Blacks hold less skilled and less high-paying jobs than whites in sectors such as construction, but they are still important. During 1981, Blacks and other minorities represented 8.1 percent of all construction workers and 12.8 percent of transportation and public utility workers, while making up 11.6 percent of the employed work force (U.S. Department of Labor March 1982).

For public employment to improve the jobs of females, expanded social service employment would be required, a policy not mentioned by most industrial policy advocates. Expanded social services are a double bonus for women, as women both work in and are the major recipients of many services such as day care and expanded education. The emphasis on better schooling thus may provide opportunities for Black college-educated women as teachers and for others as school support personnel.

Public-service employment has been roundly criticized by many, but it remains an important source of employment for Blacks and other minorities. Males would stand to benefit from expanded infrastructure repair and a vigorous affirmative action program in hiring and contracting. Females would not benefit from infrastructure construction to any great extent, although a program of repair and maintenance of public facilities might provide some jobs for them. But expanded day care, health programs, education, and other such services would have double advantages for females, especially minorities. So increasing infrastructure repair and maintenance social services in industrial policy programs would be a strong advantage for Blacks.

Affirmative Action and Direct Government Intervention

Policies to aid Blacks cannot ignore affirmative action and other forms of direct governmental intervention. Evidence suggests that affirmative action in hiring and in contracting has had positive effects on Black employment. Economist Jonathan Leonard (1984a, 1984b) has concluded that federal affirmative action programs increased the employment of minorities and also significantly raised their wages, especially when the programs were vigorously enforced and combined with skill training.

Federal antidiscrimination activity is found by some researchers to be the most likely explanation for improvement in the Black-to-white earnings ratio for some segments of the population (e.g., college-educated young) since 1964 (Brown 1981), and other analyses (summarized by Freeman) show that federal contractors, especially those facing compliance reviews, have larger increases in Black employment than other firms (1982, p. 5).

Further, this sort of protection seems necessary to protect the gains Blacks have made, even in government jobs. Although Black employment gains were substantial in an expanding public sector, the current shrinkage of government is caused rapid reversals of those gains. Federal agencies dismissed minorities at a rate 50 percent greater than that for whites in 1981; the number of layoffs doubled in 1982, hitting upper-level Black professionals hardest (*New York Times* September 17, 1983).

Of course, governmental programs might also have negative impacts. The minimum wage and welfare policies, in particular, have been examined for their negative effects on Black employment. However, economic research has found that such negative effects tend to be "moderate or modest," not overwhelming (Freeman 1982, pp. 2–3).

In recent years, conservative critics have launched a broader attack on welfare programs. Analysts such as Walter Williams, Thomas Sowell, and Charles Murray have argued that American welfare programs not only fail to assist Blacks and other economically distressed people, but also make the distressed worse off by perpetuating a culture of dependency. Murray, in *Losing Ground* (1984), advocated the elimination of virtually all federal means-tested social programs—food stamps, AFDC, and the like—arguing that the rise in these programs coincided with worsening conditions for Blacks and that the programs both fostered dependency and stifled economic growth. He stated that economic growth would aid poor Blacks more than any targeted social programs would.

In fact, critics on the left have themselves attacked poverty programs for fostering dependency (Piven and Cloward 1971). The peculiar configuration of the American welfare state has kept most transfer payment programs, especially Great Society initiatives, separate from economic development goals (Skocpol and Ikenberry 1983). But it seems a leap to conclude that poverty programs cause continued economic distress for Blacks. Proponents of such programs conclude just the opposite; if Black conditions have become worse, then more funding is required.

We are not persuaded that welfare state programs are a significant cause of America's economic troubles. Many welfare state programs in the United States should, however, be redesigned with more emphasis on economic growth, social equity, and community economic development (Jeffries and McGahey, forthcoming). As Robert Kuttner (1984) has pointed out, in the 1970s many European countries had both faster economic growth and greater social equality than the United States.

While Europe's economic growth in the 1970s is perhaps less compelling when we compare recent U.S. economic growth to current stagnation in Europe, our recent economic growth has gone hand-in-hand with growing social inequality and historically high levels of unemployment. Although economic growth is necessary for improving the lives of economically distressed Americans, it does not seem to be a sufficient reason for tolerating social inequality and high levels of Black unemployment.

But welfare programs—particularly income maintenance—have little or no political support, except in the form of unemployment insurance or Social Security, programs not usually perceived as transfer programs. Americans traditionally dislike income redistribution schemes, preferring to pin their hopes for social mobility on education (DeLeone 1979). For this reason, any expansion of income maintenance will have more support if it is tied to work or training programs, even if those programs (like many of their predecessors) do not substantially raise the labor-market returns of their participants.

Minority Participation in Industrial Policy

Thus far, industrial policy advocates have said very little about the particular problems of Black and minority workers. Most express laudable sentiments for equalizing the burdens of economic adjustment, but the major stress in industrial policy remains on restoring economic growth through increasing productivity. They encourage growth through a variety of methods, including better foreign trade policy, increased retraining of laid-off workers, closer ties between schools and training, on the one hand, and the needs of the private labor market, on the other hand, and encouragement of economic revival in central cities through a combination of incentives to private firms and construction of public works.

What strikes us about this list is not its novelty, but its familiarity. Economic policies aimed at disadvantaged workers throughout the 1960s and 1970s had exactly the same kinds of things on their agendas (although with somewhat less attention to foreign trade issues), from the earliest days of manpower policy through the War on Poverty in the 1960s, through CETA in the 1970s, and up to the latest program of the Reagan Administration.

The programs now endorsed by industrial policy proponents sound rather similar. The one goal most of them back away from, with the exception of some critics on the left, is direct income stabilization policy. Most prefer to link income stabilization to participation in the labor market or training in more extreme cases, some endorse diverting resources from the traditionally disadvantaged to the recently displaced.

It may be that human capital policies will, in fact, spur new economic growth and somehow provide enough new jobs to absorb both the disadvantaged and the displaced. But we are not optimistic about this possibility. Service and high-tech enterprises so far seem to have widened earnings gaps, not narrowed them. Many of the new jobs anticipated in the next decade will require few skills, and there may be increased competition for those jobs.

Certainly better schooling, especially in math and language basics, is a desirable social goal in itself. The goal of preparing workers to be more flexible in the labor market is in fact, more likely to be enhanced by general educational development than by occupation-specific or employer-based training. But poor economic performance for many individuals, especially minorities, is unlikely to be addressed adequately by human capital policies alone. If the scenario of raised productivity with low employment and earnings growth is played out, other sorts of policies may have to be considered. These might include a more active and permanent role for public-service employment, national wage and price policies to restrain inflation, reorganization of work with earlier retirement and work sharing, a greater role for communities in directing economic activity through such programs as producer's cooperatives in cities, and the like.

We are concerned that, once again, the employment problems of minorities (and increasing numbers of white Americans) are being defined largely in terms of a mismatch between their inadequate skills and education and the allegedly high skills demanded by a changing labor-market structure. And we fear that liberals and conservatives alike have assumed that the distinctive employment problems of minority workers will be solved by general economic growth. Growth alone did not solve the Black employment problems of the sixties and seventies (Harris 1983), and there is now depressing evidence that the Black economic situation is deteriorating even as the U.S. economy grows.

Restoring growth, especially through productivity gains driven by capital substitution, may not in and of itself provide much help for Blacks in the eighties and future decades. Adjustment policies certainly will be required in the future, but we foresee a continuing need for affirmative action programs, public employment, and income redistribution policies. It is manifestly harder to endorse such policies at a time when overall economic prospects remain so uncertain; like many industrial policy advocates, we do not minimize the political problems with such an agenda. But we hope that this article will help stimulate a dialogue between industrial policy advocates and minority leaders, for any change in the American economy that smacks of economic resources will surely require the active participation of both.

REFERENCES

BLUESTONE, BARRY, and B. HARRISON, *The Deindustrialization of America* (New York: Basic Books, 1982.

BOWLES, SAMUEL, D., GORDON, and T. E. WEISSKOPF, *Beyond the Wasteland: A Democratic Alternative to Economic Decline* (Garden City, NY: Anchor Press/ Doubleday, 1983).

BROWN, CHARLES, "The Federal Attack on Labor Market Discrimination: The Mouse That Roared," Working paper no. 664 (Cambridge, MA: National bureau of Economic Research, 1981).

Business Week, "American Rushes to High Tech for Growth" (March 28, 1983), pp. 84–98,

______. "Displaced Workers May Have to Face Lower Wages" (September 19, 1983) p. 22.

CALITRI, RONALD, *Minority Secondary Education in New York State and New York City* (New York: Aspira of New York, Inc., 1983).

Center For Public Resources, *Basic Skills in the U. S. Work Force* (New York: Center For Public Resources, 1982).

DELEONE, RICHARD H., *Small Futures: Children, Inequality, and the Limits of Liberal Reform* (New York: Harcourt Brace Jovanovich, 1979).

ELLWOOD, DAVID T., and D. WISE, "Youth Employment in the Seventies: The Changing Circumstances of Young Adults," Working paper no. 1055 (Cambridge, MA: National Bureau of Ecconomic Research, 1983).

FREEMAN, RICHARD B., "Program Report: Labor Studies," *NBER Reporter* (Summer 1982), pp. 1–6.

HARRIS, DONALD J., "Economic Growth, Structural Change, and the Relative Income Status of Blacks in the U.S. Economy, 1947–78," *Review of Black Political Economy* (Spring 1983), pp. 75–92.

JEFFRIES, JOHN M., and R. MCGAHEY, "Equity, Growth and Change: Anti-Discrimination Policy in an Era of Economic Transformation," *Review of Law and Social Change,* forthcoming.

KUTTNER, ROBERT, "THE DECLINING MIDDLE," *The Atlantic Monthly* (July 1983), pp. 60–72.

______. *The Economic Illusion: False Choices Between Prosperity and Social Justice* (Boston: Houghton-Mifflin, 1984).

LEONARD, JONATHAN S., "Employment and Occupational Advance Under Affirmative Action," Working paper no. 1270, (Cambridge, MA; National Bureau of Economic Research, 1984a).

______. "The Impact of Affirmative Action on Employment," Working paper no. 1310, (Cambridge, MA: National Bureau of Economic Research, 1984b).

Los Angeles Times, "Layoffs of Minorities Are Disportionate" (June 2, 1983).

MCGAHEY, RICHARD M. "High Tech, Low Hopes," *New York Sunday Times,* Op-Ed Section (May 15, 1983).

MARKUSEN, ANN R., "High Technology and the Future of Employment." Paper presented at the National Conference on Economic Dislocation and Job Loss, Washington, DC, April 1984.

MILLS, D. QUINN, "The Human Resource Consequences of Industrial Revitalization, in Michael L. Wachter, and Susan M. Wachter (eds.) *Toward a New U.S. Industrial Policy?* (Philadelphia: University of Pennsylvania Press, 1983), pp. 257–80.

MURRAY, CHARLES A., *Losing Ground: American Social Policy, 1950–1980* (New York: Basic Books, 1984).

New York Times, "Few on Layoffs May be Rehired; U. S. Push in World Market Urged" (May 16, 1983), p. A1.

———. "Battling to Stem the Black Layoff Tide" (September 17, 1983).

OSTERMAN, PAUL, *Getting Started: The Youth Labor Market* (Cambridge, MA: The MIT Press, 1980).

PIORE, MICHAEL J., "The Dual Labor Market: Theory and Implications," in David M. Gordon (ed.) *Problems in Political Economy: An Urban Perspective* (Lexington, MA: D.C. Health, 1977).

PIVEN, FRANCES FOX, and R. A. CLOWARD, *Regulating the Poor: The Functions of Public Welfare* (New York: Pantheon, 1971).

REICH, ROBERT B., *The Next American Frontier* (New York: The New York Times Book Co., 1983).

ROHATYN, FELIX, "Time For A Change," *New York Review of Books* (August 18, 1983), pp. 46–49.

RUMBERGER, RUSSELL W., "The Employment Impact of Government Spending," Institute for Research on Educational Finance and Governance, School of Education, Stanford University, Palo Alto, (May 1983).

———. "The Changing Skill Requirements of Jobs in the U.S. Economy," *Industrial and Labor Relations Review* (July 1981), pp. 578–90.

SKOCPOL, THEDA, and J. IKENBERRY, "The Political Formation of the American Welfare State," *Comparative Social Research* (1983), pp. 87–148.

SULLIVAN, EILEEN, "Promising Occupations and Industries," New York City Department of Employment, 1982.

U.S. Department of Labor, Bureau of Labor Statistics, *BLS Reports on Displaced Workers,* USDL 84–492, Washington, DC (November 1984).

———. *Employment and Earnings,* March 1982.

U.S. General Accounting Office, *Labor Market Problems of Teenagers Result Largely From Doing Poorly in School* (Washington, DC: U.S. Government Printing Office, 1982).

42 *Jobs: Proposals from the Democratic Left*

BOB KUTTNER

Given the long-term changes in the structure of the economy and the shape of the job market, conventional policy responses are inadequate. The mainstream program to solve unemployment is "economic growth." Growth is, of course, desirable, but if more and more output requires fewer and fewer human workers, and if a growing share of workers producing for American markets are located abroad, then growth by itself won't solve the problem. In theory, all the purchasing power generated by growth "has to go somewhere." But it can go—and is going—to provide jobs overseas, while the jobs that it provides at home are not the kind that a broad middle-class society needs.

The retraining approach is also sadly inadequate. Demand for millions of blue-collar workers with new skills to tend the machines of the high-tech economy simply does not exist. Increasingly, high-tech machines tend themselves. For example, the new field of "Computer-Assisted Design/Computer-Assisted Manufacturing" or CAD/CAM will transform the function of draftsmen. With CAD/CAM, a draftsman turns out blueprints by computer. Several junior colleges have introduced CAD/CAM technician curricula to retrain a generation of draftsmen.

This new career opportunity is, however, likely to be short-lived. Just around the technological corner is something called CIM, which stands for

Computer-*Integrated* Manufacturing. With CIM, the engineer who designs a part sits at the computer terminal himself; instead of turning out computer-generated blueprints, CIM transforms the engineer's design directly into parts, and ultimately, products. The "new" job of CAD/CAM technician will go the way of the buggy-whip maker.

By the same token, the usual liberal proposal of massive public works or public employment also fails to solve the problem. For a number of reasons, public works jobs have been short-term and relatively low-paid. Because of the complex relationship between trade unions and the Democratic party, CETA-type programs have been designed not to compete with union labor, or even with non-union full-time career workers. Public works jobs of the usual variety are not likely to replace the disappearing middle of the American labor market or to solve the long-term problem of technological unemployment.

The real solutions to the structural jobs problem, I think, include these:

A coordinated worldwide economic stimulus program, coupled with a managed-trade regime. As the unhappy experience of the first two years of the Mitterand government showed, Keynesian reflation in one country is just as difficult as "socialism in one country." Our economies today are simply too interdependent to pursue their macroeconomic policies in isolation. If one country reflates by itself, its imports rise, a balance-of-payments crisis occurs, and the attempt at reflation self-destructs.

But even a concerted recovery program does not solve the problem of imports from low-wage areas. To be sure, it is in everyone's interest to help the Third World develop; but there is more than one form of development. It would not be realistic to expect that the welfare states of Europe and North America, whose social contracts are based on decent wages, will permit their own populations to be impoverished as a result of allowing unlimited imports from low-wage countries.

An alternative to a trade free-for-all, in which different nations play by different rules and everyone employs forms of disguised protectionism, is to admit the political nature of the issue and to make the question of where production is located a matter for political bargaining. This sounds like utter heresy, but there are already numerous practical examples.

A good illustration is the trade in textiles. Beginning in the early 1960s, textile imports from low-wage countries were threatening to wipe out the U.S. textile industry. After a series of de facto moves, the United States and Europe eventually negotiated a "Multi-fiber Arrangement"—a frank exception to the General Agreement on Tariff and Trade (GATT) free-trade regime, which recognized that every industrial country wanted to retain a domestic textile industry. Under the multi-fiber agreement, imports of textiles from cheap-labor countries were permitted to grow, but at an annual rate of growth limited to 5 percent. It is worth remembering that textile production, even in the United States, is one of the lowest-wage manufacturing industries, despite a unionization rate of about 20 percent.

Armed with this shield, the American textile industry began to improve its technological advantage. During the 1960s and 1970s, with substantial union cooperation, the domestic manufacturers invested massively in advanced spinning and weaving technology. By 1980, the American industry could boast the world's highest rate of textile productivity. In this case, "protectionism"—far from leading to "subsidized stagnation"—encouraged the textile industry to modernize and stay competitive. And of course it retained jobs for more than a million U.S. workers.

Nor has the multi-fiber arrangement beggared the Third World. Textile exports have continued to grow, and new countries, such as China, have become vigorous competitors. As these newly industrializing nations become more productive, their principal customers will increasingly be their own domestic markets.

Once nations reach approximately the same general wage levels, free trade need not undermine Western living standards and Western social contracts. But as long as the comparative advantage of the newly industrializing countries is based on repression of free trade unions, authoritarian governments working with authoritarian employers, and rock-bottom industrial wages, we have every right to limit imports. Indeed, here is a case for compensating tariffs to be levied on goods from countries that prohibit collective bargaining.

A second illustration is the domestic content legislation sponsored by the United Auto Workers. The bill requires that a substantial proportion of the total value of cars sold in the United States be built with American labor. This approach is far superior to import quotas (the Reagan ad hoc remedy) because it gives the American consumer unlimited access to superior foreign-designed technologies and products and, at the same time, assures that American purchasing power will provide at least some American jobs.

Most studies suggest that the Japanese comparative advantage in automobiles is mainly the result of better management, more efficient use of technology, and superior design—not cheaper labor. Japanese wage rates are lower than American ones, but it takes fewer worker-hours to build a Japanese car. Moreover, only about 24 percent of the cost of building a car is labor cost in any case. Studies have shown that Japanese cars built in America are built with every bit as much labor productivity as those produced in Japan. And whatever advantage does result from cheaper Japanese labor is almost canceled out by shipping costs.

So building Japanese-designed and -capitalized cars in the United States would not exact a hidden tax on American consumers. Quite the contrary; it would benefit the American economy by keeping several hundred thousand high-wage production jobs at home. The domestic content bill is the exact counterpart of the development strategy long used by Japan herself. Before the Japanese Ministry of International Trade and Industry let American semiconductor manufacturers sell their products in Japan, MITI exacted a series of coproduction agreements that served to accelerate the development of Japan's domestic semiconductor industry. More recently, Japan informed Boeing that before Japan Air Lines would order a new generation of planes, Boeing would have to sign an agreement that 15 percent of the added value would be produced in Japan. And the Japanese made sure that the 15 percent represented aeronautical technology that Japan wished to acquire.

Coproduction agreements of this nature, far from restricting world trade, simply change the ground rules. They promote the transfer of technology and, more importantly, assure that products sold in a particular country generate some jobs in that country. This has been the strategy of nations like Brazil with respect to the auto industry for two decades. American and European autoworkers wishing to sell to the big Brazilian consumer market must build auto plants in Brazil. Most European countries rely on similar policies.

The multi-fiber arrangement, by its success in stimulating technological advances, suggests another counter-intuitive conclusion about the relationship of technology to trade: ultimately, *automation* is the solution to a temporary comparative advantage based on cheap labor. If textile production is labor-intensive, and Haitian labor is paid at one-tenth the rate of American labor, the ball game is over unless we are prepared to protect the American domestic market. But what happens when our textile production is heavily automated? Suddenly we can again compete with Haitian textiles on the basis of price and quality, because our capital is more efficient.

The problem, again, is the familiar productivity paradox: automation produces more output at less labor input, and thus makes "everyone" better off—but in the meantime it throws people out of work. When a machine replaces a production worker, both the firm and consumers generally benefit; the loss falls mainly on the worker who loses his job. The "gain" to the firm may be invisible, depending on how competitive the particular industry is. Replacing labor with capital may not increase profits; it may simply permit the firm to hold its own against its competitors, for in a highly competitive industry profits are held down and efficiency gains are passed along to consumers.

If productivity gains are good for the competitiveness of American industry and good for American consumers, but bad for American workers (who are, after all, the same people as consumers except between nine and five), then the $64 billion question is how to socialize productivity gains so that they truly benefit everyone.

As Wassily Leontief has pointed out, failure to redistribute the gains of automation could condemn the society to a paradoxical condition of rising productivity and rising destitution. There has been little practical experimentation with redistribution of work and leisure, especially on this side of the Atlantic; indeed, this is not even a recognized policy issue in the United States.

In Europe, trade unionists have long argued that the solution to technological unemployment is a shorter work week, earlier retirement, pay for longer vacations, and other devices to share available work. The French have applied the boldest version of this program, and it has been only a mixed success.

Under the French program, an extra week of vacation was added, the work week is gradually being reduced from forty to thirty-five hours, and firms are encouraged to enter into "solidarity contracts" with their workers and with the state, in which workers agree to take voluntary early retirement and the state bears a substantial share of the cost of supplementary pensions, so that the purchasing power of the retired workers falls only slightly.

This approach has succeeded in stabilizing the unemployment rate at 8–9 percent but has not reduced it much. What might an American version of this approach look like? Here's one variation.

Redistributing productivity gains. Some eleven million Americans are out of work, but millions more are working longer hours than they really want to. Two obvious categories are people over sixty and parents of young children.

The original Social Security Act of 1935 was motivated partly by a desire to make life more comfortable for the elderly; but it was also stimulated by the need to get older people out of the labor force to open up jobs for younger ones. Suppose we adopt a supplemental early retirement incentive program. Unlike Social Security, which is becoming a fiscal drag because it is a pay-as-you-go system, let's prepay our new early retirement program, on the model of an Individual Retirement Account. Call it an ERA, for Early Retirement Account. Under this approach, workers would have a very small deduction taken from their paychecks, which would be matched by the employer and by the government on a one-third/one-third/one-third basis. Eventually, the accumulated capital would be sufficient to permit the worker to drop to a three-day week at age sixty, at full pay. If the average workers earns $12,500 a year, two-fifths of that amount times five years is a comparatively small sum—only $25,000. The payroll deductions necessary to build up to that sum over a thirty-year period, especially given the miracle of compound interest, would be quite trivial—less than $200 a year. This system would also help the national savings rate, a favorite cause of capital-conscious conservatives. But it would do so without further enriching the rich. By enticing millions of workers over sixty to take life a little easier (but to continue productive work), this would open up millions of jobs for younger workers.

As the parent of two school-age children, I propose working parents in the two-income family as my other nominees for oppressed workers. In theory, feminism and the family with two working parents was not just supposed to give women a ticket of entry to the male rat race; they were also supposed to allow men the luxury of some female values—nurturing, caregiving, and so forth. Remember? But things haven't worked out quite that way. Despite the entry of millions of women into the labor force, wage cuts have outpaced the additional income brought into the family by working women. Average household income has actually fallen by 11 percent since 1973. The average woman with a full-time job outside the home still spends thirty hours a week on housework, according the Labor Department. And very, very few fathers have taken advantage of working spouses in order to spend more time at home. Neither the drop in family income nor the conventional structure of career ladders permits it.

A proposal: Every parent, male and female, living with a minor child shall be entitled to five years of full-time pay for half-time work. The benefit would be capped at, say, 100 percent of the first $12,000 of income, and 70 percent of the next $12,000, so that millionaires could not drop to half-time work at government subsidy.

This would accomplish two things. It would be good social policy in its own right. Parents would be able to be less frantically job-obsessed during the years when their children need them most. In the case of two-parent households, there could be a total of ten years when at least one parent was in the home at least half time. In the case of a one-parent family, this would provide a far superior alternative to welfare. Secondly, of course, this particular form of family allowance would serve to free millions of person-years of jobs for people who are now unemployed. It would be a way of sharing available work in a manner that also served the critical social goal of shoring up family life.

As the economy becomes more and more productive, and as fewer and fewer workers are needed to produce society's material goods, other variations could be added, like periodic paid sabbaticals for everyone in the workforce. The frequency, duration, and timing of these could be adjusted periodically, to provide a fine-tuning device for maintaining full employment.

There is one caveat to this approach. In macro-economic terms, a transfer is a transfer is a transfer. Somewhere, the money must exist to finance these new income transfers; otherwise, we are simply reallocating an inadequate supply of real income; income will be more equitably shared, but we will lack the money to maintain full-time living standards for part-time work. To some degree, of course, this system packs a Keynesian punch. To the extent that it reallocates purchasing power and reduces involuntary unemployment, it stimulates aggregate demand, which in turn stimulates the growth necessary to provide the real resources to finance the program. While it is unlikely that we will experience enough productivity growth to rely solely on work-sharing to solve the unemployment problem, it is nonetheless an indispensable part of an overall program.

Procurement as industrial policy. Even more fundamental than work-sharing is the need to maximize total output as fewer and fewer workers are needed for production of physical goods. Obviously, a full-employment commitment is the beginning and end of any progressive economic program. Without full employment, workers who do have jobs lose their bargaining power to demand fair wages, and the state incurs the cost of supporting all those who have lost jobs. Neither make-work nor share-work, by themselves, get the society to full employment with maximized output. We also need to devise a means to redistribute some of the increased purchasing power that results from productivity gains—and turn it into both good jobs and the production of socially desired goods and services. In other words, the state has to devise a way to create the effective demand for commodities and services that have "use-values," but for which the private market does not generate effective purchasing power.

There is a well-proven American model for accomplishing this redistribution and translation of purchasing power into jobs. It is, unfortunately, the Pentagon. To a lesser degree, it is also NASA.

Consider the World War II experience. During World War II, the U.S. government mobilized more purchasing power than the country had ever seen. During the first six months of 1942, the War Department entered more orders than the entire value of the Gross National Product of 1940! By 1944, the peak year of war production, the Keynesian prod of war production had so stimulated the economy as a whole that *civilian* output alone exceeded the economy's entire output of 1940. And we accomplished that, even though half of what we produced was made in order to be blown up. During the war, GNP expanded at an unheard-of rate of 13 percent per year. Today, we are congratulating ourselves that the 1983 "recovery" temporarily attained a growth rate of 5 percent, and economists debate whether such a rate can be sustained.

World War II was also a formidable human capital policy. People with no prior experience in factories came off the farms to take good jobs in war plants. Women also learned blue-collar production jobs. During the war we trained 300,000 skilled machinists, a training feat that has never been equaled. The United States is now facing a shortage of experienced "A" machinists, in large part because about two-thirds of their senior machinists all reached retirement age about the same time—forty years after World War II!

For thirty years, it has been fashionable in conservative circles to aver that "it wasn't the New Deal that cured the Great Depression; it was the war." True enough. But anybody who thinks that World War II was anything other than a massive dose of Keynesian demand stimulus is kidding himself and the public. The supply-side nostrums—subsidizing the cost of capital and lowering the taxes to be paid on anticipated profits—may stimulate some new productive investment (*if* the entrepreneur is reasonably sure of a market for the product). But a government contract—a guarantee to purchase what is produced—is what really works miracles. If American industry can actually design and produce *guided missiles on railroad tracks* for the sole reason that the United States government is willing to write a check to pay for them, it is safe to assume that American industry can produce almost anything.

War production generates a Keynesian "big bang" in several complementary respects. It puts the unemployed back to work, and at relatively high-paid jobs. Whether production is accomplished in the public sector or through contracts with privately owned industry is optional; it accomplishes the same Keynesian jolt to aggregate purchasing power. During World War II, a good deal of the research and development was public (or under contract to universities); much of the production was private. But there was a substantial degree of planning and coordination by the government. Not only does a World

War II-style jump in public procurement put people back to work, train them at high skills, and boost purchasing power generally; it also sets in motion a technological dynamism. We are still living off many of the technological advances seeded by World War II defense contracting—synthetic rubber, the transistor, early computer science, America's early lead in civil aviation, and of course (for better or worse), atomic energy.

Moreover, procurement contracts do not just allow industry to make profits and hire workers by meeting current orders; they also subsidize the rebuilding and improvement of capital equipment. The civil aircraft and the automobiles and the chemical fertilizer industry that roared into full production to meet the postwar demands of returning veterans and their families were not produced by plants and machines that had to be built from scratch. They were built in converted war production plants. Tank lines became car lines; ordnance production became chemical fertilizer production.

The World War II example immediately throws into relief the fallacy of the conservative claim that government allocation of capital must necessarily retard output because it deploys capital "less optimally" than the market. When output expands at 13 percent a year, the economy can afford a lot of suboptimal use of capital! Conversely, when eleven million people are out of work and industry is producing at below 70 percent of capacity, the genius of the market for allocating capital isn't worth very much.

It is axiomatic that government procurement wastes some money. But idle resources waste more money. Keynes himself, in a celebrated passage, observed that if the government buried old banknotes in bottles and paid jobless workers to dig them up, it would be preferable to keeping them idle. In a sense, paying workers to build the implements of war is tantamount to paying them to dig up old bottles. Neither process produces useful goods for the domestic civilian economy. Defense production, however, has a patriotic aura about it, which imparts to it all sorts of virtues that enable conservatives to overlook the intimate involvement of the government.

The challenge, therefore, is to find civilian economic counterparts to World War II. The closest thing to a "civilian Pentagon" proposal in the political mainstream in Senator Kennedy's call for a domestic technology agency, which would award contracts to private industry. Unlike Pentagon contracts, the purpose of these projects would be to stimulate the development of technologies and products with civilian applications.

One could go a big step further and create an agency that would enter into contracts whose purpose is not just the development of technology, but the creation of useful goods and services and the redevelopment of regional economies suffering high unemployment. If America could grow at a rate of 13 percent a year on war production (much of which was blown up), just imagine how fast we could grow if the output went for useful goods and services. Suppose the government offered to buy a minicomputer for every classroom in America, subject only to the proviso that all of the production be carried out domestically. Suppose the government guaranteed a public school teacher-pupil ratio of one to fifteen. Suppose the government determined to replace every subway car older than ten years. Suppose the government pledged to finance the development of a commercially viable system of decentralized photoelectric domestic power generation.

Make up your own wish list of socially useful products that provide jobs. If employing people to build bombs and guns and planes to blow up half of Europe could energize a rapid economic recovery, production of *anything* can energize a recovery. A civilian economic equivalent of World War II has other advantages. It is far less inflationary than a war. In war, much of the added production goes up in smoke. In the civilian equivalent a government contract translates into useful products, so there is less of a problem of too much demand chasing too little supply. Government procurement is also an excellent planning tool. The procurement can be targeted in areas of high unemployment, or to particular industries, or to create new forms of ownership.

Unfortunately, "industrial policy" has been defined largely as the subsidy or provision of *capital,* not of production. For most of its advocates, industrial policy turns out to mean, in practice, some form of development bank or equity capital agency. That is far too limiting, and also too accepting of supply-side assumptions. Subsidizing capital accomplishes little when there is no assurance of a market for the product. Government procurement, in contrast, *creates* the market. It tells the contractor exactly what to produce, but leaves it to the contractor's ingenuity to figure out *how* to produce it. Anybody who thinks this inevitably leads to lemon products has never flown in an F-15. Government contracting has created technological marvels. Unfortunately, most have been in the military sector, which is almost totally isolated from the discipline of consumer markets. But it is also possible to deploy government contracts creatively to energize new, dynamic civilian sectors—and not just by buying new generations of products. The Japanese, shrewdly, create markets for new generations of capital goods by offering to buy out old capital goods.

All of this, to be sure, will require the expenditure of tax dollars and, in the short run, will increase public deficits. But the surest way—and, in the long run, the only way—to bring deficits back to moderate proportions is to get the economy back to full employment. Deficits that reflect high unemployment and shrunken tax bases create a self-perpetuating recession that will never cure either itself or the deficits it perpetuates.

Active labor market policies. Retraining, by itself, is no remedy for 10 percent unemployment. But coupled with a commitment to full employment as a fundamental policy goal, active labor market policies are very useful. The Swedes have devised a system of local labor market boards, which report to a national labor market board (the AMS). Under this system, workers are encouraged to take advantage of retraining opportunities—not just to upgrade the quality of the workforce as a whole, to ease inflationary shortages of skilled workers, and to match manpower supply to manpower demand, but also to take up slack during periods of high unemployment. At peak periods of joblessness, as much as 5 percent of the entire Swedish labor force is in some training program under the sponsorship of a labor market board. Translated into American proportions, this would equal over five million people.

If we are serious about an industrial policy and a commitment to full employment, an entirely different approach to retraining is required. One promising variation is the use of retraining vouchers; another is wage subsidy vouchers. Under this approach, any worker laid off because of technological advances would be given a voucher that was worth, say, half of five years' pay to that worker's next employer. By subsidizing wages directly (rather than through tax gimmickry), this approach would send the dislocated worker to the head of the job line. It would also give a needed competitive advantage to cities and towns hard hit by plant relocations. In a Youngstown or a Flint, a prospective new employer would find thousands of skilled workers, all holding job vouchers that would add up to a subsidy of half of the employer's labor costs for five years.

At present, American labor market policy consists mainly of paying unemployment compensation to the unemployed, on the assumption that their joblessness is temporary. We also have a parallel system of paying means-tested (welfare) subsistence benefits to the jobless who do not have a history of gainful employment or an ability to take full-time work. All of this income transfer represents a resource that could be used for retraining and for wage subsidies, rather than simply for subsistence.

Don't mourn, organize. The single most important piece of public policy that supported the postwar, high-wage social contract was not job retraining through CETA, or even minimum wage legislation. It was the Wagner Act.

As the American workforce is radically transformed in response to changes in the mobility of capital and the technology of industry, a strong labor movement is crucial to defend decent wages, to negotiate these transformations so that they are not paid for only by workers, and to provide a long-term constituency for progressive policies generally. The present labor movement surely has its flaws, but a look at the sweep of progressive legislation over the past forty years suggests that the labor movement is, if not the only game in town, the indispensable ally for the other enterprises—the civil rights, environmental, consumerist, feminist, and other progressive movements.

During the past ten years, the postwar consensus in which big corporations grudgingly tolerated labor organization has been unilaterally abrogated by management. It is now open season on labor unions. The nearly fifty-year-old structure of labor law is in desperate need of overhaul. Government should once again be the ally of labor organizing that it was during the New Deal, and labor unions themselves should be encouraged to take on the new issues of work quality, work organization, and the influence of technology on work, as well as that most "un-American" labor issue of all: how capital is to be invested, and for what.